W9-BAQ-623

SYSTEM SOFTWARE

AN INTRODUCTION TO SYSTEMS PROGRAMMING

SECOND EDITION

LELAND L. BECK
San Diego State University

ADDISON-WESLEY PUBLISHING COMPANY
Reading, Massachusetts • Menlo Park, California • New York
Don Mills, Ontario • Wokingham, England • Amsterdam • Bonn
Sydney • Singapore • Tokyo • Madrid • San Juan

Library of Congress Cataloging-in-Publication Data

Beck, Leland L.
System software: an introduction to systems programming / by Leland L. Beck. — 2nd ed.
p. cm.
Includes bibliographical references.
ISBN 0-201-50945-8
1. Systems programming (Computer science) 1. Title.
QA76.66.B43 1990
005.4′2—dc20 89-38688
CIP

IBM® is a trademark of International Business Machines Corporation
CDC®, COMPASS™, CYBER®, and NOS™ are trademarks of Control Data Corporation.
DEC®, VAX®, PDP®, and VMS™ are trademarks of Digital Equipment Corporation.
UNIX™ is a trademark of AT&T Bell Laboratories.
UCSD Pascal® is a trademark of Regents of the University of California.

ABCDEFGHIJ-HA-89

To Christopher Lauren

PREFACE

This text is an introduction to the design and implementation of various types of system software. A central theme of the book is the relationship between machine architecture and system software. For example, the design of an assembler or an operating system is greatly influenced by the architecture of the machine on which it runs. These influences are emphasized and demonstrated through the discussion of actual pieces of system software for a variety of different machines. However, there are also similarities between software for different systems. For example, the basic structure and design of an assembler is essentially the same for most computers. These fundamental machine-independent aspects of software design are clearly identified and separated from machine-specific details.

This second edition includes a new chapter (Chapter 8) that provides an introduction to software engineering concepts and techniques. The goal of the chapter is to help students learn how to approach a large software development project (such as the semester project that is often given in a course using this text). Chapter 1 contains an expanded introduction to SIC assembler language programming, for those instructors and students who want a more detailed treatment of this topic. Chapter 3 contains a more concrete discussion of bootstrap loading, which includes the code for a SIC/XE bootstrap loader. This code also provides a concrete example of the loading process. Other minor revisions and clarifications have been made throughout the text, and there are more than 60 new exercises.

This book is intended primarily for use as a text in a junior-, senior-, or graduate-level course in system software or systems programming. It is also suitable for use as a reference or for independent study. The reader is assumed to be familiar with the assembler language for at least one machine and with common methods for representing instructions and data within a computer (for example, octal and hexadecimal notations and 1's and 2's complement representation of negative values). Good discussions and reviews of these topics can be found in Pfleeger (1982) and Gear (1981). It is also assumed that the reader is familiar with the implementation and use of basic data structures, particularly linked lists and hash tables. This material may be found in any good data structures textbook, such as Tremblay (1984), Standish (1980), and Knuth (1973a and b).

Chapter 1 contains a brief introduction to the book, and gives a description of the Simplified Instructional Computer (SIC) that is used to present fundamental software concepts. It also describes the real machines (the System/370, VAX, and CYBER) that are used as examples throughout the text. These machines have contrasting architectures and were chosen as examples in order to illustrate the variety in software and hardware systems.

Chapter 2 describes the design and implementation of assemblers. The basic concepts of program assembly are presented in Section 2.1, using the SIC machine as a teaching aid. These basic assembler functions and characteristics should remain essentially the same, regardless of what machine is being used. This gives the student a starting point from which to begin the design of an assembler for a new or unfamiliar machine. Section 2.2 discusses machine-dependent extensions to the basic structure presented in Section 2.1; this serves to emphasize the relationship between machine architecture and assembler design and implementation. Section 2.3 introduces a number of machine-independent assembler features, and Section 2.4 discusses some important alternatives for the overall structure of the assembler. These features and alternatives are not dictated by machine considerations; they are choices made by the software designer. In such areas, there is no one "right" way of doing things; a software designer needs to be aware of the available options in order to make intelligent decisions between them. Finally, Section 2.5 discusses examples of actual assemblers for a variety of real computers. This provides an illustration of both machine-dependent and machine-independent variations that reinforces the points made in previous sections of the chapter.

The same general approach is followed in discussing loaders and linkers (Chapter 3), macro processors (Chapter 4), compilers (Chapter 5), and operating systems (Chapter 6). The basic features of each type of software are described first, followed by discussions of machine-dependent and machine-independent extensions to the basic features. Design alternatives are then discussed and examples of actual pieces of software are presented.

The depth of discussion varies considerably from one chapter to another. Chapters 2–4 give reasonably complete discussions of assemblers, linkers and loaders, and macro processors. Implementation details, such as algorithms and data structures, are also included, and the student should be able to write a working assembler, loader, or macro processor from the outline supplied by the text. (I strongly recommend that such a project be assigned in a course based on this book.)

Chapters 5 and 6, on the other hand, deal with the much larger subjects of compilers and operating systems. Each of these topics has, by itself, been the subject of many entire books, so it is obviously impossible to fully discuss either one in a single chapter. Instead the goal is to give the reader a brief-but-not-superficial overview of compilers and operating systems. The most important and fundamental concepts of these types of software are introduced and illustrated with examples. More advanced topics are mentioned, and references are given for the reader who wishes to explore these areas further. Because of space limitations, most implementation details have been omitted. A similar approach is followed in Chapter 7, which discusses database management systems, text editors, and interactive debugging systems.

Chapter 8 contains an introduction to software engineering concepts and techniques. This chapter does not attempt to cover the full scope of software engineering practice. Instead, it focuses on techniques that might be most useful in designing and implementing a piece of system software such as an assembler. The goal of this chapter is to provide the student with a set of tools and methods that he or she can use in a software project based on this book. The presentation of this material is relatively independent of the rest of the text. Chapter 8 can be read at any time after the introduction to assemblers in Section 2.1.

The exercises that appear at the end of each major chapter are an important part of the text. They are intended to stimulate individual thought and class discussion; some of the questions are open-ended design problems that have no one "right answer." Many of the exercises require the reader to apply concepts that have been covered in the text, extending them to new situations. This ensures that the reader fully understands the principles involved and is able to put them into actual practice. I have purposely not included answers to the exercises because I believe that a set of answers would have the effect of stifling thought and creativity rather than stimulating it.

This book contains more material than can usually be covered in a one-semester course, which allows the instructor to place a varying degree of emphasis on different topics to suit the needs of a specific curriculum. For example, if students will later take a course that deals solely with operating systems, the instructor may wish to omit Chapter 6; the remaining chapters include enough material for a typical one-semester course. Other instructors may prefer to cover all of the major chapters, eliminating some of the more advanced sections.

One issue that deserves comment is the use of a hypothetical computer (SIC) for instructional purposes. I have used the hypothetical machine primarily because it avoids the problem of dealing with the irrelevant complexities and "quirks" found on most real computers. It also clearly separates the fundamental concepts of a piece of software from implementation details associated with a particular computer. If a real machine is used in teaching, students are often unsure about which software characteristics are truly fundamental and which are simply consequences of the particular machine used in the text.

A secondary benefit to using a hypothetical machine is that all students begin on equal footing. No student is at an unfair disadvantage because he or she happens to be unfamiliar with the hardware and software system on which the text is based. I have found this to be particularly important in my courses, which tend to attract students who have had experience on a variety of computers.

Of course, students in a course of this type need to be able to write and run programs for the machine being studied. A SIC simulator and a simple SIC assembler, which can be run on almost any computing system that supports Pascal, are available from Addison-Wesley. This enables students to develop and run system programs as though they actually had access to a SIC machine. The availability of this portable software means that it is not necessary to use any particular real computing system—SIC programs can be run on any computer that is convenient for the instructor and the students.

Finally, it should be noted that some of the reviewers of this text were initially skeptical about SIC, but changed their opinion after seeing how it could be used as an instructional aid.

Many individuals have given time and energy to help make this a better book. I particularly wish to thank Sara Baase for sharing with me, from the very beginning of the project, the benefits of her experience as an author. Substantial portions of the manuscript were reviewed by Sara Baase, Roger Bielefeld, Hedley Bond, Steve Bruell, Tim Budd, Roger King, Joan Krone, Thomas Nugent, R. Oldehoeft, Charles Pfleeger, Gerald M. Radack, Paul Ross, David Umbaugh, Vernor Vinge, Bernhard Weinberg, and Charles N. Winton. The comments and suggestions of these reviewers were extremely helpful to me in finding errors and other problems in the manuscript. Any errors which remain are, of course, entirely my responsibility, and I would be very grateful to any reader for pointing out such errors to me.

I would like to thank Matt Belshin, Jeff Blackmon, Pat Coffey, Glenn Jerpseth, Sanda Schwimmer, and Bob Swanson for their assistance in finding reference material on specific systems. I am also indebted to the students who used previous versions of this material as a text and provided many valuable suggestions.

Finally, I am deeply grateful to Marla and Kendra for providing me with encouragement and a supportive environment in which to work.

San Diego L. L. B.

CONTENTS

1 BACKGROUND

CHAPTER **1**

BACKGROUND

This chapter contains a variety of information that serves as background for the material presented later. Section 1.1 gives a brief introduction to system software and an overview of the structure of this book. Section 1.2 begins a discussion of the relationships between system software and machine structure, which continues throughout the text; Sections 1.3 through 1.7 provide an introduction to the architecture of several computers that are used as examples. Further information on most of the machine architecture topics discussed may be found in Tannenbaum (1984), Pfleeger (1982), and Gear (1981).

Most of the material in this chapter is presented at a summary level, with many details omitted. The level of detail given here is sufficient background for the remainder of the text. You should not attempt to memorize the material in this chapter, or be overly concerned with minor points. Instead, it is recommended that you read through this material, and then use it for reference as needed in later chapters. References are provided throughout the chapter for readers who want further information.

1.1 INTRODUCTION

This text is an introduction to the design of various types of system software. We also consider implementations of such software on a variety of real machines. One central theme of the book is the relationship between system software and machine architecture—the design of an assembler, operating system, etc., is influenced by the structure of the machine on which it is to run. Some of these influences are discussed in the next section; many other examples appear throughout the text.

The major topics covered in this book are assemblers, loaders and linkers, macro processors, compilers, and operating systems; each of Chapters 2 through 6 is devoted to one of these subjects. It is assumed that the reader is familiar with each of these pieces of system software from the user's point of view. This book is concerned primarily with the design and implementation of the software itself. Chapter 7 contains a survey of some other important types of system software: database management systems, text editors, and interactive debugging systems. Chapter 8 contains an introduction to software engineering concepts and techniques, focusing on the use of such methods in writing system software. This chapter can be read at any time after the introduction to assemblers in Section 2.1.

The depth of treatment in this text varies considerably from one topic to another. The chapters on assemblers, loaders and linkers, and macro processors contain enough implementation details to prepare the reader to write these types of software for a real computer. Compilers and operating systems, on the other hand, are very large topics; each has, by itself, been the subject of many complete books and courses. It is obviously impossible to provide a full coverage of these subjects in a single chapter of any reasonable size. Instead, we provide an introduction to the most important concepts and issues related to compilers and operating systems, stressing the relationships between software design and machine structure. Other subtopics are discussed as space permits, with references provided for readers who wish to explore these areas further. Our goal is to provide a good overview of these subjects that can also serve as background for students who will later take more advanced software courses. This same approach is also applied to the other topics surveyed in Chapter 7.

1.2 SYSTEM SOFTWARE AND MACHINE STRUCTURE

One characteristic in which most system software differs from application software is machine dependency. An application program is primarily concerned with the solution of some problem, using the computer as a tool; the focus is on the application, not on the computing system. System programs, on the other hand, are intended to support the operation and use of the computer itself, rather than any particular application; for this reason, they are usually related to the structure of the machine on which they are to run. For example, assemblers translate mnemonic instructions into machine code; the instruction formats, addressing modes, etc., are of direct concern in assembler design. Similarly, compilers must generate machine language code, taking into account such hardware characteristics as the number and use of registers and the machine instructions available. Operating systems are directly concerned with the management of nearly all of the resources of a computing system. Many other examples of such machine dependencies may be found throughout this book.

On the other hand, there are some aspects of system software that do not directly depend upon the type of computing system being supported. For example, the general design and logic of an assembler is basically the same on most computers. Some of the code optimization techniques used by compilers are independent of the target machine (although there are also machine-dependent optimizations). Likewise, the process of linking together independently assembled subprograms does not usually depend on the computer being used. We will also see many examples of such machine-independent features in the chapters that follow.

Because most system software is machine-dependent, we must include real machines and real pieces of software in our study. However, most real computers have certain characteristics that are unusual or even unique. It can be difficult to distinguish between those features of the software that are truly fundamental and those that depend solely on the idiosyncrasies of a particular machine. To avoid this problem, we present the fundamental functions of each piece of software through discussion of a Simplified Instructional Computer (SIC). SIC is a hypothetical computer that has been carefully designed to include the hardware features most often found on real machines, while avoiding unusual or irrelevant complexities. In this way, the central concepts of a piece of system software can be clearly separated from the implementation details associated with a particular machine. This approach provides the reader with a starting point from which to begin the design of system software for a new or unfamiliar computer.

Each major chapter in this text first introduces the basic functions of the type of system software being discussed; we then discuss machine-dependent and machine-independent extensions to these functions, and examples of implementations on actual machines. Specifically, the major chapters are divided into the following sections:

1. Features that are fundamental, and that should be found in any example of this type of software.

2. Features whose presence and character are closely related to the machine architecture.
3. Other features that are commonly found in implementations of this type of software, and that are relatively machine-independent.
4. Major design options for structuring a particular piece of software—for example, single-pass versus multi-pass processing.
5. Examples of implementations on actual machines, stressing unusual software features and those that are related to machine characteristics.

This chapter contains brief descriptions of the real machines that are used as examples. You are encouraged to read these descriptions now, and refer to them as necessary when studying the examples in each chapter.

1.3 THE SIMPLIFIED INSTRUCTIONAL COMPUTER (SIC)

In this section we describe the architecture of our Simplified Instructional Computer (SIC). This machine has been designed to illustrate the most commonly encountered hardware features and concepts, while avoiding most of the idiosyncrasies that are often found in real machines.

In many ways SIC is similar to a typical microcomputer. Like many other products, SIC comes in two versions: the standard model, and an XE version (XE stands for “extra equipment,” or perhaps “extra expensive”). The two versions have been designed to be upward compatible—that is, an object program for the standard SIC machine will also execute properly on a SIC/XE system. (Such upward compatibility is often found on real computers that are closely related to one another.) Section 1.3.1 summarizes the standard features of SIC. Section 1.3.2 describes the additional features that are included in SIC/XE. Section 1.3.3 presents simple examples of SIC and SIC/XE programming. These examples are intended to help you become more familiar with the SIC and SIC/XE instruction sets and assembler language. Practice exercises in SIC and SIC/XE programming can be found at the end of this chapter.

1.3.1 SIC Machine Structure

Memory

Memory consists of 8-bit bytes; any three consecutive bytes form a word (24 bits). All addresses on SIC are byte addresses; words are addressed by the location of their lowest numbered byte. There are a total of 32,768 (2^{15}) bytes in the computer memory.

Registers

There are five registers, all of which have special uses. Each register is 24 bits in length. The following table indicates the numbers, mnemonics, and uses of these registers. (The numbering scheme has been chosen for compatibility with the XE version of SIC.)

Mnemonic	Number	Special use
A	0	Accumulator; used for arithmetic operations
X	1	Index register; used for addressing
L	2	Linkage register; the Jump to Subroutine (JSUB) instruction stores the return address in this register
PC	8	Program counter; contains the address of the next instruction to be fetched for execution
SW	9	Status word; contains a variety of information, including a Condition Code (CC)

Data Formats

Integers are stored as 24-bit binary numbers; 2's complement representation is used for negative values. Characters are stored using their 8-bit ASCII codes (see Appendix B). There is no floating-point hardware on the standard version of SIC.

Instruction Formats

All machine instructions on the standard version of SIC have the following 24-bit format:

8	1	15
opcode	x	address

The flag bit *x* is used to indicate indexed-addressing mode.

Addressing Modes

There are two addressing modes available, indicated by the setting of the *x* bit in the instruction. The following table describes how the *target address* is calculated from the address given in the instruction. Parentheses are used to indicate the contents of a register or a memory location. For example, (X) represents the contents of register X.

Mode	Indication	Target address calculation
Direct	x = 0	TA = address
Indexed	x = 1	TA = address + (X)

Instruction Set

SIC provides a basic set of instructions that are sufficient for most simple tasks. These include instructions that load and store registers (LDA, LDX, STA, STX, etc.), as well as integer arithmetic operations (ADD, SUB, MUL, DIV). All arithmetic operations involve register A and a word in memory, with the result being left in the register. There is an instruction (COMP) that compares the value in register A with a word in memory; this instruction sets a *condition code* CC to indicate the result (<, =, or >). Conditional jump instructions (JLT, JEQ, JGT) can test the setting of CC, and jump accordingly. Two instructions are provided for subroutine linkage: JSUB jumps to the subroutine, placing the return address in register L; RSUB returns by jumping to the address contained in register L.

Appendix A gives a complete list of all SIC (and SIC/XE) instructions, with their operation codes and a specification of the function performed by each.

Input and Output

On the standard version of SIC, input and output are performed by transferring one byte at a time to or from the rightmost 8 bits of register A. Each device is assigned a unique 8-bit code. There are three I/O instructions, each of which specifies the device code as an operand.

The Test Device (TD) instruction tests whether the addressed device is ready to send or receive a byte of data. The condition code is set to indicate the result of this test. (A setting of < means ready to send or receive, and = means device busy.) A program wishing to transfer data must wait until the device is ready, then execute a Read Data (RD) or Write Data (WD). This sequence must be repeated for each byte of data to be read or written. The program shown in Fig. 2.1 (Chapter 2) illustrates this technique for performing I/O.

1.3.2 SIC/XE Machine Structure

Memory

The memory structure for SIC/XE is the same as that previously described for SIC. However, the maximum memory available on a SIC/XE system is one megabyte (2^{20} bytes). This increase leads to a change in instruction formats and addressing modes.

Registers

The following additional registers are provided by SIC/XE.

Mnemonic	Number	Special use
B	3	Base register; used for addressing
S	4	General working register—no special use
T	5	General working register—no special use
F	6	Floating-point accumulator (48 bits)

Data Formats

SIC/XE provides the same data formats as the standard version. In addition, there is a 48-bit floating-point data type with the following format:

1	11	36
s	exponent	fraction

The fraction is interpreted as a value between 0 and 1; that is, the assumed binary point is immediately before the high-order bit. For normalized floating-point numbers, the high-order bit of the fraction must be 1. The exponent is interpreted as an unsigned binary number between 0 and 2047. If the exponent has value *e* and the fraction has value *f*, the absolute value of the number represented is

$$f * 2^{(e-1024)}.$$

The sign of the floating-point number is indicated by the value of *s* (0 = positive, 1 = negative). A value of zero is represented by setting all bits (including sign, exponent, and fraction) to 0.

Instruction Formats

The larger memory available on SIC/XE means that an address will (in general) no longer fit into a 15-bit field; thus the instruction format used on the standard version of SIC is no longer suitable. There are two possible options—either use some form of relative addressing, or extend the address field to 20 bits. Both of these options are included in SIC/XE (Formats 3 and 4 in the following description). In addition, SIC/XE provides some instructions that do not reference memory at all. Formats 1 and 2 in the following description are used for such instructions.

The new set of instruction formats is as follows. The settings of the flag bits in Formats 3 and 4 are discussed under Addressing Modes. Bit *e* is used to distinguish between Formats 3 and 4 ($e = 0$ means Format 3, $e = 1$ means Format 4). Appendix A indicates the format to be used with each machine instruction.

Format 1 (1 byte):

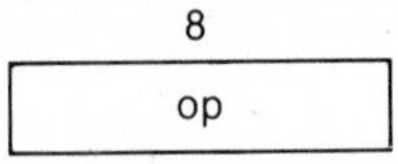

8
op

Format 2 (2 bytes):

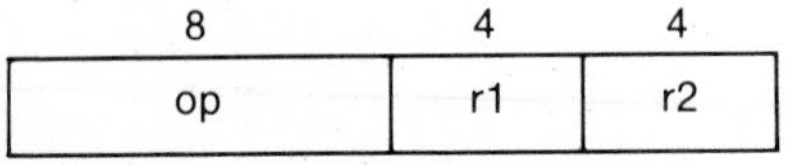

8	4	4
op	r1	r2

Format 3 (3 bytes):

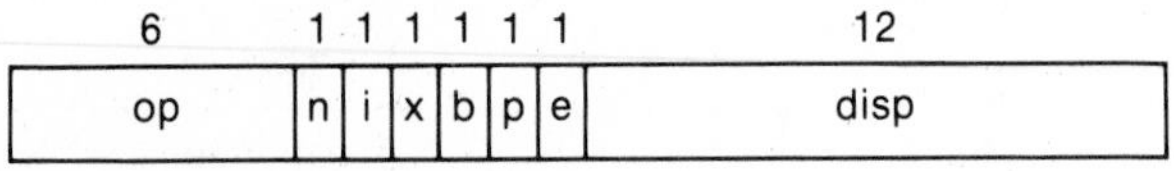

6	1	1	1	1	1	1	12
op	n	i	x	b	p	e	disp

Format 4 (4 bytes):

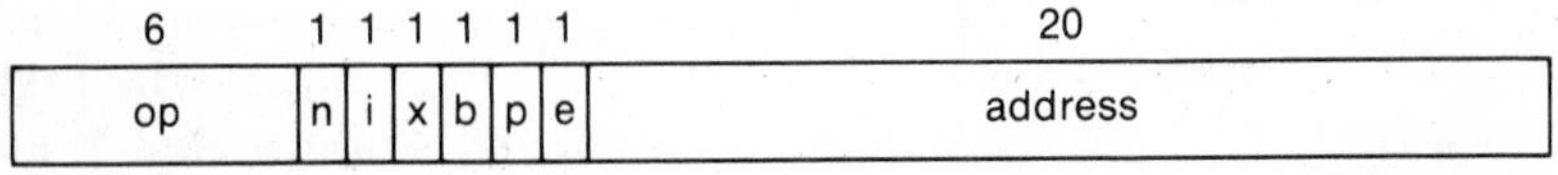

6	1	1	1	1	1	1	20
op	n	i	x	b	p	e	address

Addressing Modes

Two new relative addressing modes are available for use with instructions assembled using Format 3. These are described in the following table.

Mode	Indication	Target address calculation
Base relative	b = 1, p = 0	TA = (B) + disp ($0 \leq$ disp ≤ 4095)
Program-counter relative	b = 0, p = 1	TA = (PC) + disp ($-2048 \leq$ disp ≤ 2047)

For *base relative* addressing, the displacement field disp in a Format 3 instruction is interpreted as a 12-bit unsigned integer. For *program-counter relative* addressing, this field is interpreted as a 12-bit signed integer, with negative values represented in 2's complement notation.

If bits *b* and *p* are both set to 0, the disp field from the Format 3 instruction is taken to be the target address. For a Format 4 instruction, bits *b* and *p* must both be 0, and the target address is taken from the address field of the instruction. We will call this *direct* addressing, to distinguish it from the relative addressing modes described above.

Any of these addressing modes can also be combined with *indexed* addressing—if bit *x* is set to 1, the term (X) is added in the target address calculation. Notice that the standard version of the SIC machine uses only direct addressing (with or without indexing).

Bits *i* and *n* in Formats 3 and 4 are used to specify how the target address is used. If bit $i = 1$ and $n = 0$, the target address itself is used as the operand value; no memory reference is performed. This is termed *immediate* addressing. If bit $i = 0$ and $n = 1$, the word at the location given by the target address is fetched; the *value* contained in this word is then taken as the *address* of the operand value. This is called *indirect* addressing. If bits *i* and *n* are both 0 or both 1, the target address is taken as the location of the operand; we will refer to this as *simple* addressing. Indexing cannot be used with immediate or indirect addressing modes.

Many authors use the term *effective address* to denote what we have called the target address for an instruction. However, there is disagreement concerning the meaning of effective address when referring to an instruction that uses indirect addressing. To avoid confusion, we use the term *target address* throughout this book.

SIC/XE instructions that specify neither immediate nor indirect addressing are assembled with bits *n* and *i* both set to 1. Assemblers for the standard version of SIC will, however, set the bits in both of these positions to 0. (This is because the 8-bit binary codes for all of the SIC instructions end in 00.) All SIC/XE machines have a special hardware feature designed to provide the upward compatibility mentioned earlier. If bits *n* and *i* are both 0, then bits *b*, *p*, and *e* are considered to be part of the address field of the instruction (rather than flags indicating addressing modes). This makes Instruction Format 3 identical to the format used on the standard version of SIC, providing the desired compatibility.

Figure 1.1 gives examples of the different addressing modes available on SIC/XE. Figure 1.1(a) shows the contents of registers B, PC, and X, and of selected memory locations. (All values are given in hexadecimal.) Figure 1.1(b) gives the machine code for a series of LDA instructions. The target address generated by each instruction, and the value that is loaded into register A, are also shown. You should carefully examine these examples, being sure you understand the different addressing modes illustrated.

For ease of reference, all of the SIC/XE instruction formats and addressing modes are summarized in Appendix A.

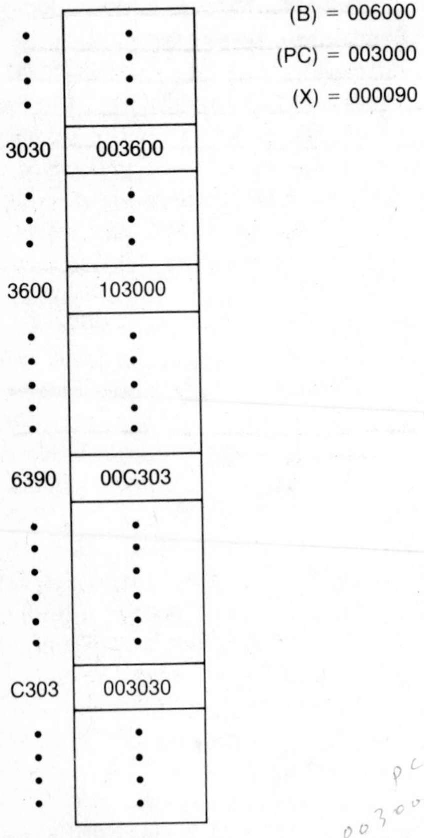

(B) = 006000

(PC) = 003000

(X) = 000090

Address	Contents
⋮	⋮
3030	003600
⋮	⋮
3600	103000
⋮	⋮
6390	00C303
⋮	⋮
C303	003030
⋮	⋮

(a)

Hex	Machine instruction: Binary								Target address	Value loaded into register A
	op	n	i	x	b	p	e	disp/address		
032600	000000	1	1	0	0	1	0	0110 0000 0000	3600	103000
03C300	000000	1	1	1	1	0	0	0011 0000 0000	6390	00C303
022030	000000	1	0	0	0	1	0	0000 0011 0000	3030	103000
010030	000000	0	1	0	0	0	0	0000 0011 0000	30	000030
003600	000000	0	0	0	0	1	1	0110 0000 0000	3600	103000
0310C303	000000	1	1	0	0	0	1	0000 1100 0011 0000 0011	C303	003030

(b)

FIGURE 1.1 Examples of SIC/XE instructions and addressing modes.

Instruction Set

SIC/XE provides all of the instructions that are available on the standard version. In addition, there are instructions to load and store the new registers (LDB, STB, etc.) and to perform floating-point arithmetic operations (ADDF, SUBF, MULF, DIVF). There are also instructions that take their operands from registers. Besides the RMO (register move) instruction, these include register–register arithmetic operations (ADDR, SUBR, MULR, DIVR). A special *supervisor call* instruction (SVC) is provided. Executing this instruction generates an interrupt that can be used for communication with the operating system. (Supervisor calls and interrupts are discussed in Chapter 6.)

There are also several other new instructions. Appendix A gives a complete list of all SIC/XE instructions, with their operation codes and a specification of the function performed by each.

Input and Output

The I/O instructions we discussed for SIC are also available on SIC/XE. In addition, there are I/O channels that can be used to perform input and output while the CPU is executing other instructions. This allows overlap of computing and I/O, resulting in more efficient system operation. The instructions SIO, TIO, and HIO are used to start, test, and halt the operation of I/O channels. (These concepts are discussed in detail in Chapter 6.)

1.3.3 SIC Programming Examples

This section presents simple examples of SIC and SIC/XE assembler language programming. These examples are intended to help you become more familiar with the SIC and SIC/XE instruction sets and assembler language. It is assumed that the reader is already familiar with the assembler language of at least one machine and with the basic ideas involved in assembly-level programming.

The primary subject of this book is systems programming, not assembler language programming. The following chapters contain discussions of various types of system software, and in some cases SIC programs are used to illustrate the points being made. This section contains material that may help you to understand these examples more easily. However, it does not contain any new material on system software or systems programming. Thus, this section can be skipped without any loss of continuity.

Figure 1.2 contains examples of data movement operations for SIC and SIC/XE. There are no memory-to-memory move instructions; thus, all data movement must be done using registers. Figure 1.2(a)

```
        LDA     FIVE            LOAD CONSTANT 5 INTO REGISTER A
        STA     ALPHA           STORE IN ALPHA
        LDCH    CHARZ           LOAD CHARACTER 'Z' INTO REGISTER A
        STCH    C1              STORE IN CHARACTER VARIABLE C1
        .
        .
        .
ALPHA   RESW    1               ONE-WORD VARIABLE
FIVE    WORD    5               ONE-WORD CONSTANT
CHARZ   BYTE    C'Z'            ONE-BYTE CONSTANT
C1      RESB    1               ONE-BYTE VARIABLE
```

(a)

```
        LDA     #5              LOAD IMMEDIATE VALUE 5 INTO REGISTER A
        STA     ALPHA           STORE IN ALPHA
        LDA     #90             LOAD ASCII CODE FOR 'Z' INTO REGISTER A
        STCH    C1              STORE IN CHARACTER VARIABLE C1
        .
        .
        .
ALPHA   RESW    1               ONE-WORD VARIABLE
C1      RESB    1               ONE-BYTE VARIABLE
```

(b)

FIGURE 1.2 Sample data movement operations for (a) SIC and (b) SIC/XE.

shows two examples of data movement. In the first, a 3-byte word is moved by loading it into register A and then storing the register at the desired destination. Exactly the same thing could be accomplished using register X (and the instructions LDX, STX) or register L (LDL, STL). In the second example, a single byte of data is moved using the instructions LDCH (Load Character) and STCH (Store Character). These instructions operate by loading or storing the rightmost 8-bit byte of register A; the other bits in register A are not affected.

Figure 1.2(a) also shows four different ways of defining storage for data items in the SIC assembler language. (These assembler directives are discussed in more detail in Section 2.1.) The statement WORD reserves one word of storage, which is initialized to a value defined in the operand field of the statement. Thus the WORD statement in Fig. 1.2(a) defines a data word labeled FIVE whose value is initialized to 5. The statement RESW reserves one or more words of storage for use by the program. For example, the RESW statement in Fig. 1.2(a) defines one word of storage labeled ALPHA, which will be used to hold a value generated by the program.

The statements BYTE and RESB perform similar storage-definition functions for data items that are characters instead of words. Thus in Fig. 1.2(a) CHARZ is a 1-byte data item whose value is initialized to the character "Z", and C1 is a 1-byte variable with no initial value.

The instructions shown in Fig. 1.2(a) would also work on SIC/XE; however, they would not take advantage of the more advanced hardware features available. Figure 1.2(b) shows the same two data-movement operations as they might be written for SIC/XE. In this example, the value 5 is loaded into register A using immediate addressing. The operand field for this instruction contains the flag # (which specifies immediate addressing) and the data value to be loaded. Similarly, the character "Z" is placed into register A by using immediate addressing to load the value 90, which is the decimal value of the ASCII code that is used internally to represent the character "Z".

Figure 1.3(a) shows examples of arithmetic instructions for SIC. All arithmetic operations are performed using register A, with the result being left in register A. Thus this sequence of instructions stores the value (ALPHA + INCR − 1) in BETA and the value (GAMMA + INCR − 1) in DELTA.

Figure 1.3(b) illustrates how the same calculations could be performed on SIC/XE. The value of INCR is loaded into register S initially, and the register-to-register instruction ADDR is used to add this value to register A when it is needed. This avoids having to fetch INCR from memory each time it is used in a calculation, which may make the program more efficient. Immediate addressing is used for the constant 1 in the subtraction operations.

Looping and indexing operations are illustrated in Fig. 1.4. Figure 1.4(a) shows a loop that copies one 11-byte character string to another. The index register (register X) is initialized to zero before the loop begins. Thus, during the first execution of the loop, the target address for the LDCH instruction will be the address of the first byte of STR1. Similarly, the STCH instruction will store the character being copied into the first byte of STR2. The next instruction, TIX, performs two functions. First it adds 1 to the value in register X, and then it compares the new value of register X to the value of the operand (in this case, the constant value 11). The condition code is set to indicate the result of this comparison. The JLT instruction jumps if the condition code is set to "Less than;" thus, the JLT causes a jump back to the beginning of the loop if the new value in register X is less than 11.

During the second execution of the loop, register X will contain the value 1. Thus, the target address for the LDCH instruction will be the second byte of STR1, and the target address for the STCH instruction will be the second byte of STR2. The TIX instruction will again add 1 to the value in register X, and the loop will continue in this way until all 11 bytes have been copied from STR1 to STR2. Notice that after the

```
          LDA      ALPHA          LOAD ALPHA INTO REGISTER A
          ADD      INCR           ADD THE VALUE OF INCR
          SUB      ONE            SUBTRACT 1
          STA      BETA           STORE IN BETA
          LDA      GAMMA          LOAD GAMMA INTO REGISTER A
          ADD      INCR           ADD THE VALUE OF INCR
          SUB      ONE            SUBTRACT 1
          STA      DELTA          STORE IN DELTA
          .
          .
          .
ONE       WORD     1              ONE-WORD CONSTANT
.                                 ONE-WORD VARIABLES
ALPHA     RESW     1
BETA      RESW     1
GAMMA     RESW     1
DELTA     RESW     1
INCR      RESW     1
```

(a)

```
          LDS      INCR           LOAD VALUE OF INCR INTO REGISTER S
          LDA      ALPHA          LOAD ALPHA INTO REGISTER A
          ADDR     S,A            ADD THE VALUE OF INCR
          SUB     #1              SUBTRACT 1
          STA      BETA           STORE IN BETA
          LDA      GAMMA          LOAD GAMMA INTO REGISTER A
          ADDR     S,A            ADD THE VALUE OF INCR
          SUB     #1              SUBTRACT 1
          STA      DELTA          STORE IN DELTA
          .
          .
          .                       ONE-WORD VARIABLES
.
ALPHA     RESW     1
BETA      RESW     1
GAMMA     RESW     1
DELTA     RESW     1
INCR      RESW     1
```

(b)

FIGURE 1.3 Sample arithmetic operations for (a) SIC and (b) SIC/XE.

TIX instruction is executed, the value in register X is equal to the number of bytes that have already been copied.

Figure 1.4(b) shows the same loop as it might be written for SIC/XE. The main difference is that the instruction TIXR is used in place of TIX. TIXR works exactly like TIX, except that the value used for comparison is taken from another register (in this case, register T), not from memory. This makes the loop more efficient, because the value does not have to be fetched from memory each time the loop is executed. Immediate addressing is used to initialize register T to the value 11 and to initialize register X to 0.

```
        LDX     ZERO            INITIALIZE INDEX REGISTER TO 0
MOVECH  LDCH    STR1,X          LOAD CHARACTER FROM STR1 INTO REGISTER A
        STCH    STR2,X          STORE CHARACTER INTO STR2
        TIX     ELEVEN          ADD 1 TO INDEX REGISTER AND COMPARE RESULT TO 11
        JLT     MOVECH          LOOP IF INDEX REGISTER IS LESS THAN 11
        .
        .
        .
STR1    BYTE    C'TEST STRING'  11-BYTE STRING CONSTANT
STR2    RESB    11              11-BYTE VARIABLE
.                               ONE-WORD CONSTANTS
ZERO    WORD    0
ELEVEN  WORD    11
```

(a)

```
        LDT     #11             INITIALIZE REGISTER T TO 11
        LDX     #0              INITIALIZE INDEX REGISTER TO 0
MOVECH  LDCH    STR1,X          LOAD CHARACTER FROM STR1 INTO REGISTER A
        STCH    STR2,X          STORE CHARACTER INTO STR2
        TIXR    T               ADD 1 TO INDEX REGISTER AND COMPARE RESULT TO 11
        JLT     MOVECH          LOOP IF INDEX REGISTER IS LESS THAN 11
        .
        .
        .
STR1    BYTE    C'TEST STRING'  11-BYTE STRING CONSTANT
STR2    RESB    11              11-BYTE VARIABLE
```

(b)

FIGURE 1.4 Sample looping and indexing operations for (a) SIC and (b) SIC/XE.

Figure 1.5 contains another example of looping and indexing operations. The variables ALPHA, BETA, and GAMMA are arrays of 100 words each. In this case, the task of the loop is to add together the corresponding elements of ALPHA and BETA, storing the results in the elements of GAMMA. The general principles of looping and indexing are the same as previously discussed. However, the value in the index register must be incremented by 3 for each iteration of this loop, because each iteration processes a 3-byte (i.e., one-word) element of the arrays. The TIX instruction always adds 1 to register X, so it is not suitable for this program fragment. Instead, we use arithmetic and comparison instructions to handle the index value.

In Fig. 1.5(a), we define a variable INDEX to hold the value to be used for indexing for each iteration of the loop. Thus, INDEX should be 0 for the first iteration, 3 for the second, and so on. INDEX is initialized to 0 before the start of the loop. The first instruction in the body of the loop loads the current value of INDEX into register X, so that it can be used for target address calculation. The next three instructions in the loop load a word from ALPHA, add the corresponding word from

BETA, and store the result in the corresponding word of GAMMA. The value of INDEX is then loaded into register A, incremented by 3, and stored back into INDEX. After being stored, the new value of INDEX is still present in register A. This value is then compared to 300 (the length of the arrays in bytes) to determine whether or not to terminate the loop. If the value of INDEX is less than 300, then all bytes of the arrays have not yet been processed. In that case, the JLT instruction causes a jump back to the beginning of the loop, where the new value of INDEX is loaded into register X.

```
            LDA     ZERO            INITIALIZE INDEX VALUE TO 0
            STA     INDEX
ADDLP       LDX     INDEX           LOAD INDEX VALUE INTO INDEX REGISTER
            LDA     ALPHA,X         LOAD A WORD FROM ALPHA INTO REGISTER A
            ADD     BETA,X          ADD A WORD FROM BETA
            STA     GAMMA,X         STORE THE RESULT IN A WORD IN GAMMA
            LDA     INDEX           ADD 3 TO INDEX VALUE
            ADD     THREE
            STA     INDEX
            COMP    K300            COMPARE NEW INDEX VALUE TO 300
            JLT     ADDLP           LOOP IF INDEX VALUE IS LESS THAN 300
            .
            .
            .
INDEX       RESW    1               ONE-WORD VARIABLE FOR INDEX VALUE
.                                   ARRAY VARIABLES -- 100 WORDS EACH
ALPHA       RESW    100
BETA        RESW    100
GAMMA       RESW    100
.                                   ONE-WORD CONSTANTS
ZERO        WORD    0
K300        WORD    300
```

(a)

```
            LDS     #3              INITIALIZE REGISTER S TO 3
            LDT     #300            INITIALIZE REGISTER T TO 300
            LDX     #0              INITIALIZE INDEX REGISTER TO 0
ADDLP       LDA     ALPHA,X         LOAD A WORD FROM ALPHA INTO REGISTER A
            ADD     BETA,X          ADD A WORD FROM BETA
            STA     GAMMA,X         STORE THE RESULT IN A WORD IN GAMMA
            ADDR    S,X             ADD 3 TO INDEX VALUE
            COMPR   X,T             COMPARE NEW INDEX VALUE TO 300
            JLT     ADDLP           LOOP IF INDEX VALUE IS LESS THAN 300
            .
            .
            .
.                                   ARRAY VARIABLES -- 100 WORDS EACH
ALPHA       RESW    100
BETA        RESW    100
GAMMA       RESW    100
```

(b)

FIGURE 1.5 Sample indexing and looping operations for (a) SIC and (b) SIC/XE.

This particular loop is cumbersome on SIC, because register A must be used for adding the array elements together and also for incrementing the index value. The loop can be written much more efficiently for SIC/XE, as shown in Fig. 1.5(b). In this example, the index value is kept permanently in register X. The amount by which to increment the index value (3) is kept in register S, and the register-to-register ADDR instruction is used to add this increment to register X. Similarly, the value 300 is kept in register T, and the instruction COMPR is used to compare registers X and T in order to decide when to terminate the loop.

Figure 1.6 shows a simple example of input and output on SIC; the same instructions would also work on SIC/XE. (The more advanced input and output facilities available on SIC/XE, such as I/O channels and interrupts, are discussed in Chapter 6.) This program fragment reads one byte of data from device F1 and copies it onto device 05. The actual input of data is performed using the RD (Read Data) instruction. The operand for the RD is a byte in memory that contains the hexadecimal code for the input device (in this case, F1). Executing the RD instruction transfers one byte of data from this device into the rightmost byte of register A. If the input device is character-oriented (for example, a keyboard or a card reader), the value placed in register A is the ASCII code for the character that was read.

Before the RD can be executed, however, the input device must be ready to transmit the data. For example, if the input device is a keyboard, the operator must have typed a character. The program checks for this by using the TD (Test Device) instruction. When the TD is executed, the status of the addressed device is tested and the condition code is set to indicate the result of this test. If the device is ready to

```
INLOOP   TD      INDEV        TEST INPUT DEVICE
         JEQ     INLOOP       LOOP UNTIL DEVICE IS READY
         RD      INDEV        READ ONE BYTE FROM INPUT DEVICE INTO REGISTER A
         STCH    DATA         STORE BYTE THAT WAS READ
         .
         .
         .
OUTLP    TD      OUTDEV       TEST OUTPUT DEVICE
         JEQ     OUTLP        LOOP UNTIL DEVICE IS READY
         LDCH    DATA         LOAD DATA BYTE INTO REGISTER A
         WD      OUTDEV       WRITE ONE BYTE TO OUTPUT DEVICE
         .
         .
         .
INDEV    BYTE    X'F1'        INPUT DEVICE NUMBER
OUTDEV   BYTE    X'05'        OUTPUT DEVICE NUMBER
DATA     RESB    1            ONE-BYTE VARIABLE
```

FIGURE 1.6 Sample input and output operations for SIC.

```
          JSUB    READ          CALL READ SUBROUTINE
          .
          .
          .
.                               SUBROUTINE TO READ 100-BYTE RECORD
READ      LDX     ZERO          INITIALIZE INDEX REGISTER TO 0
RLOOP     TD      INDEV         TEST INPUT DEVICE
          JEQ     RLOOP         LOOP IF DEVICE IS BUSY
          RD      INDEV         READ ONE BYTE OF DATA INTO REGISTER A
          STCH    RECORD,X      STORE DATA BYTE INTO RECORD
          TIX     K100          ADD 1 TO INDEX REGISTER AND COMPARE NEW VALUE TO 100
          JLT     RLOOP         LOOP IF INDEX VALUE IS LESS THAN 100
          RSUB                  EXIT FROM SUBROUTINE
          .
          .
          .
INDEV     BYTE    X'F1'         INPUT DEVICE NUMBER
RECORD    RESB    100           100-BYTE BUFFER FOR INPUT RECORD
.                               ONE-WORD CONSTANTS
ZERO      WORD    0
K100      WORD    100
```

(a)

```
          JSUB    READ          CALL READ SUBROUTINE
          .
          .
          .
.                               SUBROUTINE TO READ 100-BYTE RECORD
READ      LDX     #0            INITIALIZE INDEX REGISTER TO 0
          LDT     #100          INITIALIZE REGISTER T TO 100
RLOOP     TD      INDEV         TEST INPUT DEVICE
          JEQ     RLOOP         LOOP IF DEVICE IS BUSY
          RD      INDEV         READ ONE BYTE OF DATA INTO REGISTER A
          STCH    RECORD,X      STORE DATA BYTE INTO RECORD
          TIXR    T             ADD 1 TO INDEX REGISTER AND COMPARE NEW VALUE TO 100
          JLT     RLOOP         LOOP IF INDEX VALUE IS LESS THAN 100
          RSUB                  EXIT FROM SUBROUTINE
          .
          .
          .
INDEV     BYTE    X'F1'         INPUT DEVICE NUMBER
RECORD    RESB    100           100-BYTE BUFFER FOR INPUT RECORD
```

(b)

FIGURE 1.7 Sample subroutine call and record input operations for (a) SIC and (b) SIC/XE.

transmit data, the condition code is set to "less than;" if the device is not ready, the condition code is set to "equal." As Fig. 1.6 illustrates, the program must execute the TD instruction and then check the condition code by using a conditional jump. If the condition code is "equal" (device not ready), the program jumps back to the TD instruction. This two-instruction loop will continue until the device becomes ready; then the RD will be executed.

Output is performed in the same way. First the program uses TD to check whether the output device is ready to receive a byte of data. Then the byte to be written is loaded into the rightmost byte of register A, and the WD (Write Data) instruction is used to transmit it to the device.

Figure 1.7 shows how these instructions can be used to read a 100-byte record from an input device into memory. The read operation in this example is placed in a subroutine. This subroutine is called from the main program by using the JSUB (Jump to Subroutine) instruction. At the end of the subroutine there is an RSUB (Return from Subroutine) instruction, which returns control to the instruction that follows the JSUB.

The READ subroutine itself consists of a loop. Each execution of this loop reads one byte of data from the input device, using the same techniques illustrated in Fig. 1.6. The bytes of data that are read are stored in a 100-byte buffer area labeled RECORD. The indexing and looping techniques that are used in storing characters in this buffer are essentially the same as those illustrated in Fig. 1.4(a).

Figure 1.7(b) shows the same READ subroutine as it might be written for SIC/XE. The main differences from Fig. 1.7(a) are the use of immediate addressing and the TIXR instruction, as was illustrated in Fig. 1.4(a).

1.4 SYSTEM/370 MACHINE STRUCTURE

This section gives a brief introduction to the IBM System/370. System/370 is an architecture, rather than a specific computer. This architecture is implemented on a variety of different machines to produce the models of the 370 family. Although System/370 models differ in hardware and other physical characteristics, they are logically compatible with each other. Any program should give the same results when run on any computer in the 370 family, provided that its operation does not rely on timings or other machine-dependent properties. The 370 architecture is upward compatible with machines of the System/360® family; 360 programs should run correctly on a 370 under conditions similar to those mentioned above. Very similar architectures have also been implemented on a number of other processors by International Business Machines (IBM®) and other manufacturers.

The discussion in this section is intended to give a background that will help in understanding the examples of 370 software that we discuss later. More information can be found in IBM (1983) and Struble (1983).

Memory

Memory consists of 8-bit bytes, with each byte having a unique address. A group of two consecutive bytes beginning at an address that is a multiple of 2 is called a *halfword*. Similarly, a group of 4 bytes whose beginning address is a multiple of 4 is called a *fullword* and a group of 8 consecutive bytes whose beginning address is a multiple of 8 is called a *doubleword*. Thus a halfword contains 16 bits, a fullword contains 32 bits, and a doubleword contains 64 bits. Figure 1.8 illustrates this division of memory into bytes, halfwords, fullwords, and doublewords.

Machine instructions must be aligned on halfword boundaries (that is, addresses that are multiples of 2). In most cases, instruction operands can be located at any address in memory. For example, the 16-bit operand for an AH (Add Halfword) instruction can begin at any location—it does not have to be aligned on a halfword boundary. However, the speed of instruction execution is significantly improved when such operands are aligned on boundaries corresponding to their lengths.

The maximum main memory that is normally available on System/370 is 16 megabytes (2^{24} bytes).

Registers

System/370 has 16 general-purpose registers, numbered 0 through 15 and often designated by R0 through R15. Each register contains 32 bits. For some instructions, two adjacent registers are logically combined to form a 64-bit operand. All of the general registers can be used as accumulators in arithmetic and logical operations. All except R0 can also be used as base registers or index registers. There are four other registers that are used for floating-point operations. Each floating-point register contains 64 bits. Some instructions use two adjacent registers to store a 128-bit floating-point value.

All of the registers mentioned above are available to application programs. In addition, there are 16 *control registers,* which are in-

FIGURE 1.8 Division of IBM System/370 memory into bytes, halfwords, fullwords, and doublewords.

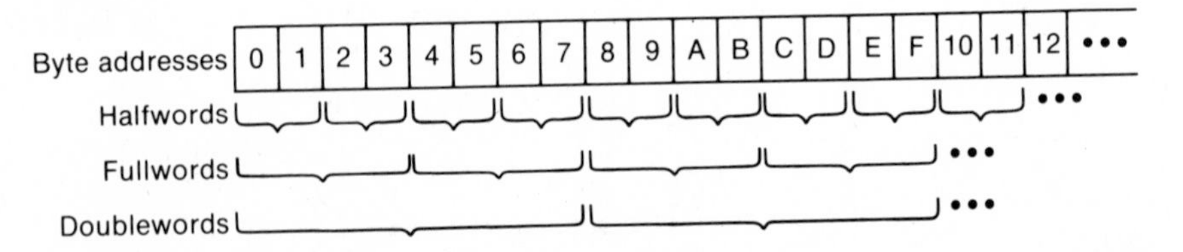

tended for use by the operating system. There is also a *program status word* (PSW), which contains a variety of system status information, including the program counter, condition code, and other control fields.

Data Formats

System/370 provides for the storage of binary and decimal integers, floating-point values, and characters. Characters are stored using their 8-bit EBCDIC codes. Binary integers are stored as 16-bit (halfword) or 32-bit (fullword) binary values. Depending upon the instruction, these values may be treated as either signed or unsigned integers. For signed integers, negative values are represented using 2's complement notation.

Decimal integers are represented using either the *zoned decimal* or the *packed decimal* format. In either case, the representation is of variable length. The number of bytes used is selected by the programmer, depending upon the largest value to be stored. In the zoned decimal format, the rightmost four bits of each byte contain the binary representation of a decimal digit (0 through 9). The leftmost four bits of all bytes except the last are normally set to 1111 (hexadecimal F); this makes the zoned decimal code for a decimal digit the same as its EBCDIC character representation. The leftmost four bits of the last byte may be interpreted as a sign—normally, these bits are hexadecimal C (positive sign), D (negative sign), or F (unsigned, assumed positive). For example, the decimal integer +53842 is represented in the zoned decimal format as hexadecimal F5F3F8F4C2 (using 5 bytes); −6071 is represented as F6F0F7D1 (4 bytes).

In the packed decimal format, each byte is divided into two 4-bit fields. In all except the last byte, each of these fields contains the binary representation of a decimal digit. In the last byte, the first field contains a decimal digit and the second field contains a sign (using the scheme described above). Thus the decimal integer +53842 is represented in packed decimal as hexadecimal 53842C (3 bytes); –6071 is represented as 06071D (3 bytes).

Floating-point values are represented using one of the following three formats.

Short format (one 32-bit word):

1	7	24
s	e	f

Long format (one 64-bit doubleword):

1	7	56
s	e	f

Extended format (two consecutive doublewords):

1	7	56
s	e1	f1

1	7	56
s	e2	f2

In each case, the value of the number is represented using a fraction *f* and an exponent *e*. The exponent is interpreted as an unsigned binary integer. The fraction is interpreted as an unsigned hexadecimal number, with the "hexadecimal point" immediately to the left of the first bit. In a normalized floating-point number, the leftmost hexadecimal digit is nonzero. For extended format floating-point numbers, *e*1 and *e*2 are concatenated to form the exponent *e*; *f*1 and *f*2 are concatenated to form the fraction *f*. The absolute value of the number being represented is given by

$$f * 16^{(e-64)}.$$

The sign of the number is given by the value of *s* (0 = positive, 1 = negative). A value of zero is represented by setting all bits (including sign, exponent, and fraction) to 0.

Notice that the exponent of a System/370 floating-point number is interpreted as a power of 16 (not a power of 2 as on most machines). This was done in order to increase the range of values that could be represented using a single 32-bit word. Because of this decision, it was necessary to interpret the fraction as a hexadecimal number (rather than a binary one). This means that as many as three of the leftmost bits of the fraction may be zero (if the leading hexadecimal digit is 1). As a result, the number of significant bits in the fraction varies according to the value of the number.

Instruction Formats

There are eight basic instruction formats on System/370. In this section, we briefly describe three of the most common formats. Details on all eight formats may be found in IBM (1983).

Most 370 machine instructions have two operands. For such instructions, there are three possibilities for operand location: both operands in registers, one in a register and one in memory, and both in memory. The following are typical instruction formats for each of these cases.

RR format:

op (8)	r1 (4)	r2 (4)

RX format:

op (8)	r1 (4)	x2 (4)	b2 (4)	d2 (12)

SS format:

op (8)	l1 (4)	l2 (4)	b1 (4)	d1 (12)	b2 (4)	d2 (12)

In the RR format, both operands are in either general-purpose or floating-point registers. The register number is encoded in the instruction (as *r*1 or *r*2). The type of register is implied by the operation code for the instruction. For RX format instructions, one operand is in a register *r*1, and the other is in memory. A full 370 memory address requires 24 bits. To save space in the instruction, the address is encoded as a base register *b*2, an (optional) index register *x*2, and a 12-bit displacement *d*2. (See the following discussion on addressing modes for a description of the target address calculation.) In the SS format, both operands are in memory. Since instructions of this type take variable-length operands, the operand lengths are encoded in the instruction (l1 and l2). The operand addresses themselves are specified with base register and displacement. No index register is allowed.

In some of the other instruction formats, the operation code field is extended to 16 bits (RRE and SSE formats). The RX format can be changed to allow a second register operand instead of the index register (RS format), or to include an 8-bit immediate operand in place of the two register numbers (SI format). One variation on the SS format specifies an 8-bit length code that is used for both operands. Another variation has only one operand specification.

Addressing Modes

On System/370, CPU instructions that refer to memory locations must use base relative addressing. The target address is obtained by adding together the contents of the specified base register, the contents of the index register (if any), and the displacement. The displacement is interpreted as an unsigned (positive) 12-bit integer; negative displacements are not allowed. General register R0 cannot be used as either a base register or an index register. If the instruction indicates a base or index register number of zero, this term is omitted in the target address calculation.

Immediate addressing on System/370 takes two forms. The Load Address (LA) instruction loads the target address itself into the specified register (rather than using this address to fetch an operand from memory). Certain other instructions such as Move Immediate (MVI) take a 1-byte immediate operand directly from the instruction. However, these are special cases in which immediate addressing is specified as part of the definition of the instruction; most instructions cannot use immediate operands.

There is no program-counter relative or indirect addressing on System/370. Direct addressing is available only in the very restricted case when both base and index register numbers are specified as zero. This case results in the 12-bit displacement being used as the actual address. In some cases, the operating system takes advantage of this ability to directly address the first 4096 bytes of memory.

Instruction Set

System/370 instructions are divided into five classes: general, decimal, floating-point, control, and I/O. Many of these instructions set the 4-bit condition code to indicate a variety of circumstances; conditional branch instructions can test the condition code setting and branch accordingly. A complete listing of System/370 instructions, and their effects on the condition code setting, may be found in IBM (1983).

General instructions include those used to load and store general-purpose registers, and to perform binary arithmetic operations and comparisons. Each of these functions may be performed by several different instructions, depending upon the types and locations of the operands involved. For example, binary addition is performed by the following machine instructions:

`A` Add fullword, memory to register

`AH` Add halfword, memory to register

`AL` Add fullword, unsigned, memory to register

`ALR` Add fullword, unsigned, register to register

`AR` Add fullword, register to register

The general instructions include conditional and unconditional branches, based on the setting of the condition code. There are also logical operations, instructions that move data from one location in memory to another, and a variety of other operations. System/370 has a supervisor call (SVC) instruction that is similar to the one found on SIC/XE.

Decimal instructions can be used to perform arithmetic operations on integers represented in the packed decimal format. There are also instructions that convert between zoned decimal and packed decimal (PACK and UNPACK), and between binary integer and packed decimal (CVD and CVB). Two other instructions provide the ability to edit data items for printing, performing such operations as inserting commas and decimal points in numeric values.

Floating-point instructions are used to load and store floating-point registers, and to perform floating-point comparisons and arithmetic operations. As with the general instructions, the operation code varies depending upon the types and locations of the operands. For example, the "add" group includes the following:

`AD` Add normalized, long format, memory to register

`ADR` Add normalized, long format, register to register

`AE` Add normalized, short format, memory to register

`AER` Add normalized, short format, register to register

`AU` Add unnormalized, short format, memory to register

`AUR` Add unnormalized, short format, register to register

`AW` Add unnormalized, long format, memory to register

`AWR` Add unnormalized, long format, register to register

`AXR` Add normalized, extended format, register to register

Control instructions include privileged operations that are primarily intended for use by the operating system. Instructions to load the PSW and the control registers are in this group, as are operations that support memory protection and a variety of other functions. We discuss some of the functions provided by such instructions when we study operating systems in Chapter 6.

Input and Output

Input and output on System/370 are performed by I/O channels similar to those we mentioned for SIC/XE. I/O instructions allow the CPU to start, stop, and test the channels, and to perform a variety of control operations on them. There is also a facility that allows the CPU to perform direct byte-by-byte transfers using a special interface, independent of the channels.

1.5 VAX MACHINE STRUCTURE

The VAX® family of computers was introduced by Digital Equipment Corporation (DEC®) in 1978. The letters VAX represent one of the most important features of this architecture—Virtual Address eXtension. Although many other computers (including System/370) have been modified to offer virtual memory, the VAX architecture was designed from the start with a virtual address space in mind. This virtual memory allows programs to operate as though they had access to an extremely large memory, regardless of the amount of memory actually present on the system. Routines in the operating system take care of the details of memory management. We discuss virtual memory in connection with our study of operating systems in Chapter 6.

The VAX architecture was designed for compatibility with the earlier PDP-11® machines. A compatibility mode is provided at the hardware level so that many PDP-11 programs can run unchanged on the VAX. It is even possible for PDP-11 programs and VAX programs to share the same machine in a multiuser environment.

This section summarizes some of the main characteristics of the VAX architecture. For further information, see DEC (1981) and Baase (1983).

Memory

The VAX memory consists of 8-bit bytes. All addresses used are byte addresses. Two consecutive bytes form a *word*; four bytes form a *longword*; eight bytes form a *quadword*; 16 bytes form an *octaword*. As on the 370, it is desirable for words, longwords, quadwords, and octawords to be aligned on appropriate boundaries in memory to improve access speed.

VAX systems may have up to 2^{23} bytes of real memory. However, all VAX programs operate in a virtual address space of 2^{32} bytes. The amount of real memory actually on the system is not ordinarily of concern to an application program. One half of the virtual address space is called *system space,* which contains the operating system, and is shared by all programs. The other half of the address space is called *process space,* and is defined separately for each program. A part of the process space contains stacks that are available to the program. Special registers and machine instructions aid in the use of these stacks.

Registers

There are 16 general-purpose registers on the VAX, denoted by R0 through R15. Some of these registers, however, have special names and uses. All general registers are 32 bits in length. Register R15 is the program counter, also called PC. It is updated during instruction execution to point to the next instruction byte to be fetched. R14 is the *stack pointer* SP, which points to the current top of the stack in the program's process space. Although it is possible to use other registers for this purpose, hardware instructions that implicitly use the stack always use SP. R13 is the *frame pointer* FP. VAX procedure call conventions build a data structure called a stack frame, and place its address in FP. R12 is the *argument pointer* AP. The procedure call convention uses AP to pass a list of arguments associated with the call.

Registers R6 through R11 have no special functions, and are available for general use by the program. Registers R0 through R5 are likewise available for general use; however, these registers are also used by some machine instructions.

In addition to the general registers, there is a *processor status longword* (PSL), which contains state variables and flags associated with a process. The PSL includes, among many other items of information, a condition code and a flag that specifies whether PDP-11 compatibility mode is being used by a process. There are also a number of control registers that are used to support various operating system functions.

Data Formats

Integers are stored as binary numbers in a byte, word, longword, quadword, or octaword; 2's complement representation is used for negative values. Characters are stored using their 8-bit ASCII codes.

There are four different floating-point data formats on the VAX, ranging in length from 4 to 16 bytes. Two of these are compatible with those found on the PDP-11, and are standard on all VAX processors. The other two are available as options, and provide for an extended range of values by allowing more bits in the exponent field. In each case, the principles are the same as those we have discussed previously: a floating point value is represented as a fraction that is to be multiplied by a specified power of 2.

VAX processors provide a packed decimal data format that is similar to that found on the System/370. There is also a *numeric* format that is used to represent numeric values with one digit per byte. In this sense it is similar to the 370 zoned decimal format, except that the digits are represented with ASCII rather than EBCDIC character codes. (This means that the high-order four bits in a byte are normally set to hexadecimal 3, rather than F.) The numeric data type, however, is more complex than the zoned decimal type—the sign may appear either in the last byte (as on the 370), or as a separate byte preceding the first digit. These two variations are called *trailing numeric* and *leading separate numeric.*

VAX also supports queues and variable-length bit strings. Data structures such as these can, of course, be implemented on any machine; however, VAX provides direct hardware support for them. There are single machine instructions that insert and remove entries in queues, and perform a variety of operations on bit strings. The existence of such powerful machine instructions and complex primitive data types is one of the more unusual features of the VAX architecture.

Instruction Formats

VAX machine instructions use a variable-length instruction format. Each instruction consists of an operation code (1 or 2 bytes) followed by up to six *operand specifiers,* depending on the type of instruction. Each operand specifier designates one of the VAX addressing modes and gives any additional information necessary to locate the operand. (See the description of addressing modes in the following section for further information.)

Addressing Modes

VAX provides a large number of addressing modes. With few exceptions, any of these addressing modes may be used with any instruction. The operand itself may be in a register (*register* mode), or its address may be specified by a register (*register deferred* mode). If the operand address is in a register, the register contents may be automatically incremented or decremented by the operand length (*autoincrement* and *autodecrement* modes). There are several base relative addressing modes, with displacement fields of different lengths; when used with register PC, these become program-counter relative modes. All of these addressing modes may also include an index register, and many of them are available in a form that specifies indirect addressing (called *deferred* modes on VAX). In addition, there are immediate operands and several special-purpose addressing modes. For further details, see DEC (1981).

Instruction Set

One of the goals of the VAX designers was to produce an instruction set that is symmetric with respect to data type. Many instruction mnemonics are formed by combining the following elements:

1. a prefix that specifies the type of operation,
2. a suffix that specifies the data type of the operands,
3. a modifier (on some instructions) that gives the number of operands involved.

For example, the instruction ADDW2 is an add operation with two operands, each a word in length. Likewise, MULL3 is a multiply operation with three longword operands, and CVTWL specifies a conversion from word to longword. (In the latter case, a two-operand instruction is assumed.) For a typical instruction, operands may be located in registers, in memory, or in the instruction itself (immediate addressing). The same machine instruction code is used, regardless of operand locations. This approach is more flexible than, for example, the System/370 architecture, which would require different instruction codes for different operand locations.

VAX provides all of the usual types of instructions for computation, data movement and conversion, comparison, branching, etc. In addition, there are a number of operations that are much more complex than the machine instructions found on most computers. These operations are, for the most part, hardware realizations of frequently occurring sequences of code. They are implemented as single instructions for efficiency and speed. For example, VAX provides instructions to load and store multiple registers, and to manipulate queues and variable-length bit fields. There are also powerful instructions for calling and returning from procedures. A single instruction saves a designated set of registers, passes a list of arguments to the procedure, maintains the stack, frame, and argument pointers, and sets a mask to enable error traps for arithmetic operations. For further information on all of the VAX instructions, see DEC (1981).

Input and Output

Input and output on the VAX are accomplished by I/O device controllers. Each controller has a set of control/status and data registers, which are assigned locations in the physical address space. The portion of the address space into which the device controller registers are mapped is called *I/O space.*

No special instructions are required to access registers in I/O space. An I/O device driver issues commands to the device controller by storing values into the appropriate registers, exactly as if they were physical memory locations. Likewise, software routines may read these registers to obtain status information. The association of an address in I/O space with a physical register in a device controller is handled by the memory management routines.

1.6 CYBER MACHINE STRUCTURE

In this section, we describe the architecture of the Control Data Corporation (CDC®) CYBER 70® and CYBER 170® series of computers. Although the different models in this series have varying hardware implementations, they are compatible from the programmer's point of view. The CYBER machine structure is essentially the same as that found on the CDC 6000 series, with a few additional features. A similar architecture has been implemented on the CYBER 180® series of computers, announced by CDC in 1984. These machines have a different word size and many improved hardware features; however, a compatibility mode allows CYBER 70 and 170 programs to run unchanged on a CYBER 180 system.

The CYBER is a multiprocessor system, with a *central processor* (CPU) and a number of *peripheral processors* (PP). The CPU normally executes user programs, while the PPs perform operating system functions. Execution of a program in the CPU can be started or interrupted by a PP to perform control functions. Each PP has access to the CPU's central memory and to the I/O devices. In addition, each PP has its own private memory.

The memory structure, registers, and machine language for the PPs are completely different and separate from those found on the CPU. However, the CPU and the PPs can exchange information by access to common memory locations. In this section, we discuss only the characteristics of the CPU. Further information on CPU and PP structures may be found in CDC (1981a) and Grishman (1974).

Memory

Central processor memory on the CYBER is composed of 60-bit words. All addresses on this machine are word addresses. Except for a small group of character-oriented instructions, it is not possible to directly reference subfields within a word. The maximum memory available is 256K ($=2^{18}$) words.

Registers

Three types of CPU registers are available to the user program: A registers, B registers, and X registers. There are eight registers of each type, designated A0 through A7, B0 through B7, and X0 through X7. The A and B registers are 18 bits in length. Generally, A registers are used for addresses, and B registers are used as index registers or to hold small integer values such as loop counters. Register B0 is permanently set to 0, so that this constant is always available. X registers are 60 bits in length. They hold the operands used for most operations, as well as the contents of words fetched from memory.

There is an unusual connection between the A registers and the X registers. When any value is placed into register A1, this value is interpreted as an address in memory; register X1 is automatically *loaded* from that memory location. The same logical connection exists between each of registers A2 through A5 and the corresponding X registers (X2 through X5). When any value (address) is placed into register A6, the contents of register X6 are automatically *stored* at that address in memory; the same applies to registers A7 and X7. No logical connection exists between registers A0 and X0.

Data Formats

Integers are stored as 60-bit binary numbers (although some integer arithmetic instructions use only the rightmost 48 bits of the word); 1's complement representation is used for negative values. Characters are stored using a 6-bit representation called CDC Display Code. Because of these 6-bit character codes, the contents of a CYBER word are normally represented using octal (rather than hexadecimal) notation.

Floating-point values are represented in the following format:

1	11	48
s	e	c

The coefficient *c* is interpreted as a 48-bit binary integer. The assumed binary point is immediately following the rightmost bit. For a normalized floating-point number, the leftmost bit of the coefficient is 1. The exponent *e* is interpreted as an unsigned 11-bit binary integer. The absolute value of a floating-point number is given by

$$c * 2^{(e-1024)}.$$

The sign *s* is 0 for a positive number. Negative values are represented by taking the 1's complement of the entire floating-point word, including sign, exponent, and coefficient. A floating-point value of zero is represented by setting all bits of the floating-point word to 0.

Certain values of the exponent *e* are reserved and are not used in any ordinary floating-point value. These reserved exponents are used to represent plus and minus infinity (the result of dividing a nonzero value by zero), and plus and minus indefinite (the result of dividing zero by zero).

Instruction Formats

Most CPU instructions on the CYBER use either a 15-bit or a 30-bit instruction format as follows.

15-bit format:

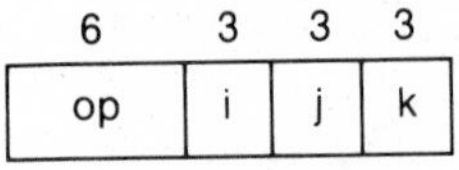

6	3	3	3
op	i	j	k

30-bit format:

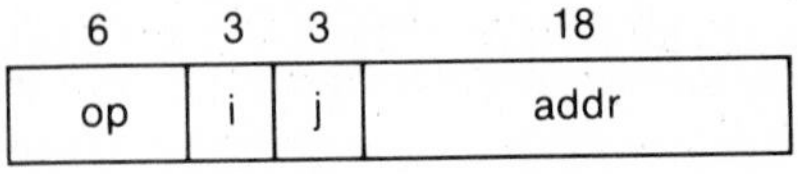

6	3	3	18
op	i	j	addr

In these formats, *op* is the machine operation code. Registers used as operands are designated by the numbers *i*, *j*, and *k*. The type of register is implied by the instruction. If one operand is a location in memory, the 30-bit format must be used; addr contains the full 18-bit memory address.

For some instructions, the operation code field is logically extended to nine bits by including *i*. Other instructions combine the *j* and *k* fields to form a single mask or shift count. There is also a special 60-bit format that is used for a small group of character-oriented instructions. The subfields in this format vary from one instruction to another. Details on all of these formats may be found in CDC (1982a).

Addressing Modes

There is only one addressing mode on the CYBER. To access memory, the full 18-bit address is placed into an A register, causing the corresponding X register to be loaded or stored. Because the actual memory address is used, this process is much like the direct addressing we have discussed. However, the Set instruction that places the address into the A register can also perform a calculation involving up to three operands. This allows the programmer to specify a large variety of target address calculations, including the equivalents of indirect, indexed, and base relative addressing. It is also possible to Set a value directly into an X register, providing an immediate addressing capability.

Instruction Set

The CYBER CPU instructions fall logically into several groups. The largest and most frequently used group contains the Set instructions, which are used to place values into A, B, or X registers. The following are a few of the many different possible forms of Set instructions:

```
SAi    Aj + Bk
SAi    address
SBi    Bj + K
SXi    Xj + Bk
```

Thus, for example, the instruction

```
SA3    A4 + B1
```

sets register A3 to the sum of the contents of registers A4 and B1. This causes register X3 to be loaded from the memory address that was placed in A3. The instruction

```
SA6    BETA
```

(where BETA is a label whose value is a memory address) places the value of BETA into register A6. This causes X6 to be stored at the location indicated by BETA. The target of the Set instruction can be any A, B, or X register. There are many possibilities for the operands.

Set instructions perform all calculations using 18-bit quantities, even when the target is an X register. There are other instruction groups that perform 60-bit integer addition and subtraction, floating-point arithmetic, shifting, and logical operations using X registers. In addition, there are four character-oriented instructions that can move strings from one location in memory to another, and perform character-by-character comparisons. (These are the instructions that use a 60-bit format.)

The CYBER provides an unconditional branch and two types of conditional branch instructions. One type tests the value in an X register, and then branches based on the result. The other type compares the values in two B registers and branches accordingly. There is no condition code on the CYBER. Comparison and branching are performed by a single instruction. There is also an instruction RJ (Return Jump) that is used for subroutine call. The return address is stored in memory (in the first instruction of the called subroutine), rather than being placed in a register.

Input and Output

All input and output on the CYBER are performed by the peripheral processors (PPs). We discuss how this is done, and the more general questions of communication between PP and CPU, in Chapter 6.

EXERCISES

Section 1.3

1. Write a sequence of instructions for SIC to set ALPHA equal to the product of BETA and GAMMA. Assume that ALPHA, BETA, and GAMMA are defined as in Fig. 1.3(a).

2. Write a sequence of instructions for SIC/XE to set ALPHA equal to 4 * BETA − 9. Assume that ALPHA and BETA are defined as in Fig. 1.3(b). Use immediate addressing for the constants.
3. Write a sequence of instructions for SIC to set ALPHA equal to the integer portion of BETA ÷ GAMMA. Assume that ALPHA and BETA are defined as in Fig. 1.3(a).
4. Write a sequence of instructions for SIC/XE to divide BETA by GAMMA, setting ALPHA to the integer portion of the quotient and DELTA to the remainder. Use register-to-register instructions to make the calculation as efficient as possible.
5. Write a sequence of instructions for SIC/XE to divide BETA by GAMMA, setting ALPHA to the value of the quotient, rounded to the nearest integer. Use register-to-register instructions to make the calculation as efficient as possible.
6. Write a sequence of instructions for SIC to clear a 20-byte string to all blanks.
7. Write a sequence of instructions for SIC/XE to clear a 20-byte string to all blanks. Use immediate addressing and register-to-register instructions to make the process as efficient as possible.
8. Suppose that ALPHA is an array of 100 words, as defined in Fig. 1.5(a). Write a sequence of instructions for SIC to set all 100 elements of the array to 0.
9. Suppose that ALPHA is an array of 100 words, as defined in Fig. 1.5(b). Write a sequence of instructions for SIC/XE to set all 100 elements of the array to 0. Use immediate addressing and register-to-register instructions to make the process as efficient as possible.
10. Suppose that RECORD contains a 100-byte record, as in Fig. 1.7(a). Write a subroutine for SIC that will write this record onto device 05.
11. Suppose that RECORD contains a 100-byte record, as in Fig. 1.7(b). Write a subroutine for SIC/XE that will write this record onto device 05. Use immediate addressing and register-to-register instructions to make the subroutine as efficient as possible.
12. Write a subroutine for SIC that will read a record into a buffer, as in Fig. 1.7(a). The record may be any length from 1 to 100 bytes. The end of the record is marked with a "null" character (ASCII code 00). The subroutine should place the length of the record read into a variable named LENGTH.
13. Write a subroutine for SIC/XE that will read a record into a buffer, as in Fig. 1.7(b). The record may be any length from 1 to 100 bytes. The end of the record is marked with a "null" character (ASCII code 00). The subroutine should place the length of the record read into a variable named LENGTH. Use immediate addressing and register-to-register instructions to make the subroutine as efficient as possible.

CHAPTER **2**

ASSEMBLERS

In this chapter we discuss the design and implementation of assemblers. There are certain fundamental functions that any assembler must perform, such as translation of mnemonic operation codes to their machine language equivalents and assignment of machine addresses to symbolic labels used by the programmer. If we consider only these fundamental functions, most assemblers are very much alike.

Beyond this most basic level, however, the features and design of an assembler depend heavily upon the source language it translates and the machine language it produces. One aspect of this dependence is, of course, the existence of different machine instruction formats and codes to accomplish (for example) an ADD operation. As we shall see, there are also many subtler ways that assemblers depend upon machine architecture. On the other hand, there are some features of an assembler language (and the corresponding assembler) that have no direct relation to machine structure—they are, in a sense, arbitrary decisions made by the designers of the language.

We begin by considering the design of a basic assembler for the standard version of our Simplified Instructional Computer. Section 2.1 introduces the most fundamental operations performed by a typical assembler, and describes common ways of accomplishing these functions. The overall logic, data structures, etc., that we describe are shared by almost all assemblers. Thus this level of presentation gives us a starting point from which to approach the study of more advanced assembler features. We can also use this basic structure as a framework from which to begin the design of an assembler for a completely new or unfamiliar machine.

In Section 2.2, we examine some typical extensions to the basic assembler structure that might be dictated by hardware considerations. We do this by discussing an assembler for the SIC/XE machine. Although this SIC/XE assembler certainly does not include all possible hardware-dependent features, it does contain some of the ones most commonly found in real machines. The principles and techniques should be easily applicable to other computers.

Section 2.3 presents a discussion of some of the most commonly encountered machine-independent assembler language features and their implementation. Once again, our purpose is not to cover all possible options, but rather to introduce concepts and techniques that can be used in new and unfamiliar situations.

Section 2.4 examines some important alternative design schemes for an assembler. These are features of an assembler that are not reflected in the assembler language. We discuss one-pass and multi-pass assemblers, and two-pass assemblers with overlay structure. We are concerned with the implementation of such assemblers, and also with the environments in which each might be useful.

Finally, in Section 2.5 we briefly consider some examples of actual assemblers for real machines. We do not attempt to discuss all aspects of these assemblers in detail. Instead, we focus on the most interesting features that are introduced by hardware or software design decisions.

2.1 BASIC ASSEMBLER FUNCTIONS

Figure 2.1 shows an assembler language program for the basic version of SIC. We use variations of this program throughout this chapter to show different assembler features. The line numbers are for reference only and are not part of the program. These numbers also help to relate corresponding parts of different versions of the program. The mnemonic instructions used are those introduced in Section 1.3.1 and Appendix A. Indexed addressing is indicated by adding the modifier ",X"

```
Line        Source statement

  5     COPY      START    1000           COPY FILE FROM INPUT TO OUTPUT
 10     FIRST     STL      RETADR         SAVE RETURN ADDRESS
 15     CLOOP     JSUB     RDREC          READ INPUT RECORD
 20               LDA      LENGTH         TEST FOR EOF (LENGTH = 0)
 25               COMP     ZERO
 30               JEQ      ENDFIL         EXIT IF EOF FOUND
 35               JSUB     WRREC          WRITE OUTPUT RECORD
 40               J        CLOOP          LOOP
 45     ENDFIL    LDA      EOF            INSERT END OF FILE MARKER
 50               STA      BUFFER
 55               LDA      THREE          SET LENGTH = 3
 60               STA      LENGTH
 65               JSUB     WRREC          WRITE EOF
 70               LDL      RETADR         GET RETURN ADDRESS
 75               RSUB                    RETURN TO CALLER
 80     EOF       BYTE     C'EOF'
 85     THREE     WORD     3
 90     ZERO      WORD     0
 95     RETADR    RESW     1
100     LENGTH    RESW     1              LENGTH OF RECORD
105     BUFFER    RESB     4096           4096-BYTE BUFFER AREA
110     .
115     .         SUBROUTINE TO READ RECORD INTO BUFFER
120     .
125     RDREC     LDX      ZERO           CLEAR LOOP COUNTER
130               LDA      ZERO           CLEAR A TO ZERO
135     RLOOP     TD       INPUT          TEST INPUT DEVICE
140               JEQ      RLOOP          LOOP UNTIL READY
145               RD       INPUT          READ CHARACTER INTO A REGISTER
150               COMP     ZERO           TEST FOR END OF RECORD (X'00')
155               JEQ      EXIT           EXIT LOOP IF EOR
160               STCH     BUFFER,X       STORE CHARACTER IN BUFFER
165               TIX      MAXLEN         LOOP UNLESS MAX LENGTH
170               JLT      RLOOP            HAS BEEN REACHED
175     EXIT      STX      LENGTH         SAVE RECORD LENGTH
180               RSUB                    RETURN TO CALLER
185     INPUT     BYTE     X'F1'          CODE FOR INPUT DEVICE
190     MAXLEN    WORD     4096
195     .
200     .         SUBROUTINE TO WRITE RECORD FROM BUFFER
205     .
210     WRREC     LDX      ZERO           CLEAR LOOP COUNTER
215     WLOOP     TD       OUTPUT         TEST OUTPUT DEVICE
220               JEQ      WLOOP          LOOP UNTIL READY
225               LDCH     BUFFER,X       GET CHARACTER FROM BUFFER
230               WD       OUTPUT         WRITE CHARACTER
235               TIX      LENGTH         LOOP UNTIL ALL CHARACTERS
240               JLT      WLOOP            HAVE BEEN WRITTEN
245               RSUB                    RETURN TO CALLER
250     OUTPUT    BYTE     X'05'          CODE FOR OUTPUT DEVICE
255               END      FIRST
```

FIGURE 2.1 Example of a SIC assembler language program.

```
Line  Loc    Source statement                  Object code

  5   1000   COPY     START   1000
 10   1000   FIRST    STL     RETADR            141033
 15   1003   CLOOP    JSUB    RDREC             482039
 20   1006            LDA     LENGTH            001036
 25   1009            COMP    ZERO              281030
 30   100C            JEQ     ENDFIL            301015
 35   100F            JSUB    WRREC             482061
 40   1012            J       CLOOP             3C1003
 45   1015   ENDFIL   LDA     EOF               00102A
 50   1018            STA     BUFFER            0C1039
 55   101B            LDA     THREE             00102D
 60   101E            STA     LENGTH            0C1036
 65   1021            JSUB    WRREC             482061
 70   1024            LDL     RETADR            081033
 75   1027            RSUB                      4C0000
 80   102A   EOF      BYTE    C'EOF'            454F46
 85   102D   THREE    WORD    3                 000003
 90   1030   ZERO     WORD    0                 000000
 95   1033   RETADR   RESW    1
100   1036   LENGTH   RESW    1
105   1039   BUFFER   RESB    4096
110          .
115          .        SUBROUTINE TO READ RECORD INTO BUFFER
120          .
125   2039   RDREC    LDX     ZERO              041030
130   203C            LDA     ZERO              001030
135   203F   RLOOP    TD      INPUT             E0205D
140   2042            JEQ     RLOOP             30203F
145   2045            RD      INPUT             D8205D
150   2048            COMP    ZERO              281030
155   204B            JEQ     EXIT              302057
160   204E            STCH    BUFFER,X          549039
165   2051            TIX     MAXLEN            2C205E
170   2054            JLT     RLOOP             38203F
175   2057   EXIT     STX     LENGTH            101036
180   205A            RSUB                      4C0000
185   205D   INPUT    BYTE    X'F1'             F1
190   205E   MAXLEN   WORD    4096              001000
195          .
200          .        SUBROUTINE TO WRITE RECORD FROM BUFFER
205          .
210   2061   WRREC    LDX     ZERO              041030
215   2064   WLOOP    TD      OUTPUT            E02079
220   2067            JEQ     WLOOP             302064
225   206A            LDCH    BUFFER,X          509039
230   206D            WD      OUTPUT            DC2079
235   2070            TIX     LENGTH            2C1036
240   2073            JLT     WLOOP             382064
245   2076            RSUB                      4C0000
250   2079   OUTPUT   BYTE    X'05'             05
255                   END     FIRST
```

FIGURE 2.2 Program from Fig. 2.1 with object code.

following the operand (see line 160). Lines beginning with "." contain comments only.

In addition to the mnemonic machine instructions, we have used the following *assembler directives:*

`START`	Specify name and starting address for the program.
`END`	Indicate the end of the source program and (optionally) specify the first executable instruction in the program.
`BYTE`	Generate character or hexadecimal constant, occupying as many bytes as needed to represent the constant.
`WORD`	Generate one-word integer constant.
`RESB`	Reserve the indicated number of bytes for a data area.
`RESW`	Reserve the indicated number of words for a data area.

The program contains a main routine that reads records from an input device (identified with device code F1) and copies them to an output device (code 05). This main routine calls subroutine RDREC to read a record into a buffer and subroutine WRREC to write the record from the buffer to the output device. Each subroutine must transfer the record one character at a time because the only I/O instructions available are RD and WD. The buffer is necessary because the I/O rates for the two devices, such as a disk and a slow printing terminal, may be very different. (In Chapter 6, we see how to use channel programs and operating system calls on a SIC/XE system to accomplish the same functions.) The end of each record is marked with a null character (hexadecimal 00). If a record is longer than the length of the buffer (4096 bytes), only the first 4096 bytes are copied. (For simplicity, the program does not deal with error recovery when a record containing 4096 bytes or more is read.) The end of the file to be copied is indicated by a zero-length record. When the end of file is detected, the program writes EOF on the output device and terminates by executing an RSUB instruction. We assume that this program was called by the operating system using a JSUB instruction; thus, the RSUB will return control to the operating system.

2.1.1 A Simple SIC Assembler

Figure 2.2 shows the same program as in Fig. 2.1, with the generated object code for each statement. The column headed Loc gives the machine address (in hexadecimal) for each part of the assembled program. We have assumed that the program starts at address 1000. (In an actual assembler listing, of course, the comments would be retained; they have been eliminated here to save space.)

The translation of source program to object code requires us to accomplish the following functions (not necessarily in the order given):

1. Convert mnemonic operation codes to their machine language equivalents—e.g., translate STL to 14 (line 10).
2. Convert symbolic operands to their equivalent machine addresses—e.g., translate RETADR to 1033 (line 10).
3. Build the machine instructions in the proper format.
4. Convert the data constants specified in the source program into their internal machine representations—e.g., translate EOF to 454F46 (line 80).
5. Write the object program and the assembly listing.

All of these functions except number 2 can easily be accomplished by simple processing of the source program, one line at a time. The translation of addresses, however, presents a problem. Consider the statement

```
10        1000      FIRST     STL       RETADR
```

If we attempt to translate the program line by line, we will be unable to process this statement because we do not know the address that will be assigned to RETADR. Because of this, most assemblers make two *passes* over the source program. The first pass does little more than scan the source program for label definitions and assign addresses (such as those in the Loc column in Fig. 2.2). The second pass performs most of the actual translation previously described.

In addition to translating the instructions of the source program, the assembler must process statements called *assembler directives* (or *pseudo-instructions*). These statements are not translated into machine instructions (although they may have an effect on the object program). Instead, they provide instructions to the assembler itself. Examples of assembler directives are statements like BYTE and WORD, which direct the assembler to generate constants as part of the object program, and RESB and RESW, which instruct the assembler to reserve memory locations without generating data values. The other assembler directives in our sample program are START, which specifies the starting memory address for the object program, and END, which marks the end of the program.

Finally, the assembler must write the generated object code onto some output device. This *object program* will later be loaded into memory for execution. The simple object program format we use contains three types of records: Header, Text, and End. The Header record contains the program name, starting address, and length; Text records contain the translated (i.e., machine code) instructions and data of the program, together with an indication of the addresses where these are to be loaded; the End record marks the end of the object program and specifies the address in the program where execution is to begin. (This is taken from the operand of the program's END statement. If no operand is specified, the address of the first executable instruction is used.)

The formats we use for these records are as follows. The details of the formats (column numbers, etc.) are arbitrary; however, the *information* contained in these records must be present (in some form) in the object program.

Header record:

Col. 1	H
Col. 2–7	Program name
Col. 8–13	Starting address of object program (hexadecimal)
Col. 14–19	Length of object program in bytes (hexadecimal)

Text record:

Col. 1	T
Col. 2–7	Starting address for object code in this record (hexadecimal)
Col. 8–9	Length of object code in this record in bytes (hexadecimal)
Col. 10–69	Object code, represented in hexadecimal (2 columns per byte of object code)

End record:

Col. 1	E
Col. 2–7	Address of first executable instruction in object program (hexadecimal)

To avoid confusion, we have used the term *column* rather than *byte* to refer to positions within object program records. This is not meant to imply the use of punched cards (or any other particular medium) for the object program.

Figure 2.3 shows the object program corresponding to Fig. 2.2, using this format. In this figure, and in the other object programs we display, the symbol ^ is used to separate fields visually. Of course, such

```
H^COPY  ^001000^00107A
T^001000^1E^141033^482039^001036^281030^301015^482061^3C1003^00102A^0C1039^00102D
T^00101E^15^0C1036^482061^081033^4C0000^454F46^000003^000000
T^002039^1E^041030^001030^E0205D^30203F^D8205D^281030^302057^549039^2C205E^38203F
T^002057^1C^101036^4C0000^F1^001000^041030^E02079^302064^509039^DC2079^2C1036
T^002073^07^382064^4C0000^05
E^001000
```

(a) Object program

FIGURE 2.3 Object program corresponding to Fig. 2.2.

symbols are not present in the actual object program. Note that there is no object code corresponding to addresses 1033–2038. This storage is simply reserved by the loader for use by the program during execution. (Chapter 3 contains a detailed discussion of the operation of the loader.)

We can now give a general description of the functions of the two passes of our simple assembler.

Pass 1 (define symbols):

1. Assign addresses to all statements in the program.
2. Save the values (addresses) assigned to all labels for use in Pass 2.
3. Perform some processing of assembler directives. (This includes processing that affects address assignment, such as determining the length of data areas defined by BYTE, RESW, etc.)

Pass 2 (assemble instructions and generate object program):

1. Assemble instructions (translating operation codes and looking up addresses).
2. Generate data values defined by BYTE, WORD, etc.
3. Perform processing of assembler directives not done during Pass 1.
4. Write the object program and the assembly listing.

In the next section we discuss these functions in more detail, describe the internal tables required by the assembler, and give an overall description of the logic flow of each pass.

2.1.2 Assembler Tables and Logic

Our simple assembler uses two major internal tables: the Operation Code Table (OPTAB) and the Symbol Table (SYMTAB). OPTAB is used to look up mnemonic operation codes and translate them to their machine language equivalents. SYMTAB is used to store values (addresses) assigned to labels.

We also need a Location Counter LOCCTR. This is a variable that is used to help in the assignment of addresses. LOCCTR is initialized to the beginning address specified in the START statement. After each source statement is processed, the length of the assembled instruction or data area to be generated is added to LOCCTR. Thus whenever we reach a label in the source program, the current value of LOCCTR gives the address to be associated with that label.

The Operation Code Table must contain (at least) the mnemonic operation code and its machine language equivalent. In more complex assemblers, this table also contains information about instruction format and length. During Pass 1, OPTAB is used to look up and validate operation codes in the source program. In Pass 2, it is used to translate the operation codes to machine language. Actually, in our simple SIC assembler, both of these processes could be done together in either Pass 1 or Pass 2. However, for a machine (such as SIC/XE) that has instructions of different lengths, we must search OPTAB in the first pass to find the instruction length for incrementing LOCCTR. Likewise, we must have the information from OPTAB in Pass 2 to tell us which instruction format to use in assembling the instruction, and any peculiarities of the object code instruction. We have chosen to retain this structure in the current discussion because it is typical of most real assemblers.

OPTAB is usually organized as a hash table, with mnemonic operation code as the key. (The information in OPTAB is, of course, predefined when the assembler itself is written, rather than being loaded into the table at execution time.) The hash table organization is particularly appropriate, since it provides fast retrieval with a minimum of searching. In most cases, OPTAB is a static table—that is, entries are not normally added to or deleted from it. In such cases it is possible to design a special hashing function or other data structure to give optimum performance for the particular set of keys being stored. Most of the time, however, a general-purpose hashing method is used. Further information about the design and construction of hash tables may be found in any good data structures text. For example, see Standish (1980) or Knuth (1973b).

The symbol table (SYMTAB) includes the name and value (address) for each label in the source program, together with flags to indicate error conditions (e.g., a symbol defined in two different places). This table may also contain information about the type, length, etc., of the data area or instruction labelled. During Pass 1 of the assembler, labels are entered into SYMTAB as they are encountered in the source program, along with their assigned addresses (from LOCCTR). During Pass 2, symbols used as operands are looked up in SYMTAB to obtain the addresses to be inserted in the assembled instructions.

SYMTAB is usually organized as a hash table for efficiency of insertion and retrieval. Since entries are rarely (if ever) deleted from this table, efficiency of deletion is not an important consideration. Because SYMTAB is used heavily throughout the assembly, care should be taken in the selection of a hashing function. Programmers often select many labels that have similar characteristics—for example, labels that start or end with the same characters (like LOOP1, LOOP2, LOOPA) or are of the same length (like A, X, Y, Z). It is important that the hashing function used perform well with such nonrandom keys. Division of the entire key by a prime table length often gives good results.

It is possible for both passes of the assembler to read the original source program as input. However, there is certain information (such as location counter values and error flags for statements) that can or should be communicated between the two passes. For this reason, Pass 1 usually writes an intermediate file containing each source statement together with its assigned address, error indicators, etc. This file is used as the input to Pass 2. This working copy of the source program can also be used to retain the results of certain operations that may be performed during Pass 1 (such as scanning the operand field for symbols and addressing flags), so that these need not be performed again during Pass 2. Similarly, pointers into OPTAB and SYMTAB may be retained for each operation code and symbol used. This avoids the need to repeat many of the table searching operations.

Figures 2.4(a) and (b) show the logic flow of the two passes of our assembler. Although described for the simple assembler we are discussing, this is also the underlying logic for more complex two-pass assemblers that we consider later. We assume for simplicity that the source lines are written in a fixed format with fields LABEL, OPCODE, and OPERAND. If one of these fields contains a character string that represents a number, we denote its numeric value with the prefix # (for example, #[OPERAND]).

At this stage, it is very important for you to understand thoroughly the algorithms in Fig. 2.4. You are strongly urged to follow through the logic in these algorithms, applying them by hand to the program in Fig. 2.1 to produce the object program of Fig. 2.3.

Much of the detail of the assembler logic has, of course, been left out to emphasize the overall structure and main concepts. You should

Pass 1:

```
begin
read first input line
if OPCODE = 'START' then
    begin
        save #[OPERAND] as starting address
        initialize LOCCTR to starting address
        write line to intermediate file
        read next input line
    end
else
    initialize LOCCTR to 0
while OPCODE < > 'END' do
    begin
        if this is not a comment line then
            begin
                if there is a symbol in the LABEL field then
                    begin
                        search SYMTAB for LABEL
                        if found then
                            set error flag (duplicate symbol)
                        else
                            insert (LABEL,LOCCTR) into SYMTAB
                    end
                search OPTAB for OPCODE
                if found then
                    add 3 {instruction length} to LOCCTR
                else if OPCODE = 'WORD' then
                    add 3 to LOCCTR
                else if OPCODE = 'RESW' then
                    add 3 * #[OPERAND] to LOCCTR
                else if OPCODE = 'RESB' then
                    add #[OPERAND] to LOCCTR
                else if OPCODE = 'BYTE' then
                    begin
                        find length of constant in bytes
                        add length to LOCCTR
                    end
                else
                    set error flag (invalid operation code)
            end {if not a comment}
        write line to intermediate file
        read next input line
    end {while}
write last line to intermediate file
save (LOCCTR - starting address) as program length
end {Pass 1}
```

FIGURE 2.4(a) Algorithm for Pass 1 of assembler.

Pass 2:

```
begin
read first input line {from intermediate file}
if OPCODE = 'START' then
    begin
        write listing line
        read next input line
    end
write Header record to object program
initialize first Text record
while OPCODE <> 'END' do
    begin
        if this is not a comment line then
            begin
                search OPTAB for OPCODE
                if found then
                    begin
                        if there is a symbol in OPERAND field then
                            begin
                                search SYMTAB for OPERAND
                                if found then
                                    store symbol value as operand address
                                else
                                    begin
                                        store 0 as operand address
                                        set error flag (undefined symbol)
                                    end
                            end {if symbol}
                        else
                            store 0 as operand address
                        assemble the object code instruction
                    end {if found}
                else if OPCODE = 'BYTE' or 'WORD' then
                    convert constant to object code
                if object code will not fit into the current Text record then
                    begin
                        write Text record to object program
                        initialize new Text record
                    end
                add object code to Text record
            end {if not comment}
        write listing line
        read next input line
    end {while}
write last Text record to object program
write End record to object program
write last listing line
end {Pass 2}
```

FIGURE 2.4(b) Algorithm for Pass 2 of assembler.

think about these details for yourself, and you should also attempt to identify those functions of the assembler that should be implemented as separate procedures or modules. (For example, the operations "search symbol table" and "read input line" might be good candidates for such implementation.) This kind of thoughtful analysis should be done *before* you make any attempt to actually implement an assembler or any other large piece of software.

Chapter 8 contains an introduction to software engineering tools and techniques, and illustrates the use of such techniques in designing and implementing a simple assembler. You may want to read this material now to gain further insight into how an assembler might be constructed.

2.2 MACHINE-DEPENDENT ASSEMBLER FEATURES

In this section, we consider the design and implementation of an assembler for the more complex XE version of SIC. In doing so, we examine the effect of the extended hardware on the structure and functions of the assembler. Many real machines have certain architectural features that are similar to those we consider here. Thus our discussion applies in large part to these machines as well as to SIC/XE.

Figure 2.5 shows the example program from Fig. 2.1 as it might be rewritten to take advantage of the SIC/XE instruction set. In our assembler language, indirect addressing is indicated by adding the prefix @ to the operand (see line 70). Immediate operands are denoted with the prefix # (lines 25, 55, 133). Instructions that refer to memory are normally assembled using either the program-counter relative or the base relative mode. The assembler directive BASE (line 13) is used in conjunction with base relative addressing. (See Section 2.2.1 for a discussion and examples.) If the displacements required for both program-counter relative and base relative addressing are too large to fit into a 3-byte instruction, then the 4-byte extended format (Format 4) must be used. The extended instruction format is specified with the prefix + added to the operation code in the source statement (see lines 15, 35, 65). It is the programmer's responsibility to specify this form of addressing when it is required.

The main differences between this version of the program and the version in Fig. 2.1 involve the use of register–register instructions (in place of register–memory instructions) wherever possible. For example, the statement on line 150 is changed from "COMP ZERO" to "COMPR A,S." Similarly, line 165 is changed from "TIX MAXLEN" to "TIXR T." In addition, immediate and indirect addressing have been used as much as possible (for example, lines 25, 55, and 70).

```
Line        Source statement

  5     COPY     START    0              COPY FILE FROM INPUT TO OUTPUT
 10     FIRST    STL      RETADR         SAVE RETURN ADDRESS
 12              LDB      #LENGTH        ESTABLISH BASE REGISTER
 13              BASE     LENGTH
 15     CLOOP   +JSUB     RDREC          READ INPUT RECORD
 20              LDA      LENGTH         TEST FOR EOF (LENGTH = 0)
 25              COMP     #0
 30              JEQ      ENDFIL         EXIT IF EOF FOUND
 35             +JSUB     WRREC          WRITE OUTPUT RECORD
 40              J        CLOOP          LOOP
 45     ENDFIL   LDA      EOF            INSERT END OF FILE MARKER
 50              STA      BUFFER
 55              LDA      #3             SET LENGTH = 3
 60              STA      LENGTH
 65             +JSUB     WRREC          WRITE EOF
 70              J        @RETADR        RETURN TO CALLER
 80     EOF      BYTE     C'EOF'
 95     RETADR   RESW     1
100     LENGTH   RESW     1              LENGTH OF RECORD
105     BUFFER   RESB     4096           4096-BYTE BUFFER AREA
110     .
115     .        SUBROUTINE TO READ RECORD INTO BUFFER
120     .
125     RDREC    CLEAR    X              CLEAR LOOP COUNTER
130              CLEAR    A              CLEAR A TO ZERO
132              CLEAR    S              CLEAR S TO ZERO
133             +LDT      #4096
135     RLOOP    TD       INPUT          TEST INPUT DEVICE
140              JEQ      RLOOP          LOOP UNTIL READY
145              RD       INPUT          READ CHARACTER INTO A REGISTER
150              COMPR    A,S            TEST FOR END OF RECORD (X'00')
155              JEQ      EXIT           EXIT LOOP IF EOR
160              STCH     BUFFER,X       STORE CHARACTER IN BUFFER
165              TIXR     T              LOOP UNLESS MAX LENGTH
170              JLT      RLOOP            HAS BEEN REACHED
175     EXIT     STX      LENGTH         SAVE RECORD LENGTH
180              RSUB                    RETURN TO CALLER
185     INPUT    BYTE     X'F1'          CODE FOR INPUT DEVICE
195     .
200     .        SUBROUTINE TO WRITE RECORD FROM BUFFER
205     .
210     WRREC    CLEAR    X              CLEAR LOOP COUNTER
212              LDT      LENGTH
215     WLOOP    TD       OUTPUT         TEST OUTPUT DEVICE
220              JEQ      WLOOP          LOOP UNTIL READY
225              LDCH     BUFFER,X       GET CHARACTER FROM BUFFER
230              WD       OUTPUT         WRITE CHARACTER
235              TIXR     T              LOOP UNTIL ALL CHARACTERS
240              JLT      WLOOP            HAVE BEEN WRITTEN
245              RSUB                    RETURN TO CALLER
250     OUTPUT   BYTE     X'05'          CODE FOR OUTPUT DEVICE
255              END      FIRST
```

FIGURE 2.5 Example of a SIC/XE program.

These changes take advantage of the more advanced SIC/XE architecture to improve the execution speed of the program. Register–register instructions are faster than the corresponding register–memory operations because they are shorter, and, more importantly, because they do not require another memory reference. (Fetching an operand from a register is much faster than retrieving it from main memory.) Likewise, when using immediate addressing, the operand is already present as part of the instruction and need not be fetched from anywhere. The use of indirect addressing often avoids the need for another instruction (as in the "return" operation on line 70). You may notice that some of the changes require the addition of other instructions to the program. For example, changing COMP to COMPR on line 150 forces us to add the CLEAR instruction on line 132. This still results in an improvement in execution speed. The CLEAR is executed only once for each record read, whereas the benefits of COMPR (as opposed to COMP) are realized for every byte of data transferred.

In Section 2.2.1, we examine the assembly of this SIC/XE program, focusing on the differences in the assembler that are required by the new addressing modes. (You may want to briefly review the instruction formats and target address calculations described in Section 1.3.2.) These changes are direct consequences of the extended hardware functions.

Section 2.2.2 discusses an indirect consequence of the change to SIC/XE. The larger main memory of SIC/XE means that we will probably have room to load and run several programs at the same time. This kind of sharing of the machine between programs is called *multiprogramming*. Such sharing often results in more productive use of the hardware. (We discuss this concept, and its implications for operating systems, in Chapter 6.) To take full advantage of this capability, however, we must be able to load programs into memory wherever there is room, rather than specifying a fixed address at assembly time. Section 2.2.2 introduces the idea of program *relocation* and discusses its implications for the assembler.

2.2.1 Instruction Formats and Addressing Modes

Figure 2.6 shows the object code generated for each statement in the program of Fig. 2.5. In this section we consider the translation of the source statements, paying particular attention to the handling of different instruction formats and different addressing modes. Note that the START statement now specifies a beginning program address of 0. As we discuss in the next section, this indicates a relocatable program. For the purposes of instruction assembly, however, the program will be translated exactly as if it were really to be loaded at machine address 0.

```
Line    Loc     Source statement                    Object code

  5     0000    COPY     START   0
 10     0000    FIRST    STL     RETADR             17202D
 12     0003             LDB     #LENGTH            69202D
 13                      BASE    LENGTH
 15     0006    CLOOP   +JSUB    RDREC              4B101036
 20     000A             LDA     LENGTH             032026
 25     000D             COMP    #0                 290000
 30     0010             JEQ     ENDFIL             332007
 35     0013            +JSUB    WRREC              4B10105D
 40     0017             J       CLOOP              3F2FEC
 45     001A    ENDFIL   LDA     EOF                032010
 50     001D             STA     BUFFER             0F2016
 55     0020             LDA     #3                 010003
 60     0023             STA     LENGTH             0F200D
 65     0026            +JSUB    WRREC              4B10105D
 70     002A             J       @RETADR            3E2003
 80     002D    EOF      BYTE    C'EOF'             454F46
 95     0030    RETADR   RESW    1
100     0033    LENGTH   RESW    1
105     0036    BUFFER   RESB    4096
110             .
115             .        SUBROUTINE TO READ RECORD INTO BUFFER
120             .
125     1036    RDREC    CLEAR   X                  B410
130     1038             CLEAR   A                  B400
132     103A             CLEAR   S                  B440
133     103C            +LDT     #4096              75101000
135     1040    RLOOP    TD      INPUT              E32019
140     1043             JEQ     RLOOP              332FFA
145     1046             RD      INPUT              DB2013
150     1049             COMPR   A,S                A004
155     104B             JEQ     EXIT               332008
160     104E             STCH    BUFFER,X           57C003
165     1051             TIXR    T                  B850
170     1053             JLT     RLOOP              3B2FEA
175     1056    EXIT     STX     LENGTH             134000
180     1059             RSUB                       4F0000
185     105C    INPUT    BYTE    X'F1'              F1
195             .
200             .        SUBROUTINE TO WRITE RECORD FROM BUFFER
205             .
210     105D    WRREC    CLEAR   X                  B410
212     105F             LDT     LENGTH             774000
215     1062    WLOOP    TD      OUTPUT             E32011
220     1065             JEQ     WLOOP              332FFA
225     1068             LDCH    BUFFER,X           53C003
230     106B             WD      OUTPUT             DF2008
235     106E             TIXR    T                  B850
240     1070             JLT     WLOOP              3B2FEF
245     1073             RSUB                       4F0000
250     1076    OUTPUT   BYTE    X'05'              05
255                      END     FIRST
```

FIGURE 2.6 Program from Fig. 2.5 with object code.

Translation of register–register instructions such as CLEAR (line 125) and COMPR (line 150) presents no new problems. The assembler must simply convert the mnemonic operation code to machine language (using OPTAB) and change each register mnemonic to its numeric equivalent. This translation is done during Pass 2, at the same point at which the other types of instructions are assembled. The conversion of register mnemonics to numbers can be done with a separate table; however, it is often convenient to use the symbol table for this purpose. To do this, SYMTAB would be preloaded with the register names (A, X, etc.) and their values (0, 1, etc.).

Most of the register–memory instructions are assembled using either program-counter relative or base relative addressing. The assembler must, in either case, calculate a displacement to be assembled as part of the object instruction. This is computed so that the correct target address results when the displacement is added to the contents of the program counter (PC) or the base register (B). Of course, the resulting displacement must be small enough to fit in the 12-bit field in the instruction. This means that the displacement must be between 0 and 4095 (for base relative mode) or between −2048 and +2047 (for program-counter relative mode).

If neither program-counter relative nor base relative addressing can be used (because the displacements are too large), then the 4-byte extended instruction format (Format 4) must be used. This 4-byte format contains a 20-bit address field, which is large enough to contain the full memory address. In this case, there is no displacement to be calculated. For example, in the instruction

```
15      0006      CLOOP    +JSUB     RDREC              4B101036
```

the operand address is 1036. This full address is stored in the instruction, with bit *e* set to 1 to indicate extended instruction format.

Note that the programmer must specify the extended format by using the prefix + (as on line 15). If neither form of relative addressing is applicable and extended format is not specified, then the instruction cannot be properly assembled. In this case, the assembler must generate an error message.

We now examine the details of the displacement calculation for program-counter relative and base relative addressing modes. The computation that the assembler needs to perform is essentially the target address calculation in reverse. You may want to review this from Section 1.3.2.

The instruction

```
10      0000      FIRST     STL      RETADR              17202D
```

is a typical example of program-counter relative assembly. During execution of instructions in SIC (as in most computers), the program counter is advanced *after* each instruction is fetched and *before* it is executed. Thus during the execution of the STL instruction, PC will contain the address of the *next* instruction (that is, 0003). From the Loc column of the listing, we see that RETADR (line 95) is assigned the address 0030. (The assembler would, of course, get this address from SYMTAB.) The displacement we need in the instruction is 30 − 3 = 2D. At execution time, the target address calculation performed will be (PC) + disp, resulting in the correct address (0030). Note that bit *p* is set to 1 to indicate program-counter relative addressing, making the last two bytes of the instruction 202D. Also note that bits *n* and *i* are both set to 1, indicating neither indirect nor immediate addressing; this makes the first byte 17 instead of 14. (See Fig. 1.1 in Section 1.3.2 for a review of the location and setting of the addressing-mode bit flags.)

Another example of program-counter relative assembly is the instruction

```
40      0017          J       CLOOP           3F2FEC
```

Here the operand address is 0006. During instruction execution, the program counter will contain the address 0001A. Thus the displacement required is 6 − 1A = −14. This is represented (using 2's complement for negative numbers) in a 12-bit field as FEC, which is the displacement assembled into the object code.

The displacement calculation process for base relative addressing is much the same as for program-counter relative mode. The main difference is that the assembler knows what the contents of the program counter will be at execution time. The base register, on the other hand, is under control of the programmer. Therefore, the programmer must tell the assembler what the base register will contain during execution of the program so that the assembler can compute displacements. This is done in our example with the assembler directive BASE. The statement "BASE LENGTH" (line 13) informs the assembler that the base register will contain the *address* of LENGTH. The preceding instruction "LDB #LENGTH" loads this value into the register during program execution. The assembler assumes for addressing purposes that register B contains this address until it encounters another BASE statement. Later in the program, it may be desirable to use register B for another purpose (for example, as temporary storage for a data value). In such a case, the programmer must use another assembler directive (perhaps NOBASE) to inform the assembler that the contents of the base register can no longer be relied upon for addressing.

It is important to understand that BASE and NOBASE are assembler directives, and produce no executable code. The programmer must provide instructions that load the proper value into the base register during execution. If this is not done properly, the target address calculation will not produce the correct operand address.

The instruction

```
160    104E        STCH    BUFFER,X     57C003
```

is a typical example of base relative assembly. According to the BASE statement, register B will contain 0033 (the address of LENGTH) during execution. The address of BUFFER is 0036. Thus the displacement in the instruction must be $36 - 33 = 3$. Notice that bits x and b are set to 1 in the assembled instruction to indicate indexed and base relative addressing. Another example is the instruction "STX LENGTH" on line 175. Here the displacement calculated is 0.

Notice the difference between the assembly of the instructions on lines 20 and 175. On line 20, "LDA LENGTH" is assembled with program-counter relative addressing. On line 175, "STX LENGTH" uses base relative addressing, as noted previously. (If you calculate the program-counter relative displacement that would be required for the statement on line 175, you will see that it is too large to fit into the 12-bit displacement field.) The statement on line 20 could also have used base relative mode. In our assembler, however, we have arbitrarily chosen to attempt program-counter relative assembly first.

The assembly of an instruction that specifies immediate addressing is simpler because no memory reference is involved. All that is necessary is to convert the immediate operand to its internal representation and insert it into the instruction. The instruction

```
55     0020        LDA     #3           010003
```

is a typical example of this, with the operand stored in the instruction as 003, and bit i set to 1 to indicate immediate addressing. Another example can be found in the instruction

```
133    103C        +LDT    #4096        75101000
```

In this case the operand (4096) is too large to fit into the 12-bit displacement field, so the extended instruction format is called for. (If the operand were too large even for this 20-bit address field, immediate addressing could not be used.)

A different way of using immediate addressing is shown in the instruction

```
12     0003        LDB     #LENGTH      69202D
```

In this statement the immediate operand is the *symbol* LENGTH.

Since the *value* of this symbol is the *address* assigned to it, this immediate instruction has the effect of loading register B with the address of LENGTH. Note here that we have combined program-counter relative addressing with immediate addressing. Although this may appear unusual, the interpretation is consistent with our previous uses of immediate operands. In general, the target address calculation is performed; then, if immediate mode is specified, the *target address* (not the *contents* stored at that address) becomes the operand. (In the LDA statement on line 55, for example, bits x, b, and p are all 0. Thus the target address is simply the displacement 003.)

The assembly of instructions that specify indirect addressing presents nothing really new. The displacement is computed in the usual way to produce the target address desired. Then bit n is set to indicate that the contents stored at this location represent the *address* of the operand, not the operand itself. Line 70 shows a statement that combines program-counter relative and indirect addressing in this way.

2.2.2 Program Relocation

As we mentioned before, it is often desirable to have more than one program at a time sharing the memory and other resources of the machine. If we knew in advance exactly which programs were to be executed concurrently in this way, we could assign addresses when the programs were assembled so that they would fit together without overlap or wasted space. Most of the time, however, it is not practical to plan program execution this closely. (We usually do not know exactly when jobs will be submitted, exactly how long they will run, etc.) Because of this, it is desirable to be able to load a program into memory wherever there is room for it. In such a situation the actual starting address of the program is not known until load time.

The program we considered in Section 2.1 is an example of an *absolute* program (or absolute assembly). This program must be loaded at address 1000 (the address that was specified at assembly time) in order to execute properly. To see this, consider the instruction

```
55      101B                LDA      THREE               00102D
```

from Fig. 2.2. In the object program (Fig. 2.3), this statement is translated as 00102D, specifying that register A is to be loaded from memory address 102D. Suppose we attempt to load and execute the program at address 2000 instead of address 1000. If we do this, address 102D will not contain the value that we expect—in fact, it will probably be part of some other user's program.

Obviously we need to make some change in the address portion of this instruction so we can load and execute our program at address 2000. On the other hand, there are parts of the program (such as the constant 3 generated from line 85) that should remain the same regardless of where the program is loaded. Looking at the object code alone, it is in general not possible to tell which values represent addresses and which represent constant data items.

Since the assembler does not know the actual location where the program will be loaded, it cannot make the necessary changes in the addresses used by the program. However, the assembler can identify for the loader those parts of the object program that need modification. An object program that contains the information necessary to perform this kind of modification is called a *relocatable* program.

To look at this in more detail, consider the program from Figs. 2.5 and 2.6. In the preceding section, we assembled this program using a starting address of 0000. In Fig. 2.7(a) we loaded this program beginning at address 0000. The JSUB instruction from line 15 is loaded at address 0006. The address field of this instruction contains 01036, which is the address of the instruction labelled RDREC. (These addresses are, of course, the same as those assigned by the assembler.)

Now suppose that we want to load this program beginning at address 5000, as shown in Fig. 2.7(b). The address of the instruction labelled RDREC is then 6036. Thus the JSUB instruction must be modified as shown to contain this new address. Likewise, if we loaded the program beginning at address 7420 (Fig. 2.7c), the JSUB instruction would need to be changed to 4B108456 to correspond to the new address of RDREC.

Note that no matter where the program is loaded, RDREC is always 1036 bytes past the starting address of the program. This means that we can solve the relocation problem in the following way:

1. When the assembler generates the object code for the JSUB instruction we are considering, it will insert the address of RDREC *relative to the start of the program*. (This is the reason we initialized the location counter to 0 for the assembly.)
2. The assembler will also produce a command for the loader, instructing it to *add* the beginning address of the program to the address field in the JSUB instruction at load time.

The command for the loader, of course, must also be a part of the object program. We can accomplish this with a Modification record having the following format.

Modification record:

Col. 1	M
Col. 2–7	Starting location of the address field to be modi-

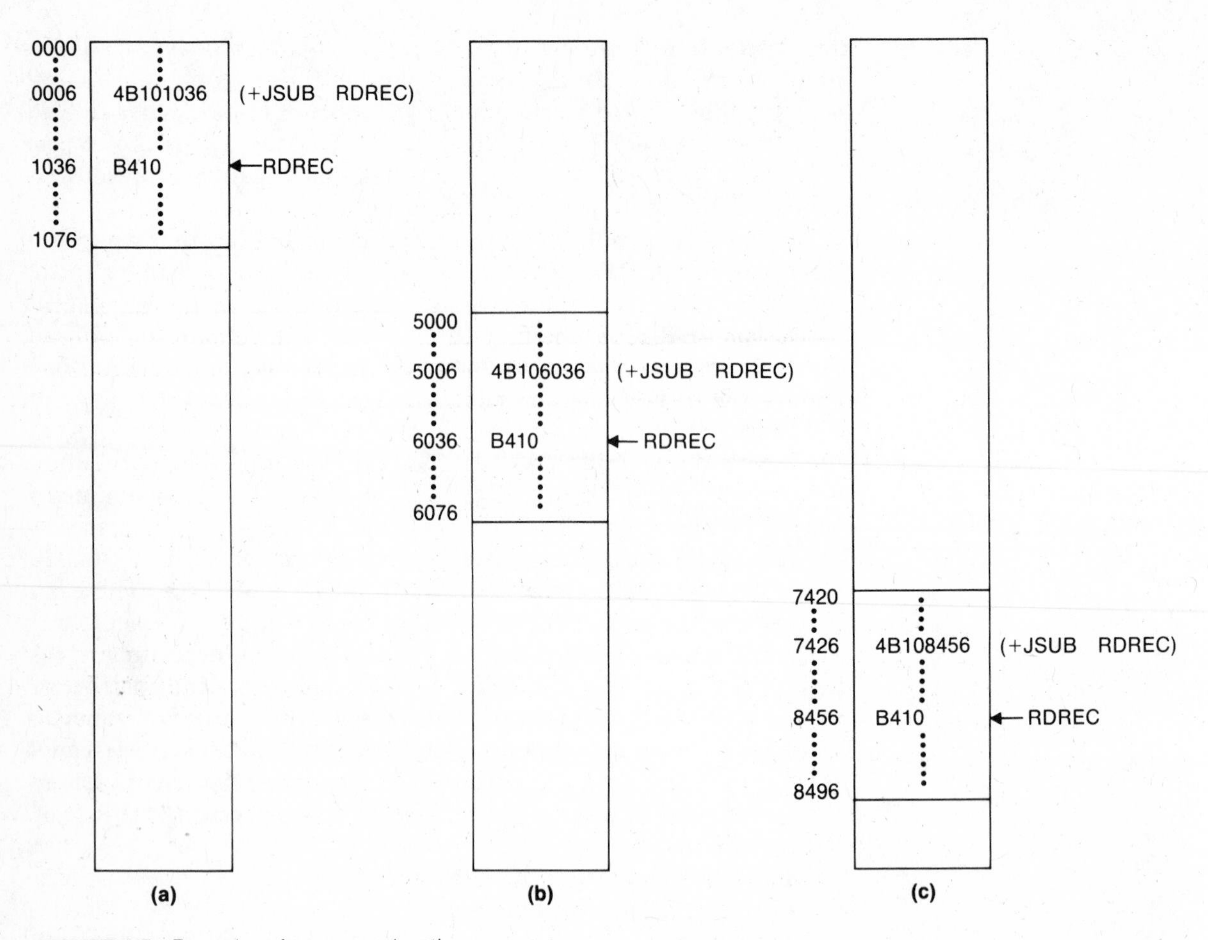

FIGURE 2.7 Examples of program relocation.

fied, relative to the beginning of the program (hexadecimal)

Col. 8–9 Length of the address field to be modified, in *half-bytes* (hexadecimal)

The length is stored in half-bytes (rather than bytes) because the address field to be modified may not occupy an integral number of bytes. (For example, the address field in the JSUB instruction we considered above occupies 20 bits, which is 5 half-bytes.) The starting location is the location of the byte containing the leftmost bits of the address field to be modified. If this field occupies an odd number of half-bytes, it is assumed to begin in the middle of the first byte at the starting location.

These conventions are, of course, closely related to the architecture of SIC/XE. For other types of machines, the half-byte approach might not be appropriate (see Exercise 2.2.8).

For the JSUB instruction we are using as an example, the Modification record would be

```
M00000705
```

This record specifies that the beginning address of the program is to be added to a field that begins at address 000007 (relative to the start of the program) and is 5 half-bytes in length. Thus in the assembled instruction 4B101036, the first 12 bits (4B1) will remain unchanged. The program load address will be added to the last 20 bits (01036) to produce the correct operand address. (You should check for yourself that this gives the results shown in Fig. 2.7.)

Exactly the same kind of relocation must be performed for the instructions on lines 35 and 65 in Fig. 2.6. The rest of the instructions in the program, however, need not be modified when the program is loaded. In some cases this is because the instruction operand is not a memory address at all (e.g., "CLEAR S" or "LDA #3"). In other cases no modification is needed because the operand is specified using program-counter relative or base relative addressing. For example, the instruction on line 10 ("STL RETADR") is assembled using program-counter relative addressing with displacement 02D. No matter where the program is loaded in memory, the word labelled RETADR will always be 2D bytes away from the STL instruction; thus no instruction modification is needed. When the STL is executed, the program counter will contain the (actual) address of the next instruction. The target address calculation process will then produce the correct (actual) operand address corresponding to RETADR.

Similarly the distance between LENGTH and BUFFER will always be 3 bytes. Thus the displacement in the base relative instruction on line 160 will be correct without modification. (The contents of the base register will, of course, depend upon where the program is loaded. However, this will be taken care of automatically when the program-counter relative instruction "LDB #LENGTH" is executed.)

By now it should be clear that the only parts of the program that require modification at load time are those that specify direct (as opposed to relative) addresses. For this SIC/XE program, the only such direct addresses are found in extended format (4-byte) instructions. This is an advantage of relative addressing—if we were to attempt to relocate the program from Fig. 2.1, we would find that almost every instruction required modification.

Figure 2.8 shows the complete object program corresponding to the source program of Fig. 2.5. Note that the Text records are exactly the

```
H^COPY  ^000000^001077
T^000000^1D^17202D^69202D^4B101036^032026^290000^332007^4B10105D^3F2FEC^032010
T^00001D^13^0F2016^010003^0F200D^4B10105D^3E2003^454F46
T^001036^1D^B410^B400^B440^75101000^E32019^332FFA^DB2013^A004^332008^57C003^B850
T^001053^1D^3B2FEA^134000^4F0000^F1^B410^774000^E32011^332FFA^53C003^DF2008^B850
T^001070^07^3B2FEF^4F0000^05
M^000007^05
M^000014^05
M^000027^05
E^000000
```

FIGURE 2.8 Object program corresponding to Fig. 2.6.

same as those that would be produced by an absolute assembler (with program starting address of 0). However, the load addresses in the Text records are interpreted as relative, rather than absolute, locations. (The same is, of course, true of the addresses in the Modification and End records.) There is one Modification record for each address field that needs to be changed when the program is relocated (in this case, the three +JSUB instructions). You should verify these Modification records yourself and make sure you understand the contents of each. In Chapter 3 we consider in detail how the loader performs the required program modification. It is important that you understand the *concepts* involved now, however, because we build on these concepts in the next section.

2.3 MACHINE-INDEPENDENT ASSEMBLER FEATURES

In this section, we discuss some common assembler features that are not closely related to machine structure. Of course, more advanced machines tend to have more complex software; therefore the features we consider are more likely to be found on larger and more complex machines. However, the presence or absence of such capabilities is much more closely related to issues such as programmer convenience and software environment than it is to machine architecture.

In Section 2.3.1 we discuss the implementation of literals within an assembler, including the required tables and processing logic. Section 2.3.2 discusses two assembler directives (EQU and ORG) whose main function is the definition of symbols. Section 2.3.3 briefly examines the use of expressions in assembler language statements, and discusses the different types of expressions and their evaluation and use.

In Sections 2.3.4 and 2.3.5 we introduce the important topics of program blocks and control sections. We discuss the reasons for providing such capabilities and illustrate some different uses with examples. We also introduce a set of assembler directives for supporting these features and discuss their implementation.

2.3.1 Literals

It is often convenient for the programmer to be able to write the value of a constant operand as a part of the instruction that uses it. This avoids having to define the constant elsewhere in the program and make up a label for it. Such an operand is called a *literal* because the value is stated "literally" in the instruction. The use of literals is illustrated by the program in Fig. 2.9. The object code generated for the statements of this program is shown in Fig. 2.10. (This program is a modification of the one in Fig. 2.5; other changes are discussed later in Section 2.3.)

In our assembler language notation, a literal is identified with the prefix =, which is followed by a specification of the literal value, using the same notation as in the BYTE statement. Thus the literal in the statement

```
45      001A      ENDFIL    LDA     =C'EOF'          032010
```

specifies a 3-byte operand whose value is the character string EOF. Likewise the statement

```
215     1062      WLOOP     TD      =X'05'           E32011
```

specifies a 1-byte literal with the hexadecimal value 05. The notation used for literals varies from assembler to assembler; however, most assemblers use some symbol (as we have used =) to make literal identification easier.

It is important to understand the difference between a literal and an immediate operand. With immediate addressing, the operand value is assembled as part of the machine instruction. With a literal, the assembler generates the specified value as a constant at some other memory location. The *address* of this generated constant is used as the target address for the machine instruction. The effect of using a literal is exactly the same as if the programmer had defined the constant explicitly and used the label assigned to the constant as the instruction operand. (In fact, the generated object code for lines 45 and 215 in Fig. 2.10 is identical to the object code for the corresponding lines in Fig. 2.6.) You should compare the object instructions generated for lines 45 and 55 in Fig. 2.10 to make sure you understand how literals and immediate operands are handled.

All of the literal operands used in a program are gathered together into one or more *literal pools*. Normally literals are placed into a pool at the end of the program. The assembly listing of a program containing literals usually includes a listing of this literal pool, which shows

```
Line        Source statement

  5    COPY    START   0               COPY FILE FROM INPUT TO OUTPUT
 10    FIRST   STL     RETADR          SAVE RETURN ADDRESS
 13            LDB     #LENGTH         ESTABLISH BASE REGISTER
 14            BASE    LENGTH
 15    CLOOP  +JSUB    RDREC           READ INPUT RECORD
 20            LDA     LENGTH          TEST FOR EOF (LENGTH = 0)
 25            COMP    #0
 30            JEQ     ENDFIL          EXIT IF EOF FOUND
 35           +JSUB    WRREC           WRITE OUTPUT RECORD
 40            J       CLOOP           LOOP
 45    ENDFIL  LDA     =C'EOF'         INSERT END OF FILE MARKER
 50            STA     BUFFER
 55            LDA     #3              SET LENGTH = 3
 60            STA     LENGTH
 65           +JSUB    WRREC           WRITE EOF
 70            J       @RETADR         RETURN TO CALLER
 93            LTORG
 95    RETADR  RESW    1
100    LENGTH  RESW    1               LENGTH OF RECORD
105    BUFFER  RESB    4096            4096-BYTE BUFFER AREA
106    BUFEND  EQU     *
107    MAXLEN  EQU     BUFEND-BUFFER   MAXIMUM RECORD LENGTH
110    .
115    .       SUBROUTINE TO READ RECORD INTO BUFFER
120    .
125    RDREC   CLEAR   X               CLEAR LOOP COUNTER
130            CLEAR   A               CLEAR A TO ZERO
132            CLEAR   S               CLEAR S TO ZERO
133           +LDT     #MAXLEN
135    RLOOP   TD      INPUT           TEST INPUT DEVICE
140            JEQ     RLOOP           LOOP UNTIL READY
145            RD      INPUT           READ CHARACTER INTO A REGISTER
150            COMPR   A,S             TEST FOR END OF RECORD (X'00')
155            JEQ     EXIT            EXIT LOOP IF EOR
160            STCH    BUFFER,X        STORE CHARACTER IN BUFFER
165            TIXR    T               LOOP UNLESS MAX LENGTH
170            JLT     RLOOP              HAS BEEN REACHED
175    EXIT    STX     LENGTH          SAVE RECORD LENGTH
180            RSUB                    RETURN TO CALLER
185    INPUT   BYTE    X'F1'           CODE FOR INPUT DEVICE
195    .
200    .       SUBROUTINE TO WRITE RECORD FROM BUFFER
205    .
210    WRREC   CLEAR   X               CLEAR LOOP COUNTER
212            LDT     LENGTH
215    WLOOP   TD      =X'05'          TEST OUTPUT DEVICE
220            JEQ     WLOOP           LOOP UNTIL READY
225            LDCH    BUFFER,X        GET CHARACTER FROM BUFFER
230            WD      =X'05'          WRITE CHARACTER
235            TIXR    T               LOOP UNTIL ALL CHARACTERS
240            JLT     WLOOP              HAVE BEEN WRITTEN
245            RSUB                    RETURN TO CALLER
255            END     FIRST
```

FIGURE 2.9 Program demonstrating additional assembler features.

```
Line    Loc     Source statement                     Object code

  5     0000    COPY    START   0
 10     0000    FIRST   STL     RETADR               17202D
 13     0003            LDB     #LENGTH              69202D
 14                     BASE    LENGTH
 15     0006    CLOOP  +JSUB    RDREC                4B101036
 20     000A            LDA     LENGTH               032026
 25     000D            COMP    #0                   290000
 30     0010            JEQ     ENDFIL               332007
 35     0013           +JSUB    WRREC                4B10105D
 40     0017            J       CLOOP                3F2FEC
 45     001A    ENDFIL  LDA     =C'EOF'              032010
 50     001D            STA     BUFFER               0F2016
 55     0020            LDA     #3                   010003
 60     0023            STA     LENGTH               0F200D
 65     0026           +JSUB    WRREC                4B10105D
 70     002A            J       @RETADR              3E2003
 93                     LTORG
        002D    *       =C'EOF'                      454F46
 95     0030    RETADR  RESW    1
100     0033    LENGTH  RESW    1
105     0036    BUFFER  RESB    4096
106     1036    BUFEND  EQU     *
107     1000    MAXLEN  EQU     BUFEND-BUFFER
110             .
115             .       SUBROUTINE TO READ RECORD INTO BUFFER
120             .
125     1036    RDREC   CLEAR   X                    B410
130     1038            CLEAR   A                    B400
132     103A            CLEAR   S                    B440
133     103C           +LDT     #MAXLEN              75101000
135     1040    RLOOP   TD      INPUT                E32019
140     1043            JEQ     RLOOP                332FFA
145     1046            RD      INPUT                DB2013
150     1049            COMPR   A,S                  A004
155     104B            JEQ     EXIT                 332008
160     104E            STCH    BUFFER,X             57C003
165     1051            TIXR    T                    B850
170     1053            JLT     RLOOP                3B2FEA
175     1056    EXIT    STX     LENGTH               134000
180     1059            RSUB                         4F0000
185     105C    INPUT   BYTE    X'F1'                F1
195             .
200             .       SUBROUTINE TO WRITE RECORD FROM BUFFER
205             .
210     105D    WRREC   CLEAR   X                    B410
212     105F            LDT     LENGTH               774000
215     1062    WLOOP   TD      =X'05'               E32011
220     1065            JEQ     WLOOP                332FFA
225     1068            LDCH    BUFFER,X             53C003
230     106B            WD      =X'05'               DF2008
235     106E            TIXR    T                    B850
240     1070            JLT     WLOOP                3B2FEF
245     1073            RSUB                         4F0000
255                     END     FIRST
        1076    *       =X'05'                       05
```

FIGURE 2.10 Program from Fig. 2.9 with object code.

the assigned addresses and the generated data values. Such a literal pool listing is shown in Fig. 2.10 immediately following the END statement. In this case, the pool consists of the single literal =X'05'.

In some cases, however, it is desirable to place literals into a pool at some other location in the object program. To allow this, we introduce the assembler directive LTORG (line 93 in Fig. 2.9). When the assembler encounters a LTORG statement, it creates a literal pool that contains all of the literal operands used since the previous LTORG (or the beginning of the program). This literal pool is placed in the object program at the location where the LTORG directive was encountered (see Fig. 2.10). Of course, literals placed in a pool by LTORG will not be repeated in the pool at the end of the program.

If we had not used the LTORG statement on line 93, the literal =C'EOF' would be placed in the pool at the end of the program. This literal pool would begin at address 1073. This means that the literal operand would be placed too far away from the instruction referencing it to allow program-counter relative addressing. The problem, of course, is the large amount of storage reserved for BUFFER. By placing the literal pool before this buffer, we avoid having to use extended format instructions when referring to the literals. The need for an assembler directive such as LTORG usually arises when it is desirable to keep the literal operand close to the instruction that uses it.

Most assemblers recognize duplicate literals—that is, the same literal used in more than one place in the program—and store only one copy of the specified data value. For example, the literal =X'05' is used in our program on lines 215 and 230. However, only one data area with this value is generated. Both instructions refer to the same address in the literal pool for their operand.

The easiest way to recognize duplicate literals is by comparison of the character strings defining them (in this case, the string =X'05'). Sometimes a slight additional saving is possible if we look at the generated data value instead of the defining expression. For example, the literals =C'EOF' and =X'454F46' would specify identical operand values. The assembler might avoid storing both literals if it recognized this equivalence. However, the benefits realized in this way are usually not great enough to justify the additional complexity in the assembler.

If we use the character string defining a literal to recognize duplicates, we must be careful of literals whose value depends upon their location in the program. Suppose, for example, that we allow literals that refer to the current value of the location counter (often denoted by the symbol *). Such literals are sometimes useful for loading base registers. For example, the statements

```
LDB    =*
BASE    *
```

as the first lines of a program would load the beginning address of the program into register B. This value would then be available for base relative addressing.

Such a notation can, however, cause a problem with the detection of duplicate literals. If a literal =* appeared on line 13 of our example program, it would specify an operand with value 0003. If the same literal appeared on line 55, it would specify an operand with value 0020. In such a case, the literal operands have identical names; however, they have different values, and both must appear in the literal pool. The same problem arises if a literal refers to any other item whose value changes between one point in the program and another.

Now we are ready to describe how the assembler handles literal operands. The basic data structure needed is a *literal table* LITTAB. For each literal used, this table contains the literal name, the operand value and length, and the address assigned to the operand when it is placed in a literal pool. LITTAB is often organized as a hash table, using the literal name or value as the key.

As each literal operand is recognized during Pass 1, the assembler searches LITTAB for the specified literal name (or value). If the literal is already present in the table, no action is needed; if it is not present, the literal is added to LITTAB (leaving the address unassigned). When Pass 1 encounters a LTORG statement or the end of the program, the assembler makes a scan of the literal table. At this time each literal currently in the table is assigned an address (unless such an address has already been filled in). As these addresses are assigned, the location counter is updated to reflect the number of bytes occupied by each literal.

During Pass 2, the operand address for use in generating object code is obtained by searching LITTAB for each literal operand encountered. The data values specified by the literals in each literal pool are inserted at the appropriate places in the object program exactly as if these values had been generated by BYTE or WORD statements. If a literal value represents an address in the program (for example, a location counter value), the assembler must also generate the appropriate Modification record.

To be sure you understand how LITTAB is created and used by the assembler, you may wish to apply the procedure we just described to the source statements in Fig. 2.9. The object code and literal pools generated should be the same as those in Fig. 2.10.

2.3.2 Symbol-Defining Statements

Up to this point the only user-defined symbols we have seen in assembler language programs have appeared as labels on instructions or data areas. The value of such a label is the address assigned to the statement on which it appears. Most assemblers provide an assembler directive that allows the programmer to define symbols and specify their values. The assembler directive generally used is EQU (for "equate"). The general form of such a statement is

```
symbol    EQU    value
```

This statement defines the given symbol (i.e., enters it into SYMTAB) and assigns to it the value specified. The value may be given as a constant or as any expression involving constants and previously defined symbols. We discuss the formation and use of expressions in the next section.

One common use of EQU is to establish symbolic names that can be used for improved readability in place of numeric values. For example, on line 133 of the program in Fig. 2.5 we used the statement

```
          +LDT    #4096
```

to load the value 4096 into register T. This value represents the maximum length record we could read with subroutine RDREC. The meaning is not, however, as clear as it might be. If we include the statement

```
MAXLEN    EQU     4096
```

in the program, we can write line 133 as

```
          +LDT    #MAXLEN
```

When the assembler encounters the EQU statement, it enters MAXLEN into SYMTAB (with value 4096). During assembly of the LDT instruction, the assembler searches SYMTAB for the symbol MAXLEN, using its value as the operand in the instruction. The resulting object code is exactly the same as in the original version of the instruction; however, the source statement is easier to understand. It is also much easier to find and change the value of MAXLEN if this becomes necessary—we would not have to search through the source code looking for places where #4096 is used.

Another common use of EQU is in defining mnemonic names for registers. We have assumed that our assembler recognizes standard mnemonics for registers—A, X, L, etc. Suppose, however, that the assembler expected register *numbers* instead of names in an instruction like RMO. This would require the programmer to write (for example) "RMO 0,1" instead of "RMO A,X." In such a case the programmer could include a sequence of EQU statements like

```
A        EQU    0
X        EQU    1
L        EQU    2
.
.
.
```

These statements cause the symbols A, X, L, . . . to be entered into SYMTAB with their corresponding values 0, 1, 2, ... An instruction like "RMO A,X" would then be allowed. The assembler would search SYMTAB, finding the values 0 and 1 for the symbols A and X, and assemble the instruction.

On a machine like SIC, there would be little point in doing this—it is just as easy to have the standard register mnemonics built into the assembler. Furthermore, the standard names (base, index, etc.) reflect the usage of the registers. Consider, however, a machine like the System/370, which has general-purpose registers. These registers are commonly designated as 0, 1, 2, . . . (or R0, R1, R2, . . .). In a particular program, however, some of these may be used as base registers, some as index registers, some as accumulators, etc. Furthermore, this usage of registers changes from one program to the next. By writing statements like

```
BASE     EQU    R1
COUNT    EQU    R2
INDEX    EQU    R3
```

the programmer can establish and use names that reflect the logical function of the registers in the program.

There is another common assembler directive that can be used to indirectly assign values to symbols. This directive is usually called ORG (for "origin"). Its form is

```
         ORG    value
```

where value is a constant or an expression involving constants and previously defined symbols. When this statement is encountered during assembly of a program, the assembler resets its location counter (LOCCTR) to the specified value. Since the values of symbols used as labels are taken from LOCCTR, the ORG statement will affect the values of all labels defined until the next ORG.

Of course the location counter is used to control assignment of storage in the object program; in most cases, altering its value would result in an incorrect assembly. Sometimes, however, ORG can be useful in label definition. Suppose that we were defining a symbol table with the following structure:

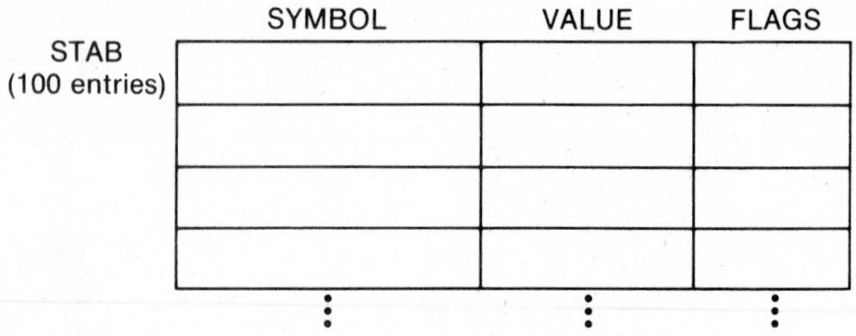

STAB (100 entries)

SYMBOL	VALUE	FLAGS
⋮	⋮	⋮

In this table, the SYMBOL field contains a 6-byte user-defined symbol; VALUE is a 1-word representation of the value assigned to the symbol; FLAGS is a 2-byte field that specifies symbol type and other information.

We could reserve space for this table with the statement

```
STAB      RESB     1100
```

We might want to refer to entries in the table using indexed addressing (placing in the index register the offset of the desired entry from the beginning of the table). Of course, we want to be able to refer to the fields SYMBOL, VALUE, and FLAGS individually, so we must also define these labels. One way of doing this would be with EQU statements:

```
SYMBOL    EQU      STAB
VALUE     EQU      STAB+6
FLAGS     EQU      STAB+9
```

This would allow us to write, for example,

```
          LDA      VALUE,X
```

to fetch the VALUE field from the table entry indicated by the contents of register X. However, this method of definition simply defines the labels; it does not make the structure of the table as clear as it might be.

We can accomplish the same symbol definition using ORG in the following way.

```
STAB      RESB    1100
          ORG     STAB
SYMBOL    RESB    6
VALUE     RESW    1
FLAGS     RESB    2
          ORG     STAB+1100
```

The first ORG resets the location counter to the value of STAB (i.e., the beginning address of the table). The label on the following RESB statement defines SYMBOL to have the current value in LOCCTR; this is the same address assigned to SYMTAB. LOCCTR is then advanced so the label on the RESW statement assigns to VALUE the address (STAB+6), and so on. The result is a set of labels with the same values as those defined with the EQU statements above. This method of definition makes it clear, however, that each entry in STAB consists of a 6-byte SYMBOL, followed by a 1-word VALUE, followed by a 2-byte FLAGS.

The last ORG statement is very important. It sets LOCCTR back to its previous value—the address of the next unassigned byte of memory after the table STAB. This is necessary so that any labels on subsequent statements, which do not represent part of STAB, are assigned the proper addresses. In some assemblers the previous value of LOCCTR is automatically remembered, so we can simply write

```
          ORG
```

(with no value specified) to return to the normal use of LOCCTR.

The descriptions of the EQU and ORG statements contain restrictions that are common to all symbol-defining assembler directives. In the case of EQU, all symbols used on the right-hand side of the statement—that is, all terms used to specify the value of the new symbol—must have been defined previously in the program. Thus, the sequence

```
ALPHA     RESW    1
BETA      EQU     ALPHA
```

would be allowed, whereas the sequence

```
BETA      EQU     ALPHA
ALPHA     RESW    1
```

would not. The reason for this is the symbol definition process. In the second example above, BETA cannot be assigned a value when it is encountered during Pass 1 of the assembly (because ALPHA does not yet have a value). However, our two-pass assembler design requires that all symbols be defined during Pass 1.

A similar restriction applies to ORG: all symbols used to specify the new location counter value must have been previously defined. Thus, for example, the sequence

```
        ORG     ALPHA
BYTE1   RESB    1
BYTE2   RESB    1
BYTE3   RESB    1
        ORG
ALPHA   RESW    1
```

could not be processed. In this case, the assembler would not know (during Pass 1) what value to assign to the location counter in response to the first ORG statement. As a result, the symbols BYTE1, BYTE2, and BYTE3 could not be assigned addresses during Pass 1.

It may appear that this restriction is a result of the particular way in which we defined the two passes of our assembler. In fact, it is a more general product of the forward-reference problem. You can easily see, for example, that the sequence of statements

```
ALPHA   EQU     BETA
BETA    EQU     DELTA
DELTA   RESW    1
```

cannot be resolved by an ordinary two-pass assembler regardless of how the work is divided between the passes. In Section 2.4.3, we briefly consider ways of handling such sequences in a more complex assembler structure.

2.3.3 Expressions

Our previous examples of assembler language statements have used single terms (labels, literals, etc.) as instruction operands. Most assemblers allow the use of expressions wherever such a single operand is permitted. Each such expression must, of course, be evaluated by the assembler to produce a single operand address or value.

Assemblers generally allow arithmetic expressions formed according to the normal rules using the operators +, −, *, and /. Division is usually defined to produce an integer result. Individual terms in the expression may be constants, user-defined symbols, or special terms. The most common such special term is the current value of the location counter (often designated by *). This term represents the value of the next unassigned memory location. Thus in Fig. 2.9 the statement

```
106     BUFEND  EQU     *
```

gives BUFEND a value that is the address of the next byte after the buffer area.

In Section 2.2 we discussed the problem of program relocation. We saw that some values in the object program are *relative* to the beginning of the program, while others are *absolute* (independent of program location). Similarly, the values of terms and expressions are either relative or absolute. A constant is, of course, an absolute term. Labels on instructions and data areas, and references to the location counter value, are relative terms. A symbol whose value is given by EQU (or some similar assembler directive) may be either an absolute term or a relative term depending upon the expression used to define its value.

Expressions are classified as either *absolute expressions* or *relative expressions* depending upon the type of value they produce. An expression that contains only absolute terms is, of course, an absolute expression. However, absolute expressions may also contain relative terms provided the relative terms occur in pairs and the terms in each such pair have opposite signs. It is not necessary that the paired terms be adjacent to each other in the expression; however, all relative terms must be capable of being paired in this way. None of the relative terms may enter into a multiplication or division operation.

A relative expression is one in which all of the relative terms except one can be paired as described above; the remaining unpaired relative term must have a positive sign. As before, no relative term may enter into a multiplication or division operation. Expressions that do not meet the conditions given for either absolute or relative expressions should be flagged by the assembler as errors.

Although the rules given above may seem arbitrary, they are actually quite reasonable. The expressions that are legal under these definitions include exactly those expressions whose value remains meaningful when the program is relocated. A relative term or expression represents some value that may be written as (S + *r*), where S is the starting address of the program and *r* is the value of the term or expression relative to the starting address. Thus a relative term usually represents some location within the program. When relative terms are paired with opposite signs, the dependency on the program starting address is cancelled out; the result is an absolute value. Consider, for example, the program of Fig. 2.9. In the statement

```
107        MAXLEN    EQU      BUFEND-BUFFER
```

both BUFEND and BUFFER are relative terms, each representing an address within the program. However, the expression represents an absolute value: the *difference* between the two addresses, which is the length of the buffer area in bytes. Notice that the assembler listing in Fig. 2.10 shows the value calculated for this expression (hexadecimal 1000) in the "Loc" column. This value does not represent an address, as do most of the other entries in that column. However, it does show the value that is associated with the symbol that appears in the source statement (MAXLEN).

Expressions such as BUFEND + BUFFER, 100 − BUFFER, or 3 * BUFFER represent neither absolute values nor locations within the program. The values of these expressions depend upon the program starting address in a way that is unrelated to anything within the program itself. Because such expressions are very unlikely to be of any use, they are considered errors.

To determine the type of an expression, we must keep track of the types of all symbols defined in the program. For this purpose we need a flag in the symbol table to indicate type of value (absolute or relative) in addition to the value itself. Thus for the program of Fig. 2.10, some of the symbol table entries might be

Symbol	Type	Value
RETADR	R	0030
BUFFER	R	0036
BUFEND	R	1036
MAXLEN	A	1000

With this information the assembler can easily determine the type of each expression used as an operand and generate Modification records in the object program for relative values.

In Section 2.3.5 we consider programs that consist of several parts that can be relocated independently of each other. As we discuss in the later section, our rules for determining the type of an expression must be modified in such instances.

2.3.4 Program Blocks

In all of the examples we have seen so far the program being assembled was treated as a unit. The source programs logically contained subroutines, data areas, etc.; however, they were handled by the assembler as one entity, resulting in a single block of object code. Within this object program the generated machine instructions and data appeared in the same order as they were written in the source program.

Many assemblers provide features that allow more flexible han-

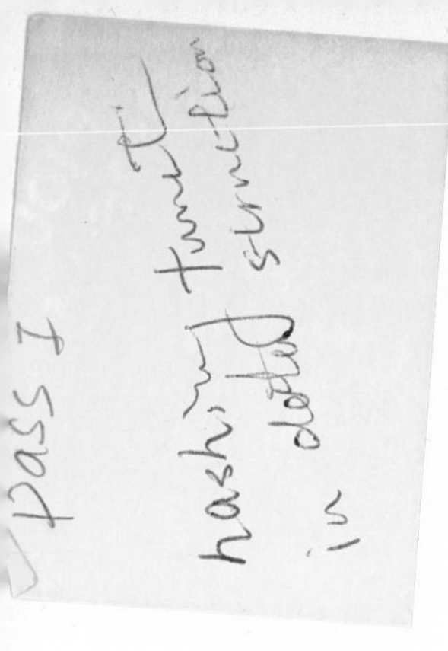

dling of the source and object programs. Some features allow the generated machine instructions and data to appear in the object program in a different order from the corresponding source statements. Other features result in the creation of several independent parts of the object program. These parts maintain their identity and are handled separately by the loader. We use the term *program blocks* to refer to segments of code that are rearranged within a single object program unit, and *control sections* to refer to segments that are translated into independent object program units. (This terminology is, unfortunately, far from uniform. As a matter of fact, in some systems the same assembler language feature is used to accomplish both of these logically different functions.) In this section we consider the use of program blocks and how they are handled by the assembler. Section 2.3.5 discusses control sections and their uses.

Figure 2.11 shows our example program as it might be written using program blocks. In this case three blocks are used. The first (unnamed) program block contains the executable instructions of the program. The second (named CDATA) contains all data areas that are a few words or less in length. The third (named CBLKS) contains all data areas that consist of larger blocks of memory. Some possible reasons for making such a division are discussed later in this section.

The assembler directive USE indicates which portions of the source program belong to the various blocks. At the beginning of the program, statements are assumed to be part of the unnamed (default) block; if no USE statements are included, the entire program belongs to this single block. The USE statement on line 92 signals the beginning of the block named CDATA. Source statements are associated with this block until the USE statement on line 103, which begins the block named CBLKS. The USE statement may also indicate a continuation of a previously begun block. Thus the statement on line 123 resumes the default block, and the statement on line 183 resumes the block named CDATA.

As we can see, each program block may actually contain several separate segments of the source program. The assembler will (logically) rearrange these segments to gather together the pieces of each block. These blocks will then be assigned addresses in the object program, with the blocks appearing in the same order in which they were first begun in the source program. The result is the same as if the programmer had physically rearranged the source statements to group together all the source lines belonging to each block.

The assembler accomplishes this logical rearrangement of code by maintaining, during Pass 1, a separate location counter for each program block. The location counter for a block is initialized to 0 when the

```
Line        Source statement

  5    COPY     START   0               COPY FILE FROM INPUT TO OUTPUT
 10    FIRST    STL     RETADR          SAVE RETURN ADDRESS
 15    CLOOP    JSUB    RDREC           READ INPUT RECORD
 20             LDA     LENGTH          TEST FOR EOF (LENGTH = 0)
 25             COMP    #0
 30             JEQ     ENDFIL          EXIT IF EOF FOUND
 35             JSUB    WRREC           WRITE OUTPUT RECORD
 40             J       CLOOP           LOOP
 45    ENDFIL   LDA     =C'EOF'         INSERT END OF FILE MARKER
 50             STA     BUFFER
 55             LDA     #3              SET LENGTH = 3
 60             STA     LENGTH
 65             JSUB    WRREC           WRITE EOF
 70             J       @RETADR         RETURN TO CALLER
 92             USE     CDATA
 95    RETADR   RESW    1
100    LENGTH   RESW    1               LENGTH OF RECORD
103             USE     CBLKS
105    BUFFER   RESB    4096            4096-BYTE BUFFER AREA
106    BUFEND   EQU     *               FIRST LOCATION AFTER BUFFER
107    MAXLEN   EQU     BUFEND-BUFFER   MAXIMUM RECORD LENGTH
110    .
115    .        SUBROUTINE TO READ RECORD INTO BUFFER
120    .
123             USE
125    RDREC    CLEAR   X               CLEAR LOOP COUNTER
130             CLEAR   A               CLEAR A TO ZERO
132             CLEAR   S               CLEAR S TO ZERO
133            +LDT     #MAXLEN
135    RLOOP    TD      INPUT           TEST INPUT DEVICE
140             JEQ     RLOOP           LOOP UNTIL READY
145             RD      INPUT           READ CHARACTER INTO A REGISTER
150             COMPR   A,S             TEST FOR END OF RECORD (X'00')
155             JEQ     EXIT            EXIT LOOP IF EOR
160             STCH    BUFFER,X        STORE CHARACTER IN BUFFER
165             TIXR    T               LOOP UNLESS MAX LENGTH
170             JLT     RLOOP             HAS BEEN REACHED
175    EXIT     STX     LENGTH          SAVE RECORD LENGTH
180             RSUB                    RETURN TO CALLER
183             USE     CDATA
185    INPUT    BYTE    X'F1'           CODE FOR INPUT DEVICE
195    .
200    .        SUBROUTINE TO WRITE RECORD FROM BUFFER
205    .
208             USE
210    WRREC    CLEAR   X               CLEAR LOOP COUNTER
212             LDT     LENGTH
215    WLOOP    TD      =X'05'          TEST OUTPUT DEVICE
220             JEQ     WLOOP           LOOP UNTIL READY
225             LDCH    BUFFER,X        GET CHARACTER FROM BUFFER
230             WD      =X'05'          WRITE CHARACTER
235             TIXR    T               LOOP UNTIL ALL CHARACTERS
240             JLT     WLOOP             HAVE BEEN WRITTEN
245             RSUB                    RETURN TO CALLER
252             USE     CDATA
253             LTORG
255             END     FIRST
```

FIGURE 2.11 Example of a program with multiple program blocks.

block is first begun. The current value of this location counter is saved when switching to another block, and the saved value is restored when resuming a previously begun block. Thus during Pass 1 each label in the program is assigned an address that is relative to the start of the block that contains it. When labels are entered into the symbol table, the block name or number is stored along with the assigned relative address. At the end of Pass 1 the latest value of the location counter for each block indicates the length of that block. The assembler can then assign to each block a starting address in the object program (beginning with relative location 0).

For code generation during Pass 2, the assembler needs the address for each symbol relative to the start of the object program (not the start of an individual program block). This is easily found from the information in SYMTAB. The assembler simply adds the location of the symbol, relative to the start of its block, to the assigned block starting address.

Figure 2.12 demonstrates this process applied to our sample program. The column headed Loc shows the relative address (within a program block) assigned to each source line and a block number indicating which program block is involved (0 = default block, 1 = CDATA, 2 = CBLKS). This is essentially the same information that is stored in SYMTAB for each symbol. Notice that the value of the symbol MAXLEN (line 107) is shown without a block number. This indicates that MAXLEN is an absolute symbol, whose value is not relative to the start of any program block.

At the end of Pass 1 the assembler constructs a working table that contains the starting addresses and lengths for all blocks. For our sample program, this table looks like

Block name	Block number	Address	Length
(default)	0	0000	0066
CDATA	1	0066	000B
CBLKS	2	0071	1000

Now consider the instruction

```
20        0006 0              LDA       LENGTH                 032060
```

SYMTAB shows the value of the operand (the symbol LENGTH) as relative location 0003 within program block 1 (CDATA). The starting address for CDATA is 0066. Thus the desired target address for this instruction is 0003 + 0066 = 0069. The instruction is to be assembled

```
Line    Loc       Source statement                    Object code

  5     0000 0    COPY      START    0
 10     0000 0    FIRST     STL      RETADR             172063
 15     0003 0    CLOOP     JSUB     RDREC              4B2021
 20     0006 0              LDA      LENGTH             032060
 25     0009 0              COMP     #0                 290000
 30     000C 0              JEQ      ENDFIL             332006
 35     000F 0              JSUB     WRREC              4B203B
 40     0012 0              J        CLOOP              3F2FEE
 45     0015 0    ENDFIL    LDA      =C'EOF'            032055
 50     0018 0              STA      BUFFER             0F2056
 55     001B 0              LDA      #3                 010003
 60     001E 0              STA      LENGTH             0F2048
 65     0021 0              JSUB     WRREC              4B2029
 70     0024 0              J        @RETADR            3E203F
 92     0000 1              USE      CDATA
 95     0000 1    RETADR    RESW     1
100     0003 1    LENGTH    RESW     1
103     0000 2              USE      CBLKS
105     0000 2    BUFFER    RESB     4096
106     1000 2    BUFEND    EQU      *
107     1000      MAXLEN    EQU      BUFEND-BUFFER
110               .
115               .         SUBROUTINE TO READ RECORD INTO BUFFER
120               .
123     0027 0              USE
125     0027 0    RDREC     CLEAR    X                  B410
130     0029 0              CLEAR    A                  B400
132     002B 0              CLEAR    S                  B440
133     002D 0             +LDT      #MAXLEN            75101000
135     0031 0    RLOOP     TD       INPUT              E32038
140     0034 0              JEQ      RLOOP              332FFA
145     0037 0              RD       INPUT              DB2032
150     003A 0              COMPR    A,S                A004
155     003C 0              JEQ      EXIT               332008
160     003F 0              STCH     BUFFER,X           57A02F
165     0042 0              TIXR     T                  B850
170     0044 0              JLT      RLOOP              3B2FEA
175     0047 0    EXIT      STX      LENGTH             13201F
180     004A 0              RSUB                        4F0000
183     0006 1              USE      CDATA
185     0006 1    INPUT     BYTE     X'F1'              F1
195               .
200               .         SUBROUTINE TO WRITE RECORD FROM BUFFER
205               .
208     004D 0              USE
210     004D 0    WRREC     CLEAR    X                  B410
212     004F 0              LDT      LENGTH             772017
215     0052 0    WLOOP     TD       =X'05'             E3201B
220     0055 0              JEQ      WLOOP              332FFA
225     0058 0              LDCH     BUFFER,X           53A016
230     005B 0              WD       =X'05'             DF2012
235     005E 0              TIXR     T                  B850
240     0060 0              JLT      WLOOP              3B2FEF
245     0063 0              RSUB                        4F0000
252     0007 1              USE      CDATA
253                         LTORG
        0007 1    *         =C'EOF'                     454F46
        000A 1    *         =X'05'                      05
255                         END      FIRST
```

FIGURE 2.12 Program from Fig. 2.11 with object code.

using program-counter relative addressing. When the instruction is executed, the program counter contains the address of the following instruction (line 25). The address of this instruction is relative location 0009 within the default block. Since the default block starts at location 0000, this address is simply 0009. Thus the required displacement is 0069 − 0009 = 60. The calculation of the other addresses during Pass 2 follows a similar pattern.

We can immediately see that the separation of the program into blocks has considerably reduced our addressing problems. Because the large buffer area is moved to the end of the object program, we no longer need to use extended format instructions on lines 15, 35, and 65. Furthermore, the base register is no longer necessary; we have deleted the LDB and BASE statements previously on lines 13 and 14. The problem of placement of literals (and literal references) in the program is also much more easily solved. We simply include a LTORG statement in the CDATA block to be sure that the literals are placed ahead of any large data areas.

Of course the use of program blocks has not accomplished anything we could not have done by rearranging the statements of the source program. For example, program readability is improved if the definitions of data areas are placed in the source program close to the statements that reference them. This is particularly true if the program is to be edited or examined from a timesharing terminal. This could be accomplished in a long subroutine (without using program blocks) by simply inserting data areas in any convenient position. However, the programmer would need to provide Jump instructions to branch around the storage thus reserved.

In the situation just discussed, machine considerations suggested that the parts of the object program appear in memory in a particular order. On the other hand, human factors suggested that the source program should be in a different order. The use of program blocks is one way of satisfying both of these requirements, with the assembler providing the required reorganization.

It is not necessary to physically rearrange the generated code in the object program to place the pieces of each program block together. The assembler can simply write the object code as it is generated during Pass 2 and insert the proper load address in each Text record. These load addresses will, of course, reflect the starting address of the block as well as the relative location of the code within the block. This process is illustrated in Fig. 2.13. The first two Text records are generated from the source program lines 5 through 70. When the USE statement on line 92 is recognized, the assembler writes out the current Text record (even though there is still room left in it). The assembler then prepares to begin a new Text record for the new program block.

```
H^COPY  ^000000^001071
T^000000^1E^172063^4B2021^032060^290000^332006^4B203B^3F2FEE^032055^0F2056^010003
T^00001E^09^0F2048^4B2029^3E203F
T^000027^1D^B410^B400^B440^75101000^E32038^332FFA^DB2032^A004^332008^57A02F^B850
T^000044^09^3B2FEA^13201F^4F0000
T^00006C^01^F1
T^00004D^19^B410^772017^E3201B^332FFA^53A016^DF2012^B850^3B2FEF^4F0000
T^00006D^04^454F46^05
E^000000
```

FIGURE 2.13 Object program corresponding to Fig. 2.11.

As it happens, the statements on lines 95 through 105 result in no generated code, so no new Text records are created. The next two Text records come from lines 125 through 180. This time the statements that belong to the next program block do result in the generation of object code. The fifth Text record contains the single byte of data from line 185. The sixth Text record resumes the default program block and the rest of the object program continues in similar fashion.

It does not matter that the Text records of the object program are not in sequence by address; the loader will simply load the object code from each record at the indicated address. When this loading is completed, the generated code from the default block will occupy relative locations 0000 through 0065; the generated code and reserved storage for CDATA will occupy locations 0066 through 0070; and the storage reserved for CBLKS will occupy locations 0071 through 1070. Figure 2.14 traces the blocks of the example program through this process of assembly and loading. Notice that the program segments marked CDATA(1) and CBLKS(1) are not actually present in the object program. Because of the way the addresses are assigned, storage will automatically be reserved for these areas when the program is loaded.

You should carefully examine the generated code in Fig. 2.12, and work through the assembly of several more instructions to be sure you understand how the assembler handles multiple program blocks. To understand how the pieces of each program block are gathered together, you may also want to simulate (by hand) the loading of the object program of Fig. 2.13.

2.3.5 Control Sections and Program Linking

In this section, we discuss the handling of programs that consist of multiple control sections. A *control section* is a part of the program that maintains its identity after assembly; each such control section can be loaded and relocated independently of the others. Different control

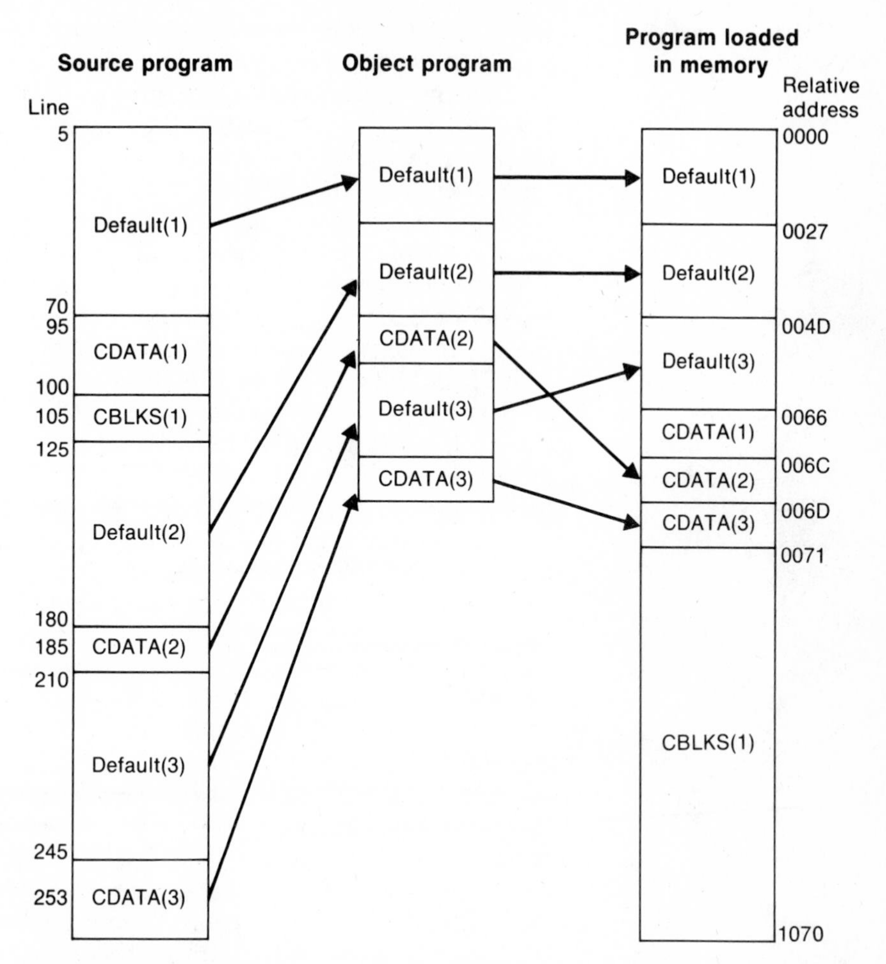

FIGURE 2.14 Program blocks from Fig. 2.11 traced through the assembly and loading processes.

sections are most often used for subroutines or other logical subdivisions of a program. The programmer can assemble, load, and manipulate each of these control sections separately. The resulting flexibility is a major benefit of using control sections. We consider examples of this when we discuss linkage editors in Chapter 3.

When control sections form logically related parts of a program, it is necessary to provide some means for *linking* them together. For example, instructions in one control section might need to refer to instructions or data located in another section. Because control sections are independently loaded and relocated, the assembler is unable to process these references in the usual way. The assembler has no idea where any other control section will be located at execution time. Such references between control sections are called *external references*. The assembler generates information for each external reference that will allow the loader to perform the required linking. In this section we describe how external references are handled by our assembler. Chapter 3 discusses in detail how the actual linking is performed.

Figure 2.15 shows our example program as it might be written using multiple control sections. In this case there are three control sections: one for the main program and one for each subroutine. The START card identifies the beginning of the assembly and gives a name (COPY) to the first control section. The first section continues until the CSECT statement on line 109. This assembler directive signals the start of a new control section named RDREC. Similarly, the CSECT statement on line 193 begins the control section named WRREC. The assembler establishes a separate location counter (beginning at 0) for each control section, just as it does for program blocks.

Control sections differ from program blocks in that they are handled separately by the assembler. (It is not even necessary for all control sections in a program to be assembled at the same time.) Symbols that are defined in one control section may not be used directly by another control section; they must be identified as external references for the loader to handle. Figure 2.15 shows the use of two assembler directives to identify such references: EXTDEF (external definition) and EXTREF (external reference). The EXTDEF statement in a control section names symbols, called *external symbols*, that are defined in this control section and may be used by other sections. Control section names (in this case COPY, RDREC, and WRREC) do not need to be named in an EXTDEF statement because they are automatically considered to be external symbols. The EXTREF statement names symbols that are used in this control section and are defined elsewhere. For example, the symbols BUFFER, BUFEND, and LENGTH are defined in the control section named COPY and made available to the other sections by the EXTDEF statement on line 6. The third control section (WRREC) uses two of these symbols, as specified in its EXTREF statement (line 207). The order in which symbols are listed in the EXTDEF and EXTREF statements is not significant.

Now we are ready to look at how external references are handled by the assembler. Figure 2.16 shows the generated object code for each statement in the program. Consider first the instruction

```
15      0003      CLOOP    +JSUB     RDREC            4B100000
```

The operand (RDREC) is named in the EXTREF statement for the control section, so this is an external reference. The assembler has no idea where the control section containing RDREC will be loaded, so it cannot assemble the address for this instruction. Instead the assembler

```
Line        Source statement

  5    COPY     START   0              COPY FILE FROM INPUT TO OUTPUT
  6             EXTDEF  BUFFER,BUFEND,LENGTH
  7             EXTREF  RDREC,WRREC
 10    FIRST    STL     RETADR         SAVE RETURN ADDRESS
 15    CLOOP   +JSUB    RDREC          READ INPUT RECORD
 20             LDA     LENGTH         TEST FOR EOF (LENGTH = 0)
 25             COMP   #0
 30             JEQ     ENDFIL         EXIT IF EOF FOUND
 35            +JSUB    WRREC          WRITE OUTPUT RECORD
 40             J       CLOOP          LOOP
 45    ENDFIL   LDA    =C'EOF'         INSERT END OF FILE MARKER
 50             STA     BUFFER
 55             LDA    #3              SET LENGTH = 3
 60             STA     LENGTH
 65            +JSUB    WRREC          WRITE EOF
 70             J      @RETADR         RETURN TO CALLER
 95    RETADR   RESW    1
100    LENGTH   RESW    1              LENGTH OF RECORD
103             LTORG
105    BUFFER   RESB    4096           4096-BYTE BUFFER AREA
106    BUFEND   EQU     *
107    MAXLEN   EQU     BUFEND-BUFFER

109    RDREC    CSECT
110    .
115    .        SUBROUTINE TO READ RECORD INTO BUFFER
120    .
122             EXTREF  BUFFER,LENGTH,BUFEND
125             CLEAR   X              CLEAR LOOP COUNTER
130             CLEAR   A              CLEAR A TO ZERO
132             CLEAR   S              CLEAR S TO ZERO
133             LDT     MAXLEN
135    RLOOP    TD      INPUT          TEST INPUT DEVICE
140             JEQ     RLOOP          LOOP UNTIL READY
145             RD      INPUT          READ CHARACTER INTO A REGISTER
150             COMPR   A,S            TEST FOR END OF RECORD (X'00')
155             JEQ     EXIT           EXIT LOOP IF EOR
160            +STCH    BUFFER,X       STORE CHARACTER IN BUFFER
165             TIXR    T              LOOP UNLESS MAX LENGTH
170             JLT     RLOOP             HAS BEEN REACHED
175    EXIT    +STX     LENGTH         SAVE RECORD LENGTH
180             RSUB                   RETURN TO CALLER
185    INPUT    BYTE    X'F1'          CODE FOR INPUT DEVICE
190    MAXLEN   WORD    BUFEND-BUFFER

193    WRREC    CSECT
195    .
200    .        SUBROUTINE TO WRITE RECORD FROM BUFFER
205    .
207             EXTREF  LENGTH,BUFFER
210             CLEAR   X              CLEAR LOOP COUNTER
212            +LDT     LENGTH
215    WLOOP    TD     =X'05'          TEST OUTPUT DEVICE
220             JEQ     WLOOP          LOOP UNTIL READY
225            +LDCH    BUFFER,X       GET CHARACTER FROM BUFFER
230             WD     =X'05'          WRITE CHARACTER
235             TIXR    T              LOOP UNTIL ALL CHARACTERS
240             JLT     WLOOP             HAVE BEEN WRITTEN
245             RSUB                   RETURN TO CALLER
255             END     FIRST
```

FIGURE 2.15 Illustration of control sections and program linking.

```
Line    Loc     Source statement                        Object code

  5     0000    COPY     START    0
  6                      EXTDEF   BUFFER,BUFEND,LENGTH
  7                      EXTREF   RDREC,WRREC
 10     0000    FIRST    STL      RETADR                 172027
 15     0003    CLOOP   +JSUB     RDREC                  4B100000
 20     0007             LDA      LENGTH                 032023
 25     000A             COMP    #0                      290000
 30     000D             JEQ      ENDFIL                 332007
 35     0010            +JSUB     WRREC                  4B100000
 40     0014             J        CLOOP                  3F2FEC
 45     0017    ENDFIL   LDA     =C'EOF'                 032016
 50     001A             STA      BUFFER                 0F2016
 55     001D             LDA     #3                      010003
 60     0020             STA      LENGTH                 0F200A
 65     0023            +JSUB     WRREC                  4B100000
 70     0027             J       @RETADR                 3E2000
 95     002A    RETADR   RESW     1
100     002D    LENGTH   RESW     1
103                      LTORG
        0030    *        =C'EOF'                         454F46
105     0033    BUFFER   RESB     4096
106     1033    BUFEND   EQU      *
107     1000    MAXLEN   EQU      BUFEND-BUFFER

109     0000    RDREC    CSECT
110             .
115             .        SUBROUTINE TO READ RECORD INTO BUFFER
120             .
122                      EXTREF   BUFFER,LENGTH,BUFEND
125     0000             CLEAR    X                      B410
130     0002             CLEAR    A                      B400
132     0004             CLEAR    S                      B440
133     0006             LDT      MAXLEN                 77201F
135     0009    RLOOP    TD       INPUT                  E3201B
140     000C             JEQ      RLOOP                  332FFA
145     000F             RD       INPUT                  DB2015
150     0012             COMPR    A,S                    A004
155     0014             JEQ      EXIT                   332009
160     0017            +STCH     BUFFER,X               57900000
165     001B             TIXR     T                      B850
170     001D             JLT      RLOOP                  3B2FE9
175     0020    EXIT    +STX      LENGTH                 13100000
180     0024             RSUB                            4F0000
185     0027    INPUT    BYTE     X'F1'                  F1
190     0028    MAXLEN   WORD     BUFEND-BUFFER          000000

193     0000    WRREC    CSECT
195             .
200             .        SUBROUTINE TO WRITE RECORD FROM BUFFER
205             .
207                      EXTREF   LENGTH,BUFFER
210     0000             CLEAR    X                      B410
212     0002            +LDT      LENGTH                 77100000
215     0006    WLOOP    TD      =X'05'                  E32012
220     0009             JEQ      WLOOP                  332FFA
225     000C            +LDCH     BUFFER,X               53900000
230     0010             WD      =X'05'                  DF2008
235     0013             TIXR     T                      B850
240     0015             JLT      WLOOP                  3B2FEE
245     0018             RSUB                            4F0000
255                      END      FIRST
        001B    *        =X'05'                          05
```

FIGURE 2.16 Program from Fig. 2.15 with object code.

inserts an address of zero and passes information to the loader, which will cause the proper address to be inserted at load time. The address of RDREC will have no predictable relationship to anything in this control section; therefore relative addressing is not possible. Thus an extended format instruction must be used to provide room for the actual address to be inserted. This is true of any instruction whose operand involves an external reference.

Similarly, the instruction

```
160     0017             +STCH    BUFFER,X         57900000
```

makes an external reference to BUFFER. The instruction is assembled using extended format with an address of zero. The *x* bit is set to 1 to indicate indexed addressing, as specified by the instruction. The statement

```
190     0028    MAXLEN   WORD     BUFEND-BUFFER    000000
```

is only slightly different. Here the value of the data word to be generated is specified by an expression involving two external references: BUFEND and BUFFER. As before, the assembler stores this value as zero. When the program is loaded, the loader will add to this data area the address of BUFEND and subtract from it the address of BUFFER, which results in the desired value.

Note the difference between the handling of the expression on line 190 and the similar expression on line 107. The symbols BUFEND and BUFFER are defined in the same control section with the EQU statement on line 107. Thus the value of the expression can be calculated immediately by the assembler. This could not be done for line 190; BUFEND and BUFFER are defined in another control section, so their values are unknown at assembly time.

As we can see from the above discussion, the assembler must remember (via entries in SYMTAB) in which control section a symbol is defined. Any attempt to refer to a symbol in another control section must be flagged as an error unless the symbol is identified (using EXTREF) as an external reference. The assembler must also allow the same symbol to be used in different control sections. For example, the conflicting definitions of MAXLEN on lines 107 and 190 should cause no problem. A reference to MAXLEN in the control section COPY would use the definition on line 107, whereas a reference to MAXLEN in RDREC would use the definition on line 190.

So far we have seen how the assembler leaves room in the object code for the values of external symbols. The assembler must also include information in the object program that will cause the loader to insert the proper values where they are required. We need two new record types in the object program and a change in a previously defined record type. As before, the exact format of these records is arbitrary; however, the same information must be passed to the loader in some form.

The two new record types are Define and Refer. A Define record gives information about external symbols that are defined in this control section—that is, symbols named by EXTDEF. A Refer record lists symbols that are used as external references by the control section—that is, symbols named by EXTREF. The formats of these records are as follows.

Define record:

Col. 1	D
Col. 2–7	Name of external symbol defined in this control section
Col. 8–13	Relative address of symbol within this control section (hexadecimal)
Col. 14–73	Repeat information in Col. 2–13 for other external symbols

Refer record:

Col. 1	R
Col. 2–7	Name of external symbol referred to in this control section
Col. 8–73	Names of other external reference symbols

The other information needed for program linking is added to the Modification record type. The new format is as follows.

Modification record (revised):

Col. 1	M
Col. 2–7	Starting address of the field to be modified, relative to the beginning of the control section (hexadecimal)
Col. 8–9	Length of the field to be modified, in *half-bytes* (hexadecimal)
Col. 10	Modification flag (+ or −)
Col. 11–16	External symbol whose value is to be added to or subtracted from the indicated field

The first three items in this record are the same as previously discussed. The two new items specify the modification to be performed:

adding or subtracting the value of some external symbol. The symbol used for modification may be defined either in this control section or in another one.

Figure 2.17 shows the object program corresponding to the source in Fig. 2.16. Notice that there is a separate set of object program records (from Header through End) for each control section. The records for each control section are exactly the same as they would be if the sections were assembled separately.

FIGURE 2.17 Object program corresponding to Fig. 2.15.

```
H^COPY ^000000^001033
D^BUFFER^000033^BUFEND^001033^LENGTH^00002D
R^RDREC ^WRREC
T^000000^1D^172027^4B100000^032023^290000^332007^4B100000^3F2FEC^032016^0F2016
T^00001D^0D^010003^0F200A^4B100000^3E2000
T^000030^03^454F46
M^000004^05^+RDREC
M^000011^05^+WRREC
M^000024^05^+WRREC
E^000000


H^RDREC ^000000^00002B
R^BUFFER^LENGTH^BUFEND
T^000000^1D^B410^B400^B440^77201F^E3201B^332FFA^DB2015^A004^332009^57900000^B850
T^00001D^0E^3B2FE9^13100000^4F0000^F1^000000
M^000018^05^+BUFFER
M^000021^05^+LENGTH
M^000028^06^+BUFEND
M^000028^06^-BUFFER
E


H^WRREC ^000000^00001C
R^LENGTH^BUFFER
T^000000^1C^B410^77100000^E32012^332FFA^53900000^DF2008^B850^3B2FEE^4F0000^05
M^000003^05^+LENGTH
M^00000D^05^+BUFFER
E
```

The Define and Refer records for each control section include the symbols named in the EXTDEF and EXTREF statements. In the case of Define, the record also indicates the relative address of each external symbol within the control section. For EXTREF symbols, no address information is available. These symbols are simply named in the Refer record.

Now let us examine the process involved in linking up external references, beginning with the source statements we discussed previously. The address field for the JSUB instruction on line 15 begins at relative address 0004. Its initial value in the object program is zero. The Modification record

```
M00000405+RDREC
```

in control section COPY specifies that the address of RDREC is to be added to this field, thus producing the correct machine instruction for execution. The other two Modification records in COPY perform similar functions for the instructions on lines 35 and 65. Likewise, the first Modification record in control section RDREC fills in the proper address for the external reference on line 160.

The handling of the data word generated by line 190 is only slightly different. The value of this word is to be BUFEND-BUFFER, where both BUFEND and BUFFER are defined in another control section. The assembler generates an initial value of zero for this word (located at relative address 0028 within control section RDREC). The last two Modification records in RDREC direct that the address of BUFEND be added to this field, and the address of BUFFER be subtracted from it. This computation, performed at load time, results in the desired value for the data word.

In Chapter 3 we discuss in detail how the required modifications are performed by the loader. At this time, however, you should be sure that you understand the concepts involved in the linking process. You should carefully examine the other Modification records in Fig. 2.17, and reconstruct for yourself how they were generated from the source program statements.

Note that the revised Modification record may still be used to perform program relocation. In the case of relocation, the modification required is adding the beginning address of the control section to certain fields in the object program. The symbol used as the name of the control section has as its value the required address. Since the control section name is automatically an external symbol, it is available for use in Modification records. Thus, for example, the Modification records from Fig. 2.8 are changed from

```
M00000705
M00001405
M00002705
```

to

```
M00000705+COPY
M00001405+COPY
M00002705+COPY
```

In this way, exactly the same mechanism can be used for program relocation and for program linking. There are more examples in the next chapter.

The existence of multiple control sections that can be relocated independently of one another makes the handling of expressions slightly more complicated. Our earlier definitions required that all of the relative terms in an expression be paired (for an absolute expression), or that all except one be paired (for a relative expression). We must now extend this restriction to specify that both terms in each pair must be relative within the same control section. The reason is simple—if the two terms represent relative locations in the same control section, their difference is an absolute value (regardless of where the control section is located). On the other hand, if they are in different control sections, their difference has a value that is unpredictable (and therefore probably useless). For example, the expression

```
BUFEND-BUFFER
```

has as its value the length of BUFFER in bytes. On the other hand, the value of the expression

```
RDREC-COPY
```

is the difference in the load addresses of the two control sections. This value depends on the way run-time storage is allocated; it is unlikely to be of any use whatsoever to an application program.

When an expression involves external references, the assembler cannot in general determine whether or not the expression is legal. The pairing of relative terms to test legality cannot be done without knowing which of the terms occur in the same control sections, and this is unknown at assembly time. In such a case, the assembler evaluates all of the terms it can, and combines these to form an initial expression value. It also generates Modification records so the loader can finish the evaluation. The loader can then check the expression for errors. We discuss this further in Chapter 3 when we examine the design of a linking loader.

2.4 ASSEMBLER DESIGN OPTIONS

In this section we discuss several important options related to the overall structure of an assembler. In many cases these options have no direct impact on the assembler language programmer, although some of them do have side effects that should be considered.

In Section 2.4.1 we describe the overlay structure commonly used for two-pass assemblers. Sections 2.4.2 and 2.4.3 discuss two alternatives to the standard two-pass assembler logic. Section 2.4.2 describes the structure and logic of one-pass assemblers. These assemblers are used when it is necessary or desirable to avoid a second pass over the source program. Section 2.4.3 introduces the notion of a multi-pass assembler, an extension to the two-pass logic that allows an assembler to handle forward references during symbol definition.

2.4.1 Two-Pass Assembler with Overlay Structure

As we have seen, most assemblers divide the processing of the source program into two passes. The internal tables and subroutines that are used only during Pass 1 are no longer needed after the first pass is completed. The routines and tables for Pass 1 and Pass 2 are never required at the same time. Of course, there are certain tables (such as SYMTAB) and certain processing subroutines (such as searching SYMTAB) that are used by both passes.

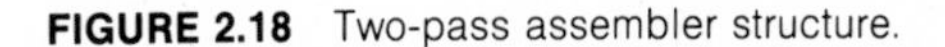

Figure 2.18 shows a representation of this overall two-pass assembler structure, which has three *segments* forming a tree structure. The Root segment contains a simple driver program whose function is to call, in turn, the other two segments (Pass 1 and Pass 2). This Root segment also includes the tables and routines needed by both passes.

Since the Pass 1 and Pass 2 segments are never needed at the same time, they can occupy the same locations in memory during execution

FIGURE 2.18 Two-pass assembler structure.

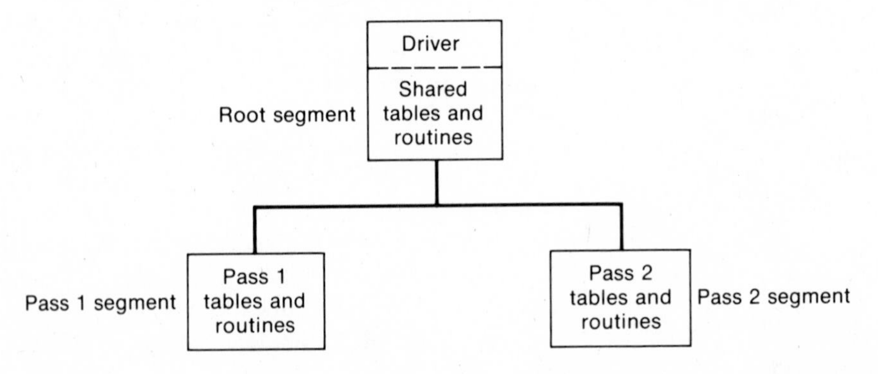

of the assembler. Initially the Root and Pass 1 segments are loaded into memory. The assembler then makes the first pass over the program being assembled. At the end of this first pass, the Pass 2 segment is loaded, replacing the Pass 1 segment. The assembler then makes its second pass of the source program (or the intermediate file) and terminates. This process is illustrated in Fig. 2.19. Note that the assembler needs much less memory to run in this way than it would if both Pass 1 and Pass 2 were loaded at the same time. Many two-pass assemblers use this technique to reduce their memory requirements.

A program that is designed to execute in this way is called an *overlay* program because some of its segments overlay others during execution. In Chapter 3 we discuss overlay programs in more detail and consider how they are implemented by the loader and operating system services.

FIGURE 2.19 Two-pass assembler with overlay execution.

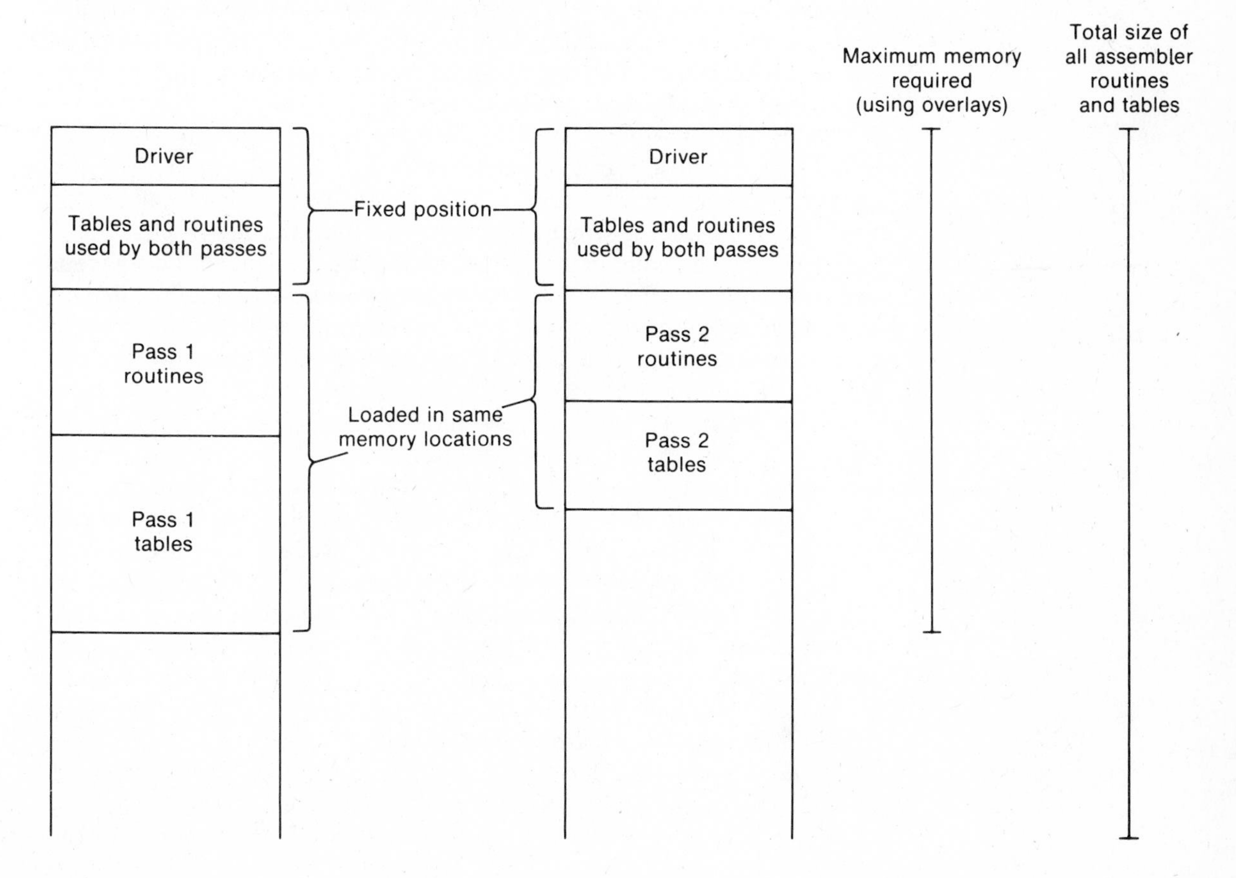

2.4.2 One-Pass Assemblers

In this section we examine the structure and design of one-pass assemblers. As we discussed in Section 2.1, the main problem in trying to assemble a program in one pass involves forward references. Instruction operands often are symbols that have not yet been defined in the source program. Thus the assembler does not know what address to insert in the translated instruction.

It is easy to eliminate forward references to data items; we can simply require that all such areas be defined in the source program before they are referenced. This restriction is not too severe. The programmer merely places all storage reservation statements at the start of the program rather than at the end. Unfortunately, forward references to labels on instructions cannot be eliminated as easily. The logic of the program often requires a forward jump—for example, in escaping from a loop after testing some condition. Requiring that the programmer eliminate all such forward jumps would be much too restrictive and inconvenient. Therefore, the assembler must make some special provision for handling forward references. To reduce the size of the problem, many one-pass assemblers do, however, prohibit (or at least discourage) forward references to data items.

There are two main types of one-pass assembler. One type produces object code directly in memory for immediate execution; the other type produces the usual kind of object program for later execution. We use the program in Fig. 2.20 to illustrate our discussion of both types. This example is the same as in Fig. 2.2, with all data item definitions placed ahead of the code that references them. The generated object code shown in Fig. 2.20 is for reference only; we will discuss how each type of one-pass assembler would actually generate the object program required.

We first discuss one-pass assemblers that generate their object code in memory for immediate execution. No object program is written out, and no loader is needed. This kind of *load-and-go* assembler is useful in a system that is oriented toward program development and testing. A university computing system for student use is a typical example of such an environment. In such a system, a large fraction of the total workload consists of program translation. Because programs are re-assembled nearly every time they are run, efficiency of the assembly process is an important consideration. A load-and-go assembler avoids the overhead of writing the object program out and reading it back in. This can be accomplished with either a one-pass or a two-pass assembler. However, a one-pass assembler also avoids the overhead of an additional pass over the source program.

```
Line    Loc        Source statement                  Object code

  0     1000    COPY      START    1000
  1     1000    EOF       BYTE     C'EOF'              454F46
  2     1003    THREE     WORD     3                   000003
  3     1006    ZERO      WORD     0                   000000
  4     1009    RETADR    RESW     1
  5     100C    LENGTH    RESW     1
  6     100F    BUFFER    RESB     4096
  9             .
 10     200F    FIRST     STL      RETADR              141009
 15     2012    CLOOP     JSUB     RDREC               48203D
 20     2015              LDA      LENGTH              00100C
 25     2018              COMP     ZERO                281006
 30     201B              JEQ      ENDFIL              302024
 35     201E              JSUB     WRREC               482062
 40     2021              J        CLOOP               3C2012
 45     2024    ENDFIL    LDA      EOF                 001000
 50     2027              STA      BUFFER              0C100F
 55     202A              LDA      THREE               001003
 60     202D              STA      LENGTH              0C100C
 65     2030              JSUB     WRREC               482062
 70     2033              LDL      RETADR              081009
 75     2036              RSUB                         4C0000
110             .
115             .         SUBROUTINE TO READ RECORD INTO BUFFER
120             .
121     2039    INPUT     BYTE     X'F1'               F1
122     203A    MAXLEN    WORD     4096                001000
124             .
125     203D    RDREC     LDX      ZERO                041006
130     2040              LDA      ZERO                001006
135     2043    RLOOP     TD       INPUT               E02039
140     2046              JEQ      RLOOP               302043
145     2049              RD       INPUT               D82039
150     204C              COMP     ZERO                281006
155     204F              JEQ      EXIT                30205B
160     2052              STCH     BUFFER,X            54900F
165     2055              TIX      MAXLEN              2C203A
170     2058              JLT      RLOOP               382043
175     205B    EXIT      STX      LENGTH              10100C
180     205E              RSUB                         4C0000
195             .
200             .         SUBROUTINE TO WRITE RECORD FROM BUFFER
205             .
206     2061    OUTPUT    BYTE     X'05'               05
207             .
210     2062    WRREC     LDX      ZERO                041006
215     2065    WLOOP     TD       OUTPUT              E02061
220     2068              JEQ      WLOOP               302065
225     206B              LDCH     BUFFER,X            50900F
230     206E              WD       OUTPUT              DC2061
235     2071              TIX      LENGTH              2C100C
240     2074              JLT      WLOOP               382065
245     2077              RSUB                         4C0000
255                       END      FIRST
```

FIGURE 2.20 Sample program for a one-pass assembler.

Because the object program is produced in memory rather than being written out on secondary storage, the handling of forward references becomes less difficult. The assembler simply generates object code instructions as it scans the source program. If an instruction operand is a symbol that has not yet been defined, the operand address is omitted when the instruction is assembled. The symbol used as an operand is entered into the symbol table (unless such an entry is already present). This entry is flagged to indicate that the symbol is undefined. The address of the operand field of the instruction that refers to the undefined symbol is added to a list of forward references associated with the symbol table entry. When the definition for a symbol is encountered, the forward reference list for that symbol is scanned (if one exists), and the proper address is inserted into any instructions previously generated.

An example should help to make this process clear. Figure 2.21(a) shows the object code and symbol table entries as they would be after scanning line 40 of the program in Fig. 2.20. The first forward reference occurred on line 15. Since the operand (RDREC) was not yet defined, the instruction was assembled with no value assigned as the

FIGURE 2.21(a) Object code in memory and symbol table entries for the program in Fig. 2.20 after scanning line 40.

Memory address	Contents			
1000	454F4600	00030000	00xxxxxx	xxxxxxxx
1010	xxxxxxxx	xxxxxxxx	xxxxxxxx	xxxxxxxx
•				
•				
•				
2000	xxxxxxxx	xxxxxxxx	xxxxxxxx	xxxxxx14
2010	100948--	--00100C	28100630	----48--
2020	--3C2012			
•				
•				
•				

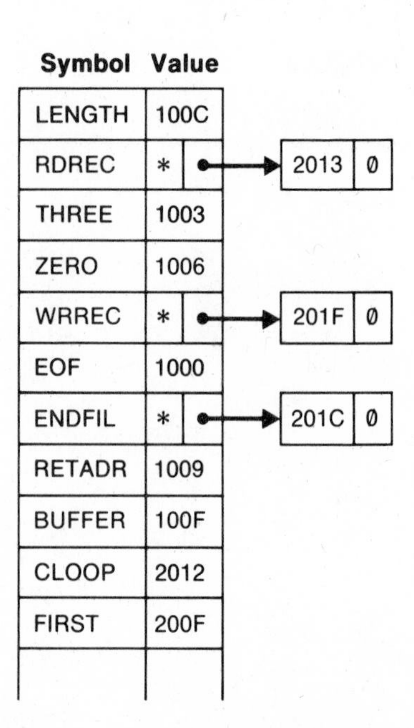

(a)

operand address (denoted in the figure by ----). RDREC was then entered into SYMTAB as an undefined symbol (indicated by *); the address of the operand field of the instruction (2013) was inserted in a list associated with RDREC. A similar process was followed with the instructions on lines 30 and 35.

Now consider Fig. 2.21(b), which corresponds to the situation after scanning line 160. Some of the forward references have been resolved by this time, while others have been added. When the symbol ENDFIL was defined (line 45), the assembler placed its value in the SYMTAB entry; it then inserted this value into the instruction operand field (at address 201C) as directed by the forward reference list. From this point on, any references to ENDFIL would not be forward references, and would not be entered into a list. Similarly, the definition of RDREC

FIGURE 2.21(b) Object code in memory and symbol table entries for the program in Fig. 2.20 after scanning line 160.

Memory address	Contents			
1000	454F4600	00030000	00xxxxxx	xxxxxxxx
1010	xxxxxxxx	xxxxxxxx	xxxxxxxx	xxxxxxxx
•				
•				
•				
2000	xxxxxxxx	xxxxxxxx	xxxxxxxx	xxxxxx14
2010	10094820	3D00100C	28100630	202448--
2020	--3C2012	0010000C	100F0010	030C100C
2030	48----08	10094C00	00F10010	00041006
2040	001006E0	20393020	43D82039	28100630
2050	----5490	0F		
•				
•				
•				

Symbol	Value	Forward reference list
LENGTH	100C	
RDREC	203D	
THREE	1003	
ZERO	1006	
WRREC	*	→ 201F → 2031 Ø
EOF	1000	
ENDFIL	2024	
RETADR	1009	
BUFFER	100F	
CLOOP	2012	
FIRST	200F	
MAXLEN	203A	
INPUT	2039	
EXIT	*	→ 2050 Ø
RLOOP	2043	

(b)

(line 125) resulted in the filling in of the operand address at location 2013. Meanwhile, two new forward references have been added: to WRREC (line 65) and EXIT (line 155). You should continue tracing through this process to the end of the program to show yourself that all of the forward references will be filled in properly. At the end of the program, any SYMTAB entries that are still marked with * indicate undefined symbols. These should be flagged by the assembler as errors.

When the end of the program is encountered, the assembly is complete. If no errors have occurred, the assembler searches SYMTAB for the value of the symbol named in the END statement (in this case, FIRST) and jumps to this location to begin execution of the assembled program.

We used an absolute program as our example because, for a load-and-go assembler, the actual address must be known at assembly time. Of course it is not necessary for this address to be specified by the programmer; it might be assigned by the system. In either case, however, the assembly process would be the same—the location counter would be initialized to the actual program starting address.

One-pass assemblers that produce object programs as output are needed on systems where external working-storage devices (for the intermediate file between the two passes) are not available. Such assemblers may also be useful when the external storage is slow or is inconvenient to use for some other reason. One-pass assemblers that produce object programs follow a slightly different procedure from that previously described. Forward references are entered into lists as before. Now, however, when the definition of a symbol is encountered, instructions that made forward references to that symbol may no longer be available in memory for modification. In general, they will already have been written out as part of a Text record in the object program. In this case the assembler must generate another Text record with the correct operand address. When the program is loaded, this address will be inserted into the instruction by the action of the loader.

Figure 2.22 illustrates this process. The second Text record contains the object code generated from lines 10 through 40 in Fig. 2.20. The operand addresses for the instructions on lines 15, 30, and 35 have been generated as 0000. When the definition of ENDFIL on line 45 is encountered, the assembler generates the third Text record. This record specifies that the value 2024 (the address of ENDFIL) is to be loaded at location 201C (the operand address field of the JEQ instruction on line 30). When the program is loaded, therefore, the value 2024 will replace the 0000 previously loaded. The other forward references in the program are handled in exactly the same way. In effect, the services of the loader are being used to complete forward references

```
H^COPY  ^001000^00107A
T^001000^09^454F46^000003^000000
T^00200F^15^141009^480000^00100C^281006^300000^480000^3C2012
T^00201C^02^2024
T^002024^19^001000^0C100F^001003^0C100C^480000^081009^4C0000^F1^001000
T^002013^02^203D
T^00203D^1E^041006^001006^E02039^302043^D82039^281006^300000^54900F^2C203A^382043
T^002050^02^205B
T^00205B^07^10100C^4C0000^05
T^00201F^02^2062
T^002031^02^2062
T^002062^18^041006^E02061^302065^50900F^DC2061^2C100C^382065^4C0000
E^00200F
```

FIGURE 2.22 Object program from one-pass assembler for program in Fig. 2.20.

that could not be handled by the assembler. Of course, the object program records must be kept in their original order when they are presented to the loader.

In this section we considered only simple one-pass assemblers that handled absolute programs. Instruction operands were assumed to be single symbols, and the assembled instructions contained the actual (not relative) addresses of the operands. More advanced assembler features such as literals were not allowed. You are encouraged to think about ways of removing some of these restrictions (see the Exercises for this section for some suggestions).

2.4.3 Multi-Pass Assemblers

In our discussion of the EQU assembler directive, we required that any symbol used on the right-hand side (i.e., in the expression giving the value of the new symbol) be defined previously in the source program. A similar requirement was imposed for ORG. As a matter of fact, such a restriction is normally applied to all assembler directives that (directly or indirectly) define symbols.

The reason for this is the symbol definition process in a two-pass assembler. Consider, for example, the sequence

```
ALPHA    EQU     BETA
BETA     EQU     DELTA
DELTA    RESW    1
```

The symbol BETA cannot be assigned a value when it is encountered during the first pass because DELTA has not yet been defined. As a result, ALPHA cannot be evaluated during the second pass. This means that any assembler that makes only two sequential passes over the source program cannot resolve such a sequence of definitions.

Restrictions such as prohibiting forward references in symbol definition are not normally a serious inconvenience for the programmer. As a matter of fact, such forward references tend to create difficulty for a person reading the program as well as for the assembler. Nevertheless, some assemblers are designed to eliminate the need for such restrictions. The general solution is a multi-pass assembler that can make as many passes as are needed to process the definitions of symbols. It is not necessary for such an assembler to make more than two passes over the entire program. Instead, the portions of the program that involve forward references in symbol definition are saved during Pass 1. Additional passes through these stored definitions are made as the assembly progresses. This process is followed by a normal Pass 2.

There are several ways of accomplishing the task outlined above. The method we describe involves storing those symbol definitions that involve forward references in the symbol table. This table also indicates which symbols are dependent on the values of others, to facilitate symbol evaluation.

Figure 2.23(a) shows a sequence of symbol-defining statements that involve forward references; the other parts of the source program are not important for our discussion, and have been omitted. The following parts of Fig. 2.23 show information in the symbol table as it might appear after processing each of the source statements shown.

Figure 2.23(b) displays symbol table entries resulting from Pass 1 processing of the statement.

```
HALFSZ    EQU    MAXLEN/2
```

MAXLEN has not yet been defined, so no value for HALFSZ can be computed. The defining expression for HALFSZ is stored in the symbol table in place of its value. The entry &1 indicates that 1 symbol in the defining expression is undefined. In an actual implementation, of course, this definition might be stored at some other location. SYMTAB would then simply contain a pointer to the defining expression. The symbol MAXLEN is also entered in the symbol table, with the flag * identifying it as undefined. Associated with this entry is a list of the symbols whose values depend on MAXLEN (in this case, HALFSZ). (Note the similarity to the way we handled forward references in a one-pass assembler.)

The same procedure is followed with the definition of MAXLEN (see Fig. 2.23c). In this case there are two undefined symbols involved in the definition: BUFEND and BUFFER. Both of these are entered into SYMTAB with lists indicating the dependence of MAXLEN upon them. Similarly, the definition of PREVBT causes this symbol to be added to the list of dependencies on BUFFER (as shown in Fig. 2.23d).

So far we have simply been saving symbol definitions for later processing. The definition of BUFFER on line 4 lets us begin evaluation of some of these symbols. Let us assume that when line 4 is read the location counter contains the hexadecimal value 1034. This address is stored as the value of BUFFER. The assembler then examines the list of symbols that are dependent on BUFFER. The symbol table entry for the first symbol in this list (MAXLEN) shows that it depends

FIGURE 2.23 Example of multi-pass assembler operation.

```
1      HALFSZ     EQU      MAXLEN/2
2      MAXLEN     EQU      BUFEND-BUFFER
3      PREVBT     EQU      BUFFER-1
         .
         .
         .
4      BUFFER     RESB     4096
5      BUFEND     EQU      *
```

(a)

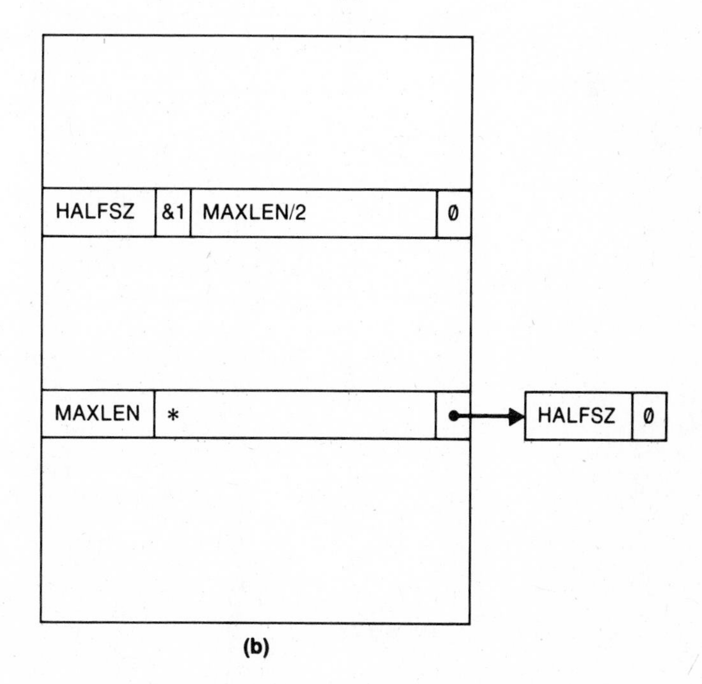

(b)

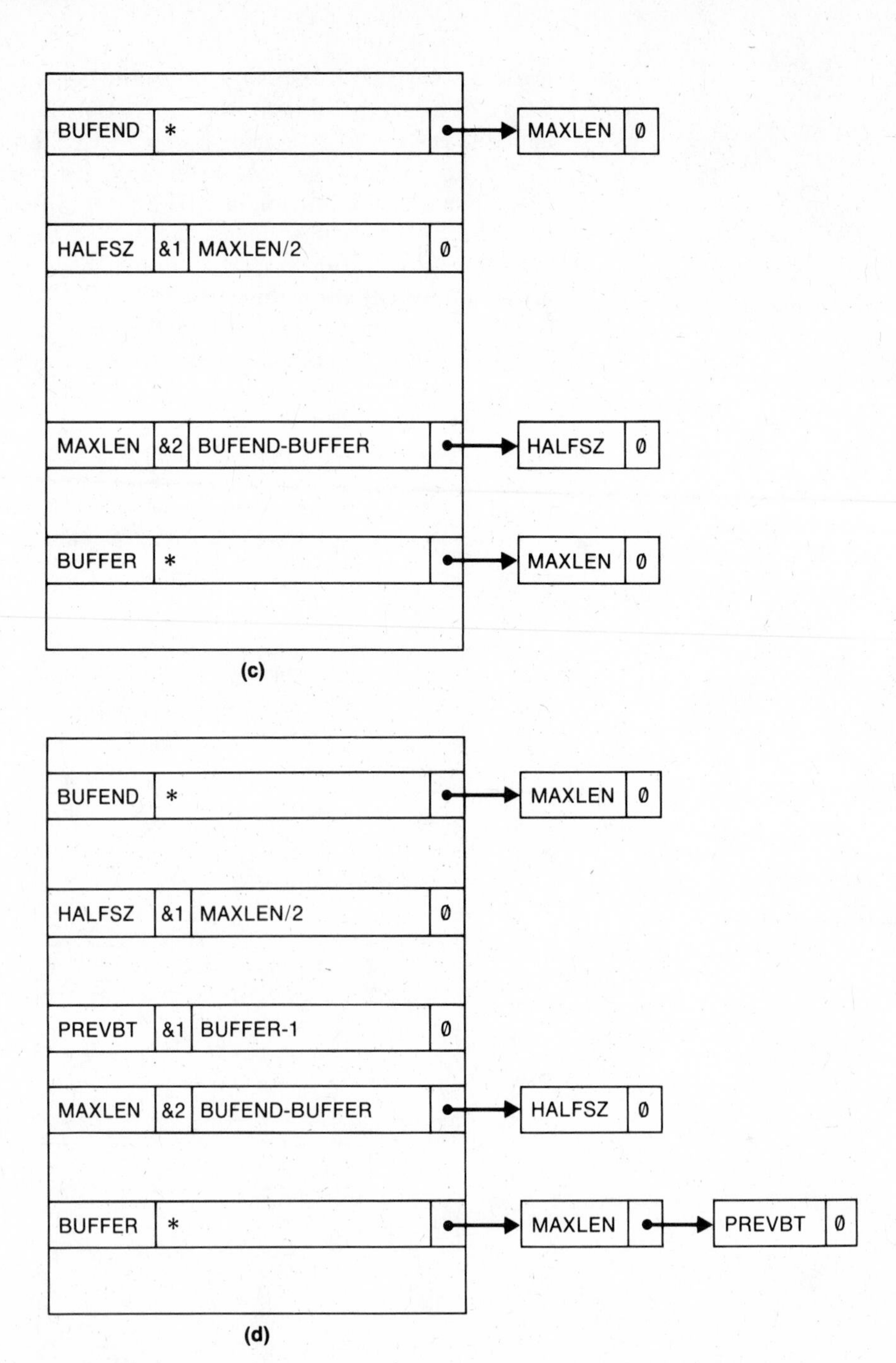

FIGURE 2.23 *(cont'd)*

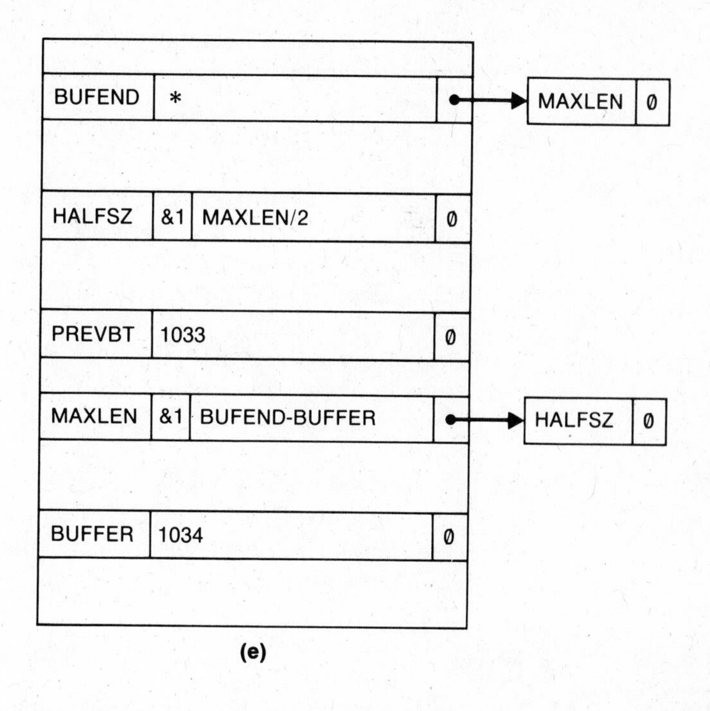

(e)

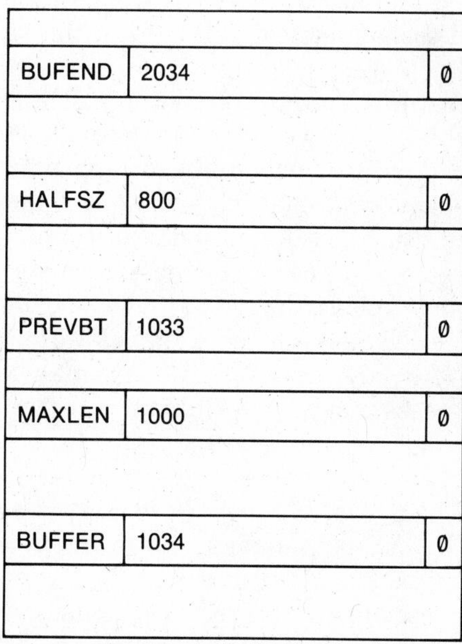

(f)

FIGURE 2.23 *(cont'd)*

on two currently undefined symbols; therefore, MAXLEN cannot be evaluated immediately. Instead, the &2 is changed to &1 to show that only one symbol in the definition (BUFEND) remains undefined. The other symbol in the list (PREVBT) can be evaluated because it depends only on BUFFER. The value of the defining expression for PREVBT is calculated and stored in SYMTAB. The result is shown in Fig. 2.23(e).

The remainder of the processing follows the same pattern. When BUFEND is defined by line 5, its value is entered into the symbol table. The list associated with BUFEND then directs the assembler to evaluate MAXLEN, and entering a value for MAXLEN causes the evaluation of the symbol in its list (HALFSZ). As shown in Fig. 2.23(f), this completes the symbol definition process. If any symbols remained undefined at the end of the program, the assembler would flag them as errors.

The procedure we have just described applies to symbols defined by assembler directives like EQU. You are encouraged to think about how this method could be modified to allow forward references in ORG statements as well.

2.5 IMPLEMENTATION EXAMPLES

We discussed many of the most common assembler features in the preceding sections. However, the variety of machines and assembler languages is very great. Most assemblers have at least some unusual features that are related to machine structure or language design. In this section we discuss three examples of assemblers for real machines. We are obviously unable to give a full description of any of these in the space available. Instead we focus on some of the most interesting or unusual features of each assembler. We are also particularly interested in areas where the assembler design differs from the basic logic and table structure described earlier.

The assembler examples we discuss are for the System/370, the VAX, and the CYBER. You may want to review the descriptions of these machines in Chapter 1 before proceeding.

2.5.1 System/370 Assembler

The System/370 uses base relative addressing for the same reason that SIC/XE did—to save space in the assembled instruction. A full 370 memory address requires 24 bits, whereas the same address can be encoded as a base register number (4 bits) and a displacement (12 bits). There is, however, a difference. The 370 does not have program-counter relative addressing, nor is there anything corresponding to the SIC/XE extended instruction format. Thus all instructions that refer to memory *must* use base relative addressing. As a result, the only parts of a 370 object program that require modification when the program is relocated are data items whose values are to be actual addresses (i.e., *address constants*).

On the 370, any of the general-purpose registers (except R0) can be used as a base register. Decisions about which registers to use in this way are left to the programmer. In a long program, it is not unusual to have several different base registers in use at the same time. The programmer specifies which registers are available for use as base registers, and the contents of these registers, with the USING assembler directive. This is similar in function to the BASE statement in our SIC/XE assembler language. Thus the statements

```
USING   LENGTH,R1
USING   BUFFER,R4
```

would identify R1 and R4 as base registers. R1 would be assumed to contain the address of LENGTH, and R4 would be assumed to contain the address of BUFFER. As in SIC/XE, the programmer must provide instructions to place these values into the registers at execution time. Additional USING statements may appear at any point in the program. If a base register is to be used later for some other purpose, the programmer indicates with the DROP statement that this register is no longer available for addressing purposes.

This additional flexibility in register usage means more work for the assembler. A *base register table* is used to remember which of the general-purpose registers are currently available as base registers, and what base addresses they contain. Processing a USING statement causes an entry to be made in this table (or an existing entry to be modified); processing a DROP statement removes the corresponding table entry. For each instruction whose operand is an address in memory, the assembler scans the table to find a base register that can be used to address that operand. If more than one register can be used, the assembler selects the base register that results in the smallest displacement. If no suitable base register is available, the instruction cannot be assembled. The process of displacement calculation is the same as we described for SIC/XE.

The System/370 assembler language also allows the programmer to write base registers and displacements explicitly in the source program. For example, the instruction

```
L     R2,8(R4)
```

specifies an operand address that is 8 bytes past the address contained in R4. This form of addressing may be useful when some register is known to contain the starting address of a table or data record, and the programmer wishes to refer to a fixed location within that table or record. The assembler simply inserts the specified values into the object code instruction: in this case base register R4 and displacement 8. The base register table is not involved, and the register used in this way need not have appeared in a USING statement.

Another issue that arises in a System/370 assembler is the *alignment* of data items and instructions. Execution of some instructions is more efficient if the instruction operands are aligned on boundaries in memory that correspond to their lengths. For example, a fullword operand (32 bits) should begin at a byte address that is a multiple of 4 and a halfword operand (16 bits) at an address that is a multiple of 2. Likewise, a machine instruction is required to begin at a byte address that is a multiple of 2.

The assembler automatically takes care of these alignment requirements by adjusting the location counter as needed. For example, suppose that the location counter contains 0015, and a fullword data item is to be defined. The location counter value would be increased at 0018 (the next higher multiple of 4) before reserving storage for this fullword. The bytes in the object program that are skipped to provide the proper alignment are called *slack bytes*. (On some 370 assemblers, the alignment of data items is an option that can be selected at assembly time.)

Slack bytes, of course, represent wasted memory in the object program. The order in which data items are defined has much to do with how much memory is lost in this way. For example, consider the sequence of storage reservation statements in Fig. 2.24(a). If, as illustrated, the location counter contains a multiple of 4 when the first statement is encountered, the assembler will need 20 bytes of memory to allocate all data items with proper alignment. Since the items themselves occupy only 14 bytes, this means that the insertion of slack bytes has increased storage requirements by nearly 50%. Grouping together items that have the same alignment requirements can greatly reduce this overhead. For example, if the same definitions were ordered as shown in Fig. 2.24(b), no slack bytes would be required.

There is little that the assembler can do about the sequence of data definitions written by the programmer. The allocation of literal operands, however, is under the assembler's direct control. Most 370 assemblers order the data items in a literal pool to minimize wasted storage. First, addresses are assigned to all doubleword literals followed by all fullwords, all halfwords, and so on.

The System/370 assembler supports control sections (called CSECTs) that are similar in function to those described in Section

Loc	Source statement				
•					
•					
•					
0024	A	DS	C	1 BYTE	
0025					(3 slack bytes)
0028	B	DS	F	FULLWORD	
002C	C	DS	3C	3 BYTES	
002F					(1 slack byte)
0030	D	DS	H	HALFWORD	
0032					(2 slack bytes)
0034	E	DS	F	FULLWORD	
0038					
•					
•					
•					

(a)

Loc	Source statement			
•				
•				
•				
0024	B	DS	F	FULLWORD
0028	E	DS	F	FULLWORD
002C	D	DS	H	HALFWORD
002E	C	DS	3C	3 BYTES
0031	A	DS	C	1 BYTE
0032				
•				
•				
•				

(b)

FIGURE 2.24 Effect of data item alignment on memory requirements.

2.3.5. However, there is no feature that directly corresponds to the program blocks discussed in Section 2.3.4. In a 370 program, CSECTs may consist of several separate segments. These are gathered together by the assembler in much the same way that program blocks were. The CSECTs, however, are not combined into a single program unit. They remain separate and are handled independently by the loader or linkage editor.

For each external symbol used in a CSECT, the programmer reserves one word of storage for an address constant. The value of this constant is inserted by the loader using a mechanism like the one described in Section 2.3.5. The programmer writes instructions to load the address constant into a base register at execution time. The assembler is informed via a USING statement that this base register can be used to accomplish the external reference.

Suppose one CSECT establishes a base register to refer to a symbol in another CSECT using the procedure just described. It is then possible to use this same base register to refer to other labels within the same external CSECT provided that the two CSECTs involved are assembled at the same time. This is because the symbol table used by the assembler contains entries for all symbols defined during the entire assembly. However, if this is done, the ability to assemble these CSECTs independently of one another is lost.

As you can see, 370 assembler language uses CSECTs for two logically different purposes: object program rearrangement (like SIC/XE program blocks) and program modularity (like SIC/XE control sections). This often results in confusion on the part of programmers first introduced to these concepts on such a system.

Another point of confusion when using CSECTs in 370 assembler language is the handling of literal pools. If there are no LTORG statements, all literals are placed at the end of the first CSECT in the assembly. In such a case, the programmer must ensure that a base register is always available for addressing this control section. The ability to assemble the CSECTs separately is lost unless LTORG statements are included at the end of each CSECT.

The System/370 assembler language also provides a special type of control section called a *dummy section* (DSECT). Any assembler language statements may appear in a DSECT; however, these statements do not become part of the object program. DSECTs are most commonly used to describe the layout of a record, table, etc. that is defined externally. The labels used within the DSECT define symbols that can be used to address fields in the record or table (after an appropriate base register is established).

Further information about a typical System/370 assembler can be found in IBM (1979), IBM (1974), and IBM (1982).

2.5.2 VAX Assembler

The VAX has a much more flexible method of encoding operands than do most machines. There are an unusually large number of addressing modes, and (with a few exceptions) any of these modes can be used with any instruction. The operand specifiers associated with the different addressing modes occupy varying amounts of storage. As we shall see, this requires more work from the assembler.

Some of the terminology used in the VAX assembler language differs from the standard set of terms we have been using. For example, there are two addressing modes in which the value of the operand is stored as a part of the instruction: immediate mode and literal mode. (In Section 2.2 we used the term *immediate addressing* for this type of operand.) These two addressing modes are functionally equivalent; in fact, they are specified by the same notation in the source program. The assembler decides which mode to use during Pass 1 by examining the operand value. If the operand is an integer between 0 and 63 (or some other value that fits in a 6-bit field), literal mode is used. If the operand does not fit into 6 bits, or if the assembler cannot yet determine its value, immediate mode is selected. The length of an operand specifier that uses immediate mode is determined by the type of operand used by that instruction (word, longword, etc.). The length of an operand specifier using literal mode is always 1 byte.

Note that VAX operands that use literal mode are handled in a very different manner from the literals discussed in Section 2.3. There is no VAX assembler language feature that corresponds to what we called literals in our SIC/XE assembler.

There are several other instances in which the length of an operand specifier (hence the length of the instruction) must be determined by the assembler. Using program-counter relative mode, for example, the displacement field may be encoded as a byte, a word, or a longword. The assembler selects the length of this field according to the displacement value required. If any of the symbols in the instruction operand are not yet defined, the assembler cannot compute the displacement during Pass 1. In such a case, the assembler selects the maximum possible displacement size (a longword). Note that this decision cannot be delayed until Pass 2 because the length of the instruction will affect the values of all labels defined later in the program.

The calculation of the displacement value for program-counter relative mode is the same as discussed in Section 2.2. However, when the target address is calculated at execution time, the program counter will contain the address of the next *operand specifier* (if there is one)—not necessarily the address of the next *instruction*. This must be taken into consideration when computing the displacement value to be inserted in the instruction.

In most assembler languages the length of an assembled instruction can be determined from the mnemonic operation code. This is not the case with VAX. Each VAX operand specifier may be from 1 to 9 bytes in length depending upon the type of operand. For example, in the instruction

```
MOVB    R1,R5
```

both operands are in registers. Each of the operand specifiers requires 1 byte; thus the assembled instruction will occupy 3 bytes. On the other hand, in the instruction

```
MOVB    1,8(R6)
```

the first operand uses literal mode (a 1-byte operand specifier), and the second uses base relative mode. Because of the size of the displacement value (8), the second operand specifier requires 2 bytes; thus the assembled instruction will occupy 4 bytes. In the instruction

```
MOVB    R1,ALPHA
```

the first operand specifier requires 1 byte. The second operand uses program-counter relative mode. The specifier may occupy from 2 to 5 bytes depending upon the size of the displacement field required. Thus this instruction will be between 4 and 7 bytes long depending upon the address assigned to ALPHA.

This means that Pass 1 of a VAX assembler must be considerably more complex than Pass 1 of, for example, a System/370 assembler. Pass 1 of a VAX assembler must analyze the operands of an instruction in addition to looking at the operation code. The operation code table must also be more complicated since it must contain information on which addressing modes are valid for each operand.

The VAX assembler language provides for control sections called PSECTs (*program sections*). Like CSECTs in 370 assembler language, PSECTs are used for the two distinct purposes of program rearrangement and program modularity. On the VAX, PSECTs are handled somewhat differently: less of the work is performed by the assembler and more by the linker. If one PSECT refers to a symbol that is defined in another PSECT within the same assembly, the reference is assembled as a program-counter relative operand. The displacement cannot be calculated by the assembler because it does not know where the PSECTs will be loaded relative to one another. The assembler does, however, generate commands that will cause the linker to calculate and insert this value. The process is similar to the one described in Section 2.3.4 for resolving references between program blocks. We consider the action of the linker in more detail in the next chapter.

A number of attributes may be specified for each PSECT, including such access properties as "read-only," "execute-only," etc. In addition, each PSECT may specify whether or not data items are to be aligned. (Alignment issues on the VAX are similar to those on the 370.) The linker gathers together all of the segments of each PSECT and all of the PSECTs with similar attributes. VAX operating system capabilities can be used to provide protection for PSECTs with certain attributes.

The VAX assembler makes the default assumption that any symbol that is used, but not defined, within an assembly represents an external reference. No explicit declaration of symbols as external references (such as the SIC/XE EXTREF statement) is required. Unfortunately, this means that errors such as misspellings of symbols are not detected until the program is linked. The programmer can, however, specify an assembler option that changes this default assumption to require that external reference symbols be declared.

Further information about the VAX assembler can be found in DEC (1982) and DEC (1979).

2.5.3 CYBER Assembler

The CYBER assembler language, called COMPASS™, differs from the other languages in several important ways. The first difference is a direct consequence of machine structure—the fact that one machine word is large enough to contain several different assembled instructions. Subject to some restrictions, the assembler must pack as many consecutive instructions as possible into each word of the object program. Of course, we could simply store one instruction per word, filling the unused space with no-op instructions, but this would be very inefficient in terms of both space and execution time.

To accomplish this packing together of instructions, the assembler maintains a *position counter*. This counter indicates the number of bits remaining in the current word of the object program. The position counter is initialized to 60 (the number of bits in a CYBER word) when a new word is begun. As each instruction is generated, the number of bits in that instruction is subtracted from the position counter. When there is no longer enough space in the current word to contain the next assembled instruction, the remaining bits (if any) are filled with no-op instructions, and a new word is started. At this time, the location counter and the origin counter (described later in this section) are incremented by 1 to point to the next word address.

The preceding description represents the normal sequence of events: instructions (or data items) are packed into a machine word until the next one will not fit, and then a new word is started. There are, however, circumstances that require us to begin a new word even though the current word has enough room for the next piece of object code. This need is a consequence of the way memory is addressed on the CYBER—by words, instead of by bytes. Consider, for example, a Jump instruction. The operand address for this instruction will be a *word* address. When the program counter is reset as a result of executing the jump, the next instruction to be fetched for execution will be the *first* instruction packed into the word at the address that is the target of the jump. This means we can jump directly to an assembled instruction *only* if that instruction appears at the beginning of a word in the object program.

The process of placing an instruction or a data item at the beginning of a word, even when there is enough room in the previous word, is called *forcing upper*. When this is done, the previous word is filled out with no-op instructions, the origin and location counters are incremented, and the position counter is set to 60 for the new word. The COMPASS assembler forces upper in the following situations:

1. if the statement being assembled has a label (because the label address will indicate the beginning of a word),
2. if the statement being assembled is a storage reservation statement that does not specify bit fields being packed into a word,
3. if the previous statement assembled is an unconditional Jump instruction (because the only way to execute the next instruction would be via a jump to it).

The programmer can specify a force upper operation for any other statement by coding a + in the label field of the source statement. An automatic force upper can be cancelled by coding a − in this field.

In addition to the position counter, the assembler maintains a *location counter* and an *origin counter*. Together the location and origin counters serve the same purposes as the LOCCTR described in Section 2.1. The main function of the origin counter is the same as for our LOCCTR: the origin counter, together with the position counter, indicates the relative location of the next piece of object code to be assembled. The location counter is used by the assembler to assign addresses to labels within the program. In our earlier description of assembler function, we also used LOCCTR for this purpose.

Normally, the location counter contains the same value as the origin counter. If necessary, however, the programmer may use assembler directives to change the location counter value. This is useful only in very special circumstances when code is to be assembled at one location and subsequently moved to another location before execution.

In this case the programmer resets the location counter to correspond to the address where the code will be located during execution.

Figure 2.25 shows an example of such a use of the location counter. This program contains a large buffer area that will be used for input and output during execution. There are three different options that may be selected when the program is called; however, only one option will be used during each execution. The executable instructions for the three options are coded in separate portions of the source program. The code for each option occupies a considerable amount of space.

If the object program were loaded with the buffer area and all three options, it would require a large amount of storage. We can save memory with the technique shown in Fig. 2.25(a). The code for Option 1, the largest option, is assembled and loaded in the usual way; however, the code for Options 2 and 3 is loaded into the buffer area. This area is not used until input is read, so the initial placement of instructions here is no problem. If either Option 2 or Option 3 is selected, the appropriate section of object code is moved by the program from the buffer area, replacing the previously loaded code for Option 1. (See Fig. 2.25b.) Now the proper option is selected simply by calling whatever code is in the option area, and the buffer is available to receive input.

The programmer must be sure, however, that labels are properly defined if this kind of object code movement is to be done. Consider, for example, the instruction labelled SKIP2 in Fig. 2.25(c) (line 11). The

FIGURE 2.25 Movement of object code during program execution.

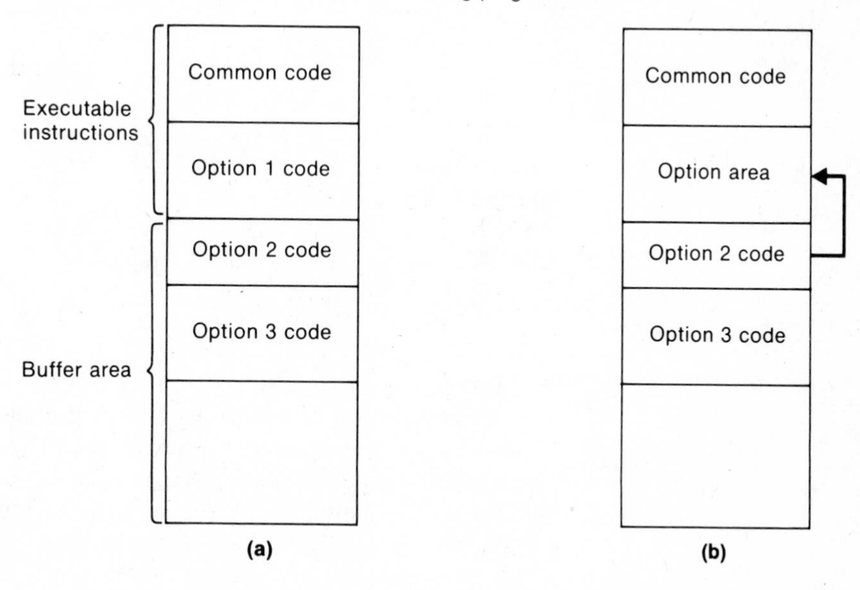

```
Line          Source statement

  1                  IDENT
                       :
  2        OPT1      SA1       ...              OPTION 1 CODE
                       :
  3                  ZR        B4,SKIP1
                       :
  4        SKIP1     SA6       ...
                       :
  5                  EQ        ...              EXIT
  6        BUFFER    BSS       1000             1000-WORD BUFFER
  7                  ORG       BUFFER
  8                  LOC       OPT1
  9        OPT2      SA2       ...              OPTION 2 CODE
                       :
 10                  NZ        B3,SKIP2
                       :
 11        SKIP2     SA7       ...
                       :
 12                  EQ        ...              EXIT
 13                  LOC       OPT1
 14        OPT3      SA6       ...              OPTION 3 CODE
                       :
 15                  EQ        B1,B3,SKIP3
                       :
 16        SKIP3     SA6       ...
                       :
 17                  EQ        ...              EXIT
 18                  ORG       BUFFER+1000
 19                  LOC       *O
                       :
 20                  END
```

(c)

FIGURE 2.25 *(cont'd)*

origin counter has been reset (line 7) so the code for Option 2 and Option 3 will be assigned to addresses within the buffer. If the program were assembled in the usual way, the value of SKIP2 would therefore be an address in the buffer area. During execution of the code for this option, however, the instructions will actually be located in the option area. The target of the Jump instruction on line 10 should be an address in this option area, not in the buffer.

This problem is solved by resetting the location counter to contain the beginning address of the option area (line 8). Recall that the values for labels in this assembler are taken from the location counter, not the origin counter. Thus the value assigned to SKIP2 will reflect the address where it will be located during execution, not the address where it is first assembled. The same procedure is followed for the Option 3 code. After the three options are assembled, the ORG statement on line 18 sets the origin counter back to the next address after the buffer. The statement on line 19 then resets the location counter to contain the same value as the origin counter, and normal assembly is resumed.

COMPASS provides for program blocks (called *subprogram blocks*) and control sections (called *subprograms*). The processing of these is much the same as described in Section 2.3, except that all literals are assigned to a separate block within each subprogram. The assembler maintains a separate set of counters (origin, location, and position) for each subprogram block.

The last unusual COMPASS feature we discuss arises from the nature of the assembler language. On most machines it is possible to determine the machine operation code for an object instruction from the mnemonic code in the source statement. This is not the case on the CYBER. Consider the statements

```
SA1  A2
SA1  B2+TAB
SA1  X1+B1
SA1  X1-B1
```

The first of these would be assembled as a 15-bit instruction with machine opcode 54; the second would be a 30-bit instruction with opcode 51; the third would be a 15-bit instruction with opcode 53; the fourth is illegal.

Mnemonics such as SA1 (and most of the others in CYBER assembler language) must be interpreted by examining the instruction operands. Source program instructions such as these are known as *syntactically identified instructions* because they are recognized through identification of their *syntax* (i.e., the form of the statement). The COMPASS assembler processes such instructions by scanning the mnemonic operation code and the operands, and by extracting a character string that describes the syntax. For the third instruction given above, this syntax descriptor would be SAX+B, which specifies a Set A instruction whose operand is an X register plus a B register. There are 22 different possible subfield descriptors that can be combined to produce such a syntax description.

Syntactically identified instructions are stored in the operation code table according to their syntax descriptors. The table entry provides additional information that specifies how the different parts of the source statement are to be inserted into the object instruction. Before processing a source language instruction, the assembler first searches the operation code table for the mnemonic instruction code. For some instructions, this search succeeds. However, for syntactically identified instructions such as SA1, it will fail. If the search fails, the assembler scans the source statement, extracting a syntax descriptor as previously outlined. It then performs a second table search. If this second search fails, the instruction cannot be recognized.

Further information about the CYBER assembler can be found in CDC (1982a)

EXERCISES

Section 2.1

1. Apply the algorithm described in Fig. 2.4 to assemble the source program in Fig. 2.1. Your results should be the same as those shown in Figs. 2.2 and 2.3.

2. Apply the algorithm described in Fig. 2.4 to assemble the following SIC source program:

```
SUM      START   4000
FIRST    LDX     ZERO
         LDA     ZERO
LOOP     ADD     TABLE,X
         TIX     COUNT
         JLT     LOOP
         STA     TOTAL
         RSUB
TABLE    RESW    2000
COUNT    RESW    1
ZERO     WORD    0
TOTAL    RESW    1
         END     FIRST
```

3. As mentioned in the text, a number of operations in the algorithm of Fig. 2.4 are not explicitly spelled out. (One example would be scanning the instruction operand field for the modifier ",X".) List as many of these implied operations as you can, and think about how they might be implemented.

4. Suppose that you are to write a "disassembler"—that is, a system program that takes an ordinary program as input and produces a listing of the source version of the program. What tables and data structures would be

required, and how would they be used? How many passes would be needed? What problems would arise in recreating the source program?

5. Many assemblers use free-format input. Labels must start in Column 1 of the source statement, but other fields (opcode, operands, comments) may begin in any column. The various fields are separated by blanks. How could our assembler logic be modified to allow this?

6. The algorithm in Fig. 2.4 provides for the detection of some assembly errors; however, there are many more such errors that might occur. List error conditions that might arise during the assembly of a SIC program. When and how would each type of error be detected, and what action should the assembler take for each?

7. Suppose that the SIC assembler language is changed to include a new form of the RESB statement, such as

```
          RESB     n'c'
```

which reserves *n* bytes of memory and initializes all of these bytes to the character 'c'. For example, Line 105 in Fig. 2.5 could be changed to

```
BUFFER    RESB     4096' '
```

This feature could be implemented by simply generating the required number of bytes in Text records. However, this could lead to a large increase in the size of the object program—for example, the object program in Fig. 2.8 would be about 40 times its previous size. Propose a way to implement this new form of RESB without such a large increase in object program size.

Section 2.2

1. Could the assembler decide for itself which instructions need to be assembled using extended format? (This would avoid the necessity for the programmer to code + in such instructions.)

2. As we have described it, the BASE statement simply gives information to the assembler. The programmer must also write an instruction like LDB to load the correct value into the base register. Could the assembler automatically generate the LDB instruction from the BASE statement? If so, what would be the advantages and disadvantages of doing this?

3. Generate the object code for each statement in the following SIC/XE program:

```
SUM       START    0
FIRST     LDX      #0
          LDA      #0
          +LDB     #TABLE2
          BASE     TABLE2
LOOP      ADD      TABLE,X
          ADD      TABLE2,X
```

```
                TIX       COUNT
                JLT       LOOP
               +STA       TOTAL
                RSUB
      COUNT     RESW      1
      TABLE     RESW      2000
      TABLE2    RESW      2000
      TOTAL     RESW      1
                END       FIRST
```

4. Generate the complete object program for the source program in Exercise 3.

5. Modify the algorithm described in Fig. 2.4 to handle all of the SIC/XE addressing modes discussed. How would these modifications be reflected in the modular structure discussed in Chapter 8?

6. Modify the algorithm described in Fig. 2.4 to handle relocatable programs. How would these modifications be reflected in the modular structure discussed in Chapter 8?

7. Suppose that you are writing a disassembler for SIC/XE (see Exercise 2.1.4.) How would your disassembler deal with the various addressing modes and instruction formats?

8. Our Modification record format is well suited for SIC/XE programs because all address fields in instructions and data words fall neatly into half-bytes. What sort of Modification record could we use if this were not the case (that is, if address fields could begin anywhere within a byte and could be of any length)?

9. Suppose that we made the program in Fig. 2.1 a relocatable program. This program is written for the *standard* version of SIC, so all operand addresses are actual addresses, and there is only one instruction format. Nearly every instruction in the object program would need to have its operand address modified at load time. This would mean a large number of Modification records (more than doubling the size of the object program). How could we include the required relocation information without this large increase in object program size?

10. Suppose that you are writing an assembler for a machine that has *only* program-counter relative addressing. (That is, there are no direct-addressing instruction formats and no base relative addressing.) Suppose that you wish to assemble an instruction whose operand is an absolute address in memory—for example,

```
LDA      100
```

to load Register A from address (hexadecimal) 100 in memory. How might such an instruction be assembled in a relocatable program? What relocation operations would be required?

Section 2.3

1. Modify the algorithm described in Fig. 2.4 to handle literals.

2. In the program of Fig. 2.9, could we have used literals on lines 135 and 145? Why might we prefer *not* to use a literal here?

3. With a minor extension to our literal notation, we could write the instruction on line 55 of Fig. 2.9 as

```
LDA     =W'3'
```

   specifying as the literal operand a word with value 3. Would this be a good idea?

4. Immediate operands and literals are both ways of specifying an operand value in a source statement. What are the advantages and disadvantages of each? When might each be preferable to the other?

5. Suppose we made the following changes to the program in Fig. 2.9:

   a) Delete the LTORG statement on line 93.

   b) Change the statement on line 45 to +LDA. . . .

   c) Change the operands on lines 135 and 145 to use literals (and delete line 185).

   Show the resulting object code for lines 45, 135, 145, 215, and 230. Also show the literal pool with addresses and data values. Note: you do not need to retranslate the entire program to do this.

6. Assume that the symbols ALPHA and BETA are labels in a source program. What is the difference between the following two sequences of statements?

   a)
```
LDA   ALPHA-BETA
```

   b)
```
LDA   ALPHA
SUB   BETA
```

7. What is the difference between the following sequences of statements?

   a)
```
        LDA   #3
```

   b)
```
THREE   EQU    3
          .
          .
          .
        LDA   #THREE
```

   c)
```
THREE   EQU    3
          .
          .
          .
        LDA    THREE
```

8. Modify the algorithm described in Fig. 2.4 to handle multiple program blocks.

9. Modify the algorithm described in Fig. 2.4 to handle multiple control sections.

10. Suppose all the features we described in Section 2.3 were to be implemented in an assembler. How would the symbol table required be different from the one discussed in Section 2.1?

11. Which of the features described in Section 2.3 would create additional problems in the writing of a disassembler (see Exercise 2.1.4)? Describe these problems, and discuss possible solutions.

12. When different control sections are assembled together, some references between them could be handled by the assembler (instead of being passed on to the loader). In the program of Fig. 2.15, for example, the expression on line 190 could be evaluated directly by the assembler because its symbol table contains all of the required information. What would be the advantages and disadvantages of doing this?

13. In the program of Fig. 2.11, suppose we used only two program blocks: the default block and CBLKS. Assume that the data items in CDATA are to be included in the default block. What changes in the source program would accomplish this? Show the object program (corresponding to Fig. 2.13) that would result.

14. Suppose LENGTH is defined as in the program of Fig. 2.9. What would be the difference between the following sequences of statements?

    a)
    ```
    LDA    LENGTH
    SUB    #1
    ```

    b)
    ```
    LDA    LENGTH-1
    ```

15. Referring to the definitions of symbols in Fig. 2.10, give the value, type, and intuitive meaning (if any) of each of the following expressions:

    a) `BUFFER-FIRST`

    b) `BUFFER+4095`

    c) `MAXLEN-1`

    d) `BUFFER+MAXLEN-1`

    e) `BUFFER-MAXLEN`

    f) `2*LENGTH`

    g) `2*MAXLEN-1`

    h) `MAXLEN-BUFFER`

    i) `FIRST+BUFFER`

    j) `FIRST-BUFFER+BUFEND`

16. In the program of Fig. 2.9, what is the advantage of writing (on line 107)

    ```
    MAXLEN    EQU    BUFEND-BUFFER
    ```

    instead of

    ```
    MAXLEN    EQU    4096  ?
    ```

17. In the program of Fig. 2.15, could we change line 190 to

    ```
    MAXLEN    EQU    BUFEND-BUFFER
    ```

and line 133 to

```
+LDT    #MAXLEN
```

as we did in Fig. 2.9?

18. The assembler could simply assume that any reference to a symbol not defined within a control section is an external reference. This change would eliminate the need for the EXTREF statement. Would this be a good idea?

19. How could an assembler that allows external references avoid the need for an EXTDEF statement? What would be the advantages and disadvantages of doing this?

20. The assembler could automatically use extended format for instructions whose operands involve external references. This would eliminate the need for the programmer to code + in such statements. What would be the advantages and disadvantages of doing this?

21. On some systems, control sections can be composed of several different parts, just as program blocks can. What problems does this pose for the assembler? How might these problems be solved?

22. Assume that the symbols RDREC and COPY are defined as in Fig. 2.15. According to our rules, the expression

```
RDREC-COPY
```

would be illegal (that is, the assembler and/or the loader would reject it). Suppose that for some reason the program really needs the value of this expression. How could such a thing be accomplished without changing the rules for expressions?

23. We discussed a large number of assembler directives, and many more could be implemented in an actual assembler. Checking for them one at a time using comparisons might be quite inefficient. How could we use a table, perhaps similar to OPTAB, to speed recognition and handling of assembler directives? (Hint: the answer to this problem may depend upon the language in which the assembler itself is written.)

24. Other than the listing of the source program with generated object code, what assembler outputs might be useful to the programmer? Suggest some optional listings that might be generated and discuss any data structures or algorithms involved in producing them.

Section 2.4

1. Consider a basic assembler as described by the algorithm in Fig. 2.4. What tables and routines would be in the Root segment of such an assembler?

2. What tables and routines would be in the Root segment of a more advanced assembler that implements all the features described in Section 2.3?

3. The process of fixing up a few forward references should involve less overhead than making a complete second pass of the source program. Why don't all assemblers use the one-pass technique for efficiency?

4. Suppose we wanted our assembler to produce a cross-reference listing for all symbols used in the program. For the program of Fig. 2.5, such a listing might look like

Symbol	Defined on line	Used on lines
COPY	5	
FIRST	10	255
CLOOP	15	40
ENDFIL	45	30
EOF	80	45
RETADR	95	10,70
LENGTH	100	12,13,20,60,175,212
.		
.		
.		

How might this be done by the assembler? Indicate changes to the logic and tables discussed in Section 2.1 that would be required.

5. Could a one-pass assembler produce a relocatable object program and handle external references? Describe the processing logic that would be involved and identify any potential difficulties.

6. How could literals be implemented in a one-pass assembler?

7. We discussed one-pass assemblers as though instruction operands could only be single symbols. How could a one-pass assembler handle an instruction like

```
JEQ     ENDFIL+3
```

where ENDFIL has not yet been defined?

8. Outline the logic flow for a simple one-pass load-and-go assembler.

9. Using the methods outlined in Chapter 8, develop a modular design for a one-pass assembler that produces object code in memory.

10. Suppose that an instruction involving a forward reference is to be assembled using program-counter relative addressing. How might this be handled by a one-pass assembler?

11. The process of fixing up forward references in a one-pass assembler that produces an object program is very similar to the linking process described in Section 2.3.5. Why didn't we just use Modification records to fix up the forward references?

12. How could we extend the methods of Section 2.4.3 to handle forward references in ORG statements?

CHAPTER **3**

LOADERS AND LINKERS

As we have seen, an object program contains translated instructions and data values from the source program, and specifies addresses in memory where these items are to be loaded. Our discussions in Chapter 2 introduced the following three processes:

1. *Loading,* which brings the object program into memory for execution.
2. *Relocation,* which modifies the object program so that it can be loaded at an address different from the location originally specified (see Section 2.2.2).
3. *Linking,* which combines two or more separate object programs and supplies the information needed to allow references between them (see Section 2.3.5).

A *loader* is a system program that performs the loading function. Many loaders also support relocation and linking. Some systems have a *linker* (or *linkage editor*) to perform the linking operations and a separate loader to handle relocation and loading. In most cases all the program translators (i.e., assemblers and compilers) on a particular system produce object programs in the same format. Thus one system loader or linker can be used regardless of the original source programming language.

In this chapter we study the design and implementation of loaders and linkers. For simplicity we often use the term "loader" in place of "loader and/or linker." Because the processes of assembly and loading are closely related, this chapter is similar in structure to the preceding one. Many of the same examples used in our study of assemblers are carried forward in this chapter. During our discussion of assemblers, we studied a number of features and capabilities that are of concern to both the assembler and the loader. In the present chapter we encounter many of the same concepts again. This time, of course, we are primarily concerned with the operation of the loader; however, it is important to remember the close connections between program translation and loading.

As in the preceding chapter, we begin by discussing the most basic software function—in this case, loading an object program into memory for execution. Section 3.1 presents the design of an *absolute loader* and illustrates its operation. Such a loader might be found on a simple SIC machine that uses the sort of assembler described in Section 2.1.

Section 3.2 examines the issues of relocation and linking from a loader's point of view. We consider some possible alternatives for object program representation and examine how these are related to issues of machine structure. We also present the design of a *linking loader*, a more advanced type of loader that is typical of those found on most modern computing systems.

Section 3.3 presents a selection of commonly encountered loader features that are not directly related to machine structure. As before, our purpose is not to cover all possible options, but to introduce some of the concepts and techniques most frequently found in loaders.

Section 3.4 discusses alternative ways of accomplishing loader functions. We consider the various times at which relocation and linking can be performed, and the advantages and disadvantages associated with each. In this context we study linkage editors (which perform linking before loading) and dynamic linking schemes (which delay linking until execution time).

Finally, in Section 3.5 we briefly discuss some examples of actual loaders and linkers. We consider the same machines for which we described assemblers in Section 2.5 and point out relationships between the assembler and loader designs. As before, we are primarily concerned with aspects of each piece of software that are related to hardware or software design decisions.

3.1 BASIC LOADER FUNCTIONS

In this section we discuss the most fundamental functions of a loader—bringing an object program into memory and starting its execution. You are probably already familiar with how these basic functions are performed. This section is intended as a review to set the stage for our later discussion of more advanced loader functions. Section 3.1.1 discusses the functions and design of an absolute loader and gives the outline of an algorithm for such a loader. Section 3.1.2 presents an example of a very simple absolute loader for SIC/XE, to clarify the coding techniques that are involved.

3.1.1 Design of an Absolute Loader

We consider the design of an absolute loader that might be used with the sort of assembler described in Section 2.1. The object program format used is the same as that described in Section 2.1.1. An example of such an object program is shown in Fig. 3.1(a).

Because our loader does not need to perform such functions as linking and program relocation, its operation is very simple. All functions are accomplished in a single pass. The Header record is checked to verify that the correct program has been presented for loading (and that it will fit into the available memory). As each Text record is read, the object code it contains is moved to the indicated address in memory. When the End record is encountered, the loader jumps to the specified address to begin execution of the loaded program. Figure 3.1(b) shows a representation of the program from Fig. 3.1(a) after loading. The contents of memory locations for which there is no Text record are shown as *xxxx*. This indicates that the previous contents of these locations remain unchanged.

Figure 3.2 shows an algorithm for the simple loader logic we have discussed. Although this process is extremely simple, there is one aspect that deserves comment. In our object program, each byte of assembled code is given using its hexadecimal representation in character form. For example, the machine operation code for an STL instruction would be represented by the *pair of characters* 14. When these are read by the loader (as part of the object program), they will occupy *two* bytes of memory. In the instruction as loaded for execution, however, this operation code must be stored in a *single* byte with *hexadecimal value* 14. Thus each pair of bytes from the object program record must be packed together into one byte during loading. It is very important to realize that in Fig. 3.1(a), each printed character represents one *byte* of the object program record. In Fig. 3.1(b), on the other hand, each printed character represents one *hexadecimal digit* in memory (i.e., a half-byte).

This method of representing an object program is inefficient in

```
H^COPY  ^001000^00107A
T^001000^1E^141033^482039^001036^281030^301015^482061^3C1003^00102A^0C1039^00102D
T^00101E^15^0C1036^482061^081033^4C0000^454F46^000003^000000
T^002039^1E^041030^001030^E0205D^30203F^D8205D^281030^302057^549039^2C205E^38203F
T^002057^1C^101036^4C0000^F1^001000^041030^E02079^302064^509039^DC2079^2C1036
T^002073^07^382064^4C0000^05
E^001000
```

(a) Object program

Memory address	Contents			
0000	xxxxxxxx	xxxxxxxx	xxxxxxxx	xxxxxxxx
0010	xxxxxxxx	xxxxxxxx	xxxxxxxx	xxxxxxxx
⋮	⋮	⋮	⋮	⋮
0FF0	xxxxxxxx	xxxxxxxx	xxxxxxxx	xxxxxxxx
1000	14103348	20390010	36281030	30101548
1010	20613C10	0300102A	0C103900	102D0C10
1020	36482061	0810334C	0000454F	46000003
1030	000000xx	xxxxxxxx	xxxxxxxx	xxxxxxxx
⋮	⋮	⋮	⋮	⋮
2030	xxxxxxxx	xxxxxxxx	xx041030	001030E0
2040	205D3020	3FD8205D	28103030	20575490
2050	392C205E	38203F10	10364C00	00F10010
2060	00041030	E0207930	20645090	39DC2079
2070	2C103638	20644C00	0005xxxx	xxxxxxxx
2080	xxxxxxxx	xxxxxxxx	xxxxxxxx	xxxxxxxx
⋮	⋮	⋮	⋮	⋮

← COPY

(b) Program loaded in memory

FIGURE 3.1 Loading of an absolute program.

FIGURE 3.2 Algorithm for an absolute loader.

```
begin
read Header record
verify program name and length
read first Text record
while record type <> 'E' do
    begin
    {if object code is in character form, convert into internal representation}
    move object code to specified location in memory
    read next object program record
    end
jump to address specified in End record
end
```

terms of both space and execution time. Therefore, most machines store object programs in a *binary* form, with each byte of object code stored as a single byte in the object program. In this type of representation, of course, a byte may contain any binary value. We must be sure that our file and device conventions do not cause some of the object program bytes to be interpreted as control characters. For example, the convention described in Section 2.1—indicating the end of a record with a byte containing hexadecimal 00—would obviously be unsuitable for use with a binary object program.

Obviously object programs stored in binary form do not lend themselves well to printing or to reading by human beings. Therefore, we continue to use character representations of object programs in our examples in this book.

3.1.2 A Simple Bootstrap Loader

When a computer is first turned on or restarted, a special type of absolute loader, called a *bootstrap loader,* is executed. This bootstrap loads the first program to be run by the computer—usually an operating system. (Bootstrap loaders are discussed in more detail in Section 3.4.3.) In this section, we examine a very simple bootstrap loader for SIC/XE. In spite of its simplicity, this program illustrates almost all of the logic and coding techniques that are used in an absolute loader.

Figure 3.3 shows the source code for our bootstrap loader. The bootstrap itself begins at address 0 in the memory of the machine. It loads the operating system (or some other program) starting at address 80. Because this loader is used in a unique situation (the initial program load for the system), the program to be loaded can be represented in a very simple format. Each byte of object code to be loaded is represented on device F1 as two hexadecimal digits (just as it is in a Text record of a SIC object program). However, there is no Header record, End record, or control information (such as addresses or lengths). The object code from device F1 is always loaded into consecutive bytes of memory, starting at address 80. After all of the object code from device F1 has been loaded, the bootstrap jumps to address 80, which begins the execution of the program that was loaded.

Much of the work of the bootstrap loader is performed by the subroutine GETC. This subroutine reads one character from device F1 and converts it from the ASCII character code to the value of the hexadecimal digit that is represented by that character. For example, the ASCII code for the character "0" (hexadecimal 30) is converted to the numeric value 0. Likewise, the ASCII codes for "1" through "9" (hexadecimal 31 through 39) are converted to the numeric values 1 through 9, and the codes for "A" through "F" (hexadecimal 41 through

```
BOOT      START    0                  BOOTSTRAP LOADER FOR SIC/XE
.
. THIS BOOTSTRAP READS OBJECT CODE FROM DEVICE F1 AND ENTERS IT INTO
. MEMORY STARTING AT ADDRESS 80 (HEXADECIMAL). AFTER ALL OF THE CODE
. FROM DEVF1 HAS BEEN ENTERED INTO MEMORY, THE BOOTSTRAP EXECUTES A
. JUMP TO ADDRESS 80 TO BEGIN EXECUTION OF THE PROGRAM JUST LOADED.
.
.          REGISTER X = NEXT ADDRESS TO BE LOADED
.
LOOP      CLEAR    A                  CLEAR REGISTER A TO ZERO
          LDX      #128               INITIALIZE REGISTER X TO HEX 80
          JSUB     GETC               READ HEX DIGIT FROM PROGRAM BEING LOADED
          RMO      A,S                SAVE IN REGISTER S
          SHIFTL   S,4                MOVE TO HIGH-ORDER 4 BITS OF BYTE
          JSUB     GETC               GET NEXT HEX DIGIT
          ADDR     S,A                COMBINE DIGITS TO FORM ONE BYTE
          STCH     0,X                STORE AT ADDRESS IN REGISTER X
          TIXR     X,X                ADD 1 TO MEMORY ADDRESS BEING LOADED
          J        LOOP               LOOP UNTIL END OF INPUT IS REACHED
.
. SUBROUTINE TO READ ONE CHARACTER FROM INPUT DEVICE AND CONVERT IT
. FROM ASCII CODE TO HEXADECIMAL DIGIT VALUE. THE CONVERTED HEX DIGIT
. VALUE IS RETURNED IN REGISTER A. WHEN AN END-OF-FILE IS READ,
. CONTROL IS TRANSFERRED TO THE STARTING ADDRESS (HEX 80).
.
GETC      TD       INPUT              TEST INPUT DEVICE
          JEQ      GETC               LOOP UNTIL READY
          RD       INPUT              READ CHARACTER
          COMP     #4                 IF CHARACTER IS HEX 04 (END OF FILE),
          JEQ      80                    JUMP TO START OF PROGRAM JUST LOADED
          COMP     #48                COMPARE TO HEX 30 (CHARACTER '0')
          JLT      GETC               SKIP CHARACTERS LESS THAN '0'
          SUB      #48                SUBTRACT HEX 30 FROM ASCII CODE
          COMP     #10                IF RESULT IS LESS THAN 10, CONVERSION IS
          JLT      RETURN                COMPLETE. OTHERWISE, SUBTRACT 7 MORE
          SUB      #7                    (FOR HEX DIGITS 'A' THROUGH 'F')
RETURN    RSUB                        RETURN TO CALLER
INPUT     BYTE     X'F1'              CODE FOR INPUT DEVICE
          END      LOOP
```

FIGURE 3.3 Bootstrap loader for SIC/XE.

46) are converted to the values 10 through 15. This is accomplished by subtracting 48 (hexadecimal 30) from the character codes for "0" through "9", and subtracting 55 (hexadecimal 37) from the codes for "A" through "F". The subroutine GETC jumps to address 80 when an end-of-file (hexadecimal 04) is read from device F1. It skips all other input characters that have ASCII codes less than hexadecimal 30. This causes the bootstrap to ignore any control bytes (such as end-of-line) that are read.

The main loop of the bootstrap keeps the address of the next memory location to be loaded in register X. GETC is used to read and convert a pair of characters from device F1 (representing one byte of object code to be loaded). These two hexadecimal digit values are combined into a single byte by shifting the first one left 4 bit positions and adding the second to it. The resulting byte is stored at the address currently in register X, by a STCH instruction that refers to location 0

using indexed addressing. The TIXR instruction is then used to add 1 to the value in register X. (Because we are not interested in the result of the comparison performed by TIXR, register X is also used as the second operand for this instruction.)

You should work through the execution of this bootstrap routine by hand with several bytes of sample input, keeping track of the exact contents of all registers and memory locations as you go. This will help you to become familiar with the machine-level details of how loading is performed.

For simplicity, the bootstrap routine in Fig. 3.3 does not do any error checking—it assumes that its input is correct. You are encouraged to think about the different kinds of error conditions that might arise during the loading, and how these could be handled.

3.2 MACHINE-DEPENDENT LOADER FEATURES

The absolute loader described in Section 2.1 is certainly simple and efficient; however, this scheme has several potential disadvantages. One of the most obvious is the need for the programmer to specify (when the program is assembled) the actual address at which it will be loaded into memory. If we are considering a very simple computer with a small memory (such as the standard version of SIC), this does not create much difficulty. There is only room to run one program at a time, and the starting address for this single user program is known in advance. On a larger and more advanced machine (such as SIC/XE), the situation is not quite as easy. We would often like to run several independent programs together, sharing memory (and other system resources) between them. This means that we do not know in advance where a program will be loaded. Efficient sharing of the machine requires that we write relocatable programs instead of absolute ones.

Writing absolute programs also makes it difficult to use subroutine libraries efficiently. Most such libraries (for example, scientific or mathematical packages) contain many more subroutines than will be used by any one program. To make efficient use of memory, it is important to be able to select and load exactly those routines that are needed. This could not be done effectively if all of the subroutines had preassigned absolute addresses.

In this section we consider the design and implementation of a more complex loader. The loader we present is one that is suitable for use on a SIC/XE system and is typical of those that are found on most modern computers. This loader provides for program relocation and linking, as well as for the simple loading functions described in the preceding section. As part of our discussion, we examine the effect of machine structure on the design of the loader.

The need for program relocation is an indirect consequence of the change to larger and more powerful computers. The way relocation is implemented in a loader is also dependent upon machine characteristics. Section 3.2.1 discusses these dependencies by examining different implementation techniques and the circumstances in which they might be used.

Section 3.2.2 examines program linking from the loader's point of view. Linking is not a machine-dependent function in the sense that relocation is; however, the same implementation techniques are often used for these two functions. In addition, the process of linking usually involves relocation of some of the routines being linked together. (See, for example, the previous discussion concerning the use of subroutine libraries.) For these reasons we discuss linking together with relocation in this section.

Section 3.2.3 discusses the data structures used by a typical linking (and relocating) loader, and gives a description of the processing logic involved. The algorithm presented here serves as a starting point for discussion of some of the more advanced loader features in the following sections.

3.2.1 Relocation

Loaders that allow for program relocation are called *relocating loaders* or *relative loaders*. The concept of program relocation was introduced in Section 2.2.2; you may want to briefly review that discussion before reading further. In this section we discuss two methods for specifying relocation as part of the object program.

The first method we discuss is essentially the same as that introduced in Chapter 2: A Modification record is used to describe each part of the object code that must be changed when the program is relocated. (The format of the Modification record is given in Section 2.3.5.) Figure 3.4 shows a SIC/XE program we use to illustrate this first method of specifying relocation. The program is the same as the one in Fig. 2.6; it is reproduced here for convenience. Most of the instructions in this program use relative or immediate addressing. The only portions of the assembled program that contain actual addresses are the extended format instructions on lines 15, 35, and 65. Thus these are the only items whose values are affected by relocation.

Figure 3.5 displays the object program corresponding to the source in Fig. 3.4. Notice that there is one Modification record for each value that must be changed during relocation (in this case, the three instructions previously mentioned). Each Modification record specifies the

```
Line   Loc    Source statement                     Object code

  5    0000   COPY     START   0
 10    0000   FIRST    STL     RETADR              17202D
 12    0003            LDB     #LENGTH             69202D
 13                    BASE    LENGTH
 15    0006   CLOOP   +JSUB    RDREC               4B101036
 20    000A            LDA     LENGTH              032026
 25    000D            COMP    #0                  290000
 30    0010            JEQ     ENDFIL              332007
 35    0013           +JSUB    WRREC               4B10105D
 40    0017            J       CLOOP               3F2FEC
 45    001A   ENDFIL   LDA     EOF                 032010
 50    001D            STA     BUFFER              0F2016
 55    0020            LDA     #3                  010003
 60    0023            STA     LENGTH              0F200D
 65    0026           +JSUB    WRREC               4B10105D
 70    002A            J       @RETADR             3E2003
 80    002D   EOF      BYTE    C'EOF'              454F46
 95    0030   RETADR   RESW    1
100    0033   LENGTH   RESW    1
105    0036   BUFFER   RESB    4096
110           .
115           .        SUBROUTINE TO READ RECORD INTO BUFFER
120           .
125    1036   RDREC    CLEAR   X                   B410
130    1038            CLEAR   A                   B400
132    103A            CLEAR   S                   B440
133    103C           +LDT     #4096               75101000
135    1040   RLOOP    TD      INPUT               E32019
140    1043            JEQ     RLOOP               332FFA
145    1046            RD      INPUT               DB2013
150    1049            COMPR   A,S                 A004
155    104B            JEQ     EXIT                332008
160    104E            STCH    BUFFER,X            57C003
165    1051            TIXR    T                   B850
170    1053            JLT     RLOOP               3B2FEA
175    1056   EXIT     STX     LENGTH              134000
180    1059            RSUB                        4F0000
185    105C   INPUT    BYTE    X'F1'               F1
195           .
200           .        SUBROUTINE TO WRITE RECORD FROM BUFFER
205           .
210    105D   WRREC    CLEAR   X                   B410
212    105F            LDT     LENGTH              774000
215    1062   WLOOP    TD      OUTPUT              E32011
220    1065            JEQ     WLOOP               332FFA
225    1068            LDCH    BUFFER,X            53C003
230    106B            WD      OUTPUT              DF2008
235    106E            TIXR    T                   B850
240    1070            JLT     WLOOP               3B2FEF
245    1073            RSUB                        4F0000
250    1076   OUTPUT   BYTE    X'05'               05
255                    END     FIRST
```

FIGURE 3.4 Example of a SIC/XE program (from Fig. 2.6).

```
H^COPY  ^000000^001077
T^000000^1D^17202D^69202D^4B101036^032026^290000^332007^4B10105D^3F2FEC^032010
T^00001D^13^0F2016^010003^0F200D^4B10105D^3E2003^454F46
T^001036^1D^B410^B400^B440^75101000^E32019^332FFA^DB2013^A004^332008^57C003^B850
T^001053^1D^3B2FEA^134000^4F0000^F1^B410^774000^E32011^332FFA^53C003^DF2008^B850
T^001070^07^3B2FEF^4F0000^05
M^000007^05+COPY
M^000014^05+COPY
M^000027^05+COPY
E^000000
```

FIGURE 3.5 Object program with relocation by Modification records.

starting address and length of the field whose value is to be altered. It then describes the modification to be performed. In this example, all modifications add the value of the symbol COPY, which represents the starting address of the program. The actual logic the loader uses to perform these modifications is discussed in Section 3.2.3. More examples of relocation specified in this manner appear in the next section when we examine the relationship between relocation and linking.

The Modification record scheme is a convenient means for specifying program relocation; however, it is not well suited for use with all machine structures. Consider, for example, the program in Fig. 3.6. This is a relocatable program written for the standard version of SIC. The important difference between this example and the one in Fig. 3.4 is that the standard SIC machine does not use relative addressing. In this program the addresses in all the instructions except RSUB must be modified when the program is relocated. This would require 31 Modification records, which results in an object program more than twice as large as the one in Fig. 3.5.

On a machine that primarily uses direct addressing and has a fixed instruction format, it is often more efficient to specify relocation using a different technique. Figure 3.7 shows this method applied to our SIC program example. There are no Modification records. The Text records are the same as before except that there is a *relocation bit* associated with each word of object code. Since all SIC instructions occupy one word, this means that there is one relocation bit for each possible instruction. The relocation bits are gathered together into a *bit mask* following the length indicator in each Text record. In Fig. 3.7 this mask is represented (in character form) as three hexadecimal digits. These characters are underlined for easier identification in the figure.

```
Line    Loc     Source statement              Object code

  5     0000    COPY    START   0
 10     0000    FIRST   STL     RETADR          140033
 15     0003    CLOOP   JSUB    RDREC           481039
 20     0006            LDA     LENGTH          000036
 25     0009            COMP    ZERO            280030
 30     000C            JEQ     ENDFIL          300015
 35     000F            JSUB    WRREC           481061
 40     0012            J       CLOOP           3C0003
 45     0015    ENDFIL  LDA     EOF             00002A
 50     0018            STA     BUFFER          0C0039
 55     001B            LDA     THREE           00002D
 60     001E            STA     LENGTH          0C0036
 65     0021            JSUB    WRREC           481061
 70     0024            LDL     RETADR          080033
 75     0027            RSUB                    4C0000
 80     002A    EOF     BYTE    C'EOF'          454F46
 85     002D    THREE   WORD    3               000003
 90     0030    ZERO    WORD    0               000000
 95     0033    RETADR  RESW    1
100     0036    LENGTH  RESW    1
105     0039    BUFFER  RESB    4096
110             .
115             .       SUBROUTINE TO READ RECORD INTO BUFFER
120             .
125     1039    RDREC   LDX     ZERO            040030
130     103C            LDA     ZERO            000030
135     103F    RLOOP   TD      INPUT           E0105D
140     1042            JEQ     RLOOP           30103F
145     1045            RD      INPUT           D8105D
150     1048            COMP    ZERO            280030
155     104B            JEQ     EXIT            301057
160     104E            STCH    BUFFER,X        548039
165     1051            TIX     MAXLEN          2C105E
170     1054            JLT     RLOOP           38103F
175     1057    EXIT    STX     LENGTH          100036
180     105A            RSUB                    4C0000
185     105D    INPUT   BYTE    X'F1'           F1
190     105E    MAXLEN  WORD    4096            001000
195             .
200             .       SUBROUTINE TO WRITE RECORD FROM BUFFER
205             .
210     1061    WRREC   LDX     ZERO            040030
215     1064    WLOOP   TD      OUTPUT          E01079
220     1067            JEQ     WLOOP           301064
225     106A            LDCH    BUFFER,X        508039
230     106D            WD      OUTPUT          DC1079
235     1070            TIX     LENGTH          2C0036
240     1073            JLT     WLOOP           381064
245     1076            RSUB                    4C0000
250     1079    OUTPUT  BYTE    X'05'           05
255                     END     FIRST
```

FIGURE 3.6 Relocatable program for a standard SIC machine.

```
H^COPY  ^000000^00107A
T^000000^1E^FFC^140033^481039^000036^280030^300015^481061^3C0003^00002A^0C0039^00002D
T^00001E^15^E00^0C0036^481061^080033^4C0000^454F46^000003^000000
T^001039^1E^FFC^040030^000030^E0105D^30103F^D8105D^280030^301057^548039^2C105E^38103F
T^001057^0A^800^100036^4C0000^F1^001000
T^001061^19^FE0^040030^E01079^301064^508039^DC1079^2C0036^381064^4C0000^05
E^000000
```

FIGURE 3.7 Object program with relocation by bit mask.

If the relocation bit corresponding to a word of object code is set to 1, the program's starting address is to be added to this word when the program is relocated. A bit value of 0 indicates that no modification is necessary. If a Text record contains fewer than 12 words of object code, the bits corresponding to unused words are set to 0. Thus the bit mask FFC (representing the bit string 111111111100) in the first Text record specifies that all 10 words of object code are to be modified during relocation. These words contain the instructions corresponding to lines 10 through 55 in Fig. 3.6. The mask E00 in the second Text record specifies that the first three words are to be modified. The remainder of the object code in this record represents data constants (and the RSUB instruction) and thus does not require modification.

The other Text records follow the same pattern. Note that the object code generated from the LDX instruction on line 210 begins a new Text record even though there is room for it in the preceding record. This occurs because each relocation bit is associated with a 3-byte segment of object code in the Text record. Any value that is to be modified during relocation must coincide with one of these 3-byte segments so that it corresponds to a relocation bit. The assembled LDX instruction does require modification because of the direct address. However, if it were placed in the preceding Text record, it would not be properly aligned to correspond to a relocation bit because of the 1-byte data value generated from line 185. Therefore, this instruction must begin a new Text record in the object program.

You should carefully examine the remainder of the object program in Fig. 3.7. Make sure you understand how the relocation bits are generated by the assembler and used by the loader.

Some computers provide a hardware relocation capability that eliminates some of the need for the loader to perform program relocation. For example, some such machines consider all memory references to be relative to the beginning of the user's assigned area of memory. The conversion of these relative addresses to actual addresses is performed as the program is executed. (We discuss this further when we study memory management in Chapter 6.) As the next section illustrates, however, the loader must still handle relocation of subprograms in connection with linking.

3.2.2 Program Linking

The basic concepts involved in program linking were introduced in Section 2.3.5. Before proceeding you may want to review that discussion and the examples in that section. In this section we consider more complex examples of external references between programs and examine the relationship between relocation and linking. The next section gives an algorithm for a linking and relocating loader.

Figure 2.15 in Section 2.3.5 showed a program made up of three control sections. These control sections could be assembled together (that is, in the same invocation of the assembler), or they could be assembled independently of one another. In either case, however, they would appear as separate segments of object code after assembly (see Fig. 2.17). The programmer has a natural inclination to think of a program as a logical entity that combines all of the related control sections. From the loader's point of view, however, there is no such thing as a program in this sense—there are only control sections that are to be linked, relocated, and loaded. The loader has no way of knowing (and no need to know) which control sections were assembled at the same time.

Consider the three (separately assembled) programs in Fig. 3.8, each of which consists of a single control section. Each program contains a list of items (LISTA, LISTB, LISTC); the ends of these lists are marked by the labels ENDA, ENDB, ENDC. The labels on the beginnings and ends of the lists are external symbols (that is, they are available for use in linking). Note that each program contains exactly the same set of references to these external symbols. Three of these are instruction operands (REF1 through REF3), and the others are the values of data words (REF4 through REF8). In considering this example, we examine the differences in the way these identical expressions are handled within the three programs. This emphasizes the relationship between the relocation and linking processes. To focus on these issues, we have not attempted to make these programs appear realistic. All portions of the programs not involved in the relocation and linking process are omitted. The same applies to the generated object programs shown in Fig. 3.9.

Consider first the reference marked REF1. For the first program (PROGA), REF1 is simply a reference to a label within the program. It is assembled in the usual way as a program-counter relative instruc-

```
Loc          Source statement                                  Object code

0000    PROGA      START    0
                   EXTDEF   LISTA,ENDA
                   EXTREF   LISTB,ENDB,LISTC,ENDC
                     .
                     .
0020    REF1       LDA      LISTA                              03201D
0023    REF2      +LDT      LISTB+4                            77100004
0027    REF3       LDX     #ENDA-LISTA                         050014
                     .
                     .
0040    LISTA      EQU      *
                     .
0054    ENDA       EQU      *
0054    REF4       WORD     ENDA-LISTA+LISTC                   000014
0057    REF5       WORD     ENDC-LISTC-10                      FFFFF6
005A    REF6       WORD     ENDC-LISTC+LISTA-1                 00003F
005D    REF7       WORD     ENDA-LISTA-(ENDB-LISTB)            000014
0060    REF8       WORD     LISTB-LISTA                        FFFFC0
                   END      REF1
```

```
Loc          Source statement                                  Object code

0000    PROGB      START    0
                   EXTDEF   LISTB,ENDB
                   EXTREF   LISTA,ENDA,LISTC,ENDC
                     .
                     .
0036    REF1      +LDA      LISTA                              03100000
003A    REF2       LDT      LISTB+4                            772027
003D    REF3      +LDX     #ENDA-LISTA                         05100000
                     .
                     .
0060    LISTB      EQU      *
                     .
0070    ENDB       EQU      *
0070    REF4       WORD     ENDA-LISTA+LISTC                   000000
0073    REF5       WORD     ENDC-LISTC-10                      FFFFF6
0076    REF6       WORD     ENDC-LISTC+LISTA-1                 FFFFFF
0079    REF7       WORD     ENDA-LISTA-(ENDB-LISTB)            FFFFF0
007C    REF8       WORD     LISTB-LISTA                        000060
                   END
```

FIGURE 3.8 Sample programs illustrating linking and relocation.

```
Loc          Source statement                                   Object code

0000    PROGC     START    0
                  EXTDEF   LISTC,ENDC
                  EXTREF   LISTA,ENDA,LISTB,ENDB
                    .
                    .
                    .
0018    REF1     +LDA      LISTA                                03100000
001C    REF2     +LDT      LISTB+4                              77100004
0020    REF3     +LDX     #ENDA-LISTA                           05100000
                    .
                    .
                    .
0030    LISTC     EQU      *
                    .
                    .
0042    ENDC      EQU      *
0042    REF4      WORD     ENDA-LISTA+LISTC                     000030
0045    REF5      WORD     ENDC-LISTC-10                        000008
0048    REF6      WORD     ENDC-LISTC+LISTA-1                   000011
004B    REF7      WORD     ENDA-LISTA-(ENDB-LISTB)              000000
004E    REF8      WORD     LISTB-LISTA                          000000
                  END
```

FIGURE 3.8 *(cont'd)*

FIGURE 3.9 Object programs corresponding to Fig. 3.8.

```
H^PROGA ^000000^000063
D^LISTA ^000040^ENDA  ^000054
R^LISTB ^ENDB  ^LISTC ^ENDC
.
.
T^000020^0A^03201D^77100004^050014
.
.
T^000054^0F^000014^FFFFF6^00003F^000014^FFFFC0
M^000024^05^+LISTB
M^000054^06^+LISTC
M^000057^06^+ENDC
M^000057^06^-LISTC
M^00005A^06^+ENDC
M^00005A^06^-LISTC
M^00005A^06^+PROGA
M^00005D^06^-ENDB
M^00005D^06^+LISTB
M^000060^06^+LISTB
M^000060^06^-PROGA
E^000020
```

```
H^PROGB ^000000^00007F
D^LISTB ^000060^ENDB  ^000070
R^LISTA ^ENDA  ^LISTC ^ENDC
⋮
T^000036^0B^03100000^772027^05100000
⋮
T^000070^0F^000000^FFFFF6^FFFFFF^FFFFF0^000060
M^000037^05^+LISTA
M^00003E^05^+ENDA
M^00003E^05^-LISTA
M^000070^06^+ENDA
M^000070^06^-LISTA
M^000070^06^+LISTC
M^000073^06^+ENDC
M^000073^06^-LISTC
M^000076^06^+ENDC
M^000076^06^-LISTC
M^000076^06^+LISTA
M^000079^06^+ENDA
M^000079^06^-LISTA
M^00007C^06^+PROGB
M^00007C^06^-LISTA
E
```

```
H^PROGC ^000000^000051
D^LISTC ^000030^ENDC  ^000042
R^LISTA ^ENDA  ^LISTB ^ENDB
⋮
T^000018^0C^03100000^77100004^05100000
⋮
T^000042^0F^000030^000008^000011^000000^000000
M^000019^05^+LISTA
M^00001D^05^+LISTB
M^000021^05^+ENDA
M^000021^05^-LISTA
M^000042^06^+ENDA
M^000042^06^-LISTA
M^000042^06^+PROGC
M^000048^06^+LISTA
M^00004B^06^+ENDA
M^00004B^06^-LISTA
M^00004B^06^-ENDB
M^00004B^06^+LISTB
M^00004E^06^+LISTB
M^00004E^06^-LISTA
E
```

FIGURE 3.9 *(cont'd)*

tion. No modification for relocation or linking is necessary. In PROGB, on the other hand, the same operand refers to an external symbol. The assembler uses an extended-format instruction with address field set to 000000. The object program for PROGB (see Fig. 3.9) contains a Modification record instructing the loader to add the value of the symbol LISTA to this address field when the program is linked. This reference is handled in exactly the same way for PROGC.

The reference marked REF2 is processed in a similar manner. For PROGA, the operand expression consists of an external reference plus a constant. The assembler stores the value of the constant in the address field of the instruction and a Modification record directs the loader to add to this field the value of LISTB. In PROGB, the same expression is simply a local reference and is assembled using a program-counter relative instruction with no relocation or linking required.

REF3 is an immediate operand whose value is to be the difference between ENDA and LISTA (that is, the length of the list in bytes). In PROGA, the assembler has all of the information necessary to compute this value. For the assembly of PROGB (and PROGC), however, the values of the labels are unknown. In these programs, the expression must be assembled as an external reference (with two Modification records) even though the final result will be an absolute value independent of the locations at which the programs are loaded.

The remaining references illustrate a variety of other possibilities. The general approach taken is for the assembler to evaluate as much of the expression as it can. The remaining terms are passed on to the loader via Modification records. To see this, consider REF4. The assembler for PROGA can evaluate all of the expression in REF4 except for the value of LISTC. This results in an initial value of (hexadecimal) 000014 and one Modification record. However, the same expression in PROGB contains no terms that can be evaluated by the assembler. The object code therefore contains an initial value of 000000 and three Modification records. For PROGC, the assembler can supply the value of LISTC relative to the beginning of the program (but not the actual address, which is not known until the program is loaded). The initial value of this data word contains the relative address of LISTC (hexadecimal 000030). Modification records instruct the loader to add the beginning address of the program (i.e., the value of PROGC), to add the value of ENDA, and to subtract the value of LISTA. Thus the expression in REF4 represents a simple external reference for PROGA, a more complicated external reference for PROGB, and a combination of relocation and external references for PROGC.

You should work through references REF5 through REF8 for yourself to be sure you understand how the object code and Modification records in Fig. 3.9 were generated.

Figure 3.10(a) shows these three programs as they might appear in memory after loading and linking. PROGA has been loaded starting at address 4000, with PROGB and PROGC immediately following. Note that each of REF4 through REF8 has resulted (after relocation and linking is performed) in the same value in each of the three programs. This is as it should be, since the same source expression appeared in each program.

For example, the value for reference REF4 in PROGA is located at address 4054 (the beginning address of PROGA plus 0054, the relative address of REF4 within PROGRA). Figure 3.10(b) shows the details of how this value is computed. The initial value (from the Text record) is 000014. To this is added the address assigned to LISTC, which is 4112 (the beginning address of PROGC plus 30). In PROGB, the value for

FIGURE 3.10(a) Programs from Fig. 3.8 after linking and loading.

Memory address	Contents				
0000	xxxxxxxx	xxxxxxxx	xxxxxxxx	xxxxxxxx	
⋮	⋮	⋮	⋮	⋮	
3FF0	xxxxxxxx	xxxxxxxx	xxxxxxxx	xxxxxxxx	
4000	....	....	....	....	
4010	....	....	....	....	
4020	03201D77	1040C705	0014....	....	←PROGA
4030	....	....	....	....	
4040	....	....	....	....	
4050	....	00412600	00080040	51000004	
4060	000083..	....	....	....	
4070	....	....	....	....	
4080	....	....	....	....	
4090	....	....	..031040	40772027	←PROGB
40A0	05100014	....	....	....	
40B0	....	....	....	....	
40C0	....	....	....	....	
40D0	00	41260000	08004051	00000400	
40E0	0083....	....	....	....	
40F0	....	....	0310	40407710	
4100	40C70510	0014....	....	....	←PROGC
4110	....	....	....	....	
4120	....	00412600	00080040	51000004	
4130	000083xx	xxxxxxxx	xxxxxxxx	xxxxxxxx	
4140	xxxxxxxx	xxxxxxxx	xxxxxxxx	xxxxxxxx	
⋮	⋮	⋮	⋮	⋮	

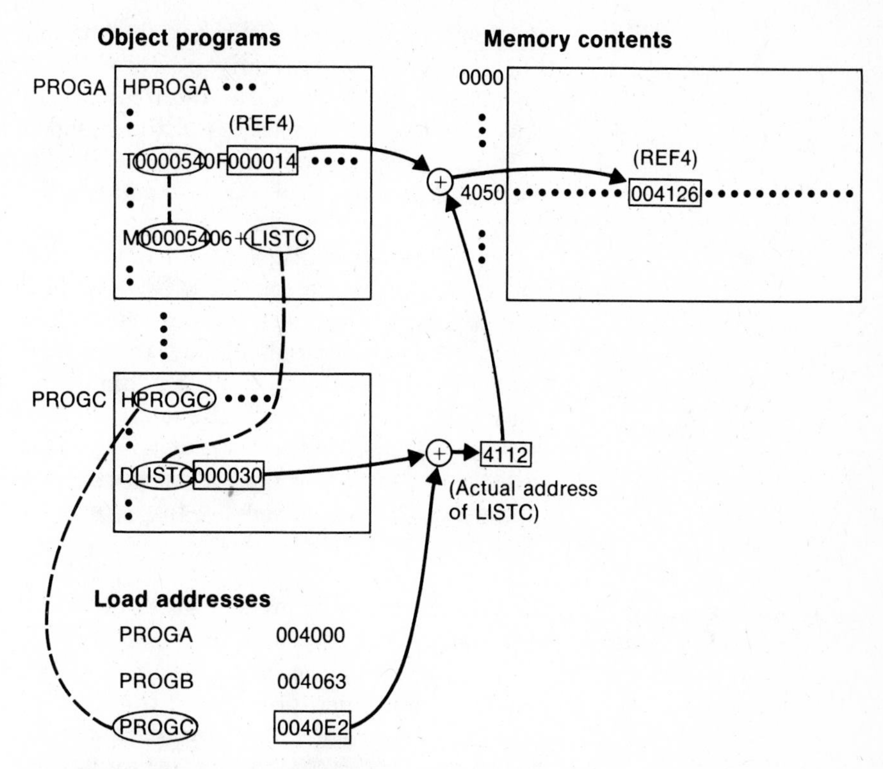

FIGURE 3.10(b) Relocation and linking operations performed on REF4 from PROGA.

REF4 is located at relative address 70 (actual address 40D3). To the initial value (000000), the loader adds the values of ENDA (4054) and LISTC (4112), and subtracts the value of LISTA (4040). The result, 004126, is the same as was obtained in PROGA. Similarly, the computation for REF4 in PROGC results in the same value. The same is also true for each of the other references REF5 through REF8.

For the references that are instruction operands, the calculated values after loading do not always appear to be equal. This is because there is an additional address calculation step involved for program-counter relative (or base relative) instructions. In these cases it is the *target addresses* that are the same. For example, in PROGA the reference REF1 is a program-counter relative instruction with displacement 01D. When this instruction is executed, the program counter contains the value 4023 (the actual address of the next instruction). The resulting target address is 4040. No relocation is necessary for this instruction since the program counter will always contain the actual (not relative) address of the next instruction. We could also think of this process as automatically providing the needed relocation at execution time through the target address calculation. In PROGB, on the other hand, reference REF1 is an extended format instruction that contains a direct (actual) address. This address, after linking, is 4040—the same as the target address for the same reference in PROGA.

You should work through the details of the other references to see that the target addresses (for REF2 and REF3) or the data values (for REF5 through REF8) are the same in each of the three programs. You do not need to worry about how these calculations are actually performed by the loader because the algorithm and data structures for doing this are discussed in the next section. It is important, however, that you *understand* the calculations to be performed, and that you are able to carry out the computations by hand (following the instructions that are contained in the object programs).

3.2.3 Tables and Logic for a Linking Loader

Now we are ready to present an algorithm for a linking (and relocating) loader. We use Modification records for relocation so that the linking and relocation functions are performed using the same mechanism. As mentioned previously, this type of loader is often found on machines (like SIC/XE) whose relative addressing makes the relocation of most instruction addresses unnecessary.

The algorithm for a linking loader is considerably more complicated than the absolute loader algorithm discussed in Section 3.1. The input to such a loader consists of a set of object programs (i.e., control sections) that are to be linked together. It is possible (and common) for a control section to make an external reference to a symbol whose definition does not appear until later in this input stream. In such a case the required linking operation cannot be performed until an address is assigned to the external symbol involved (that is, until the later control section is read). Thus a linking loader usually makes two passes over its input, just as an assembler does. In terms of general function, the two passes of a linking loader are quite similar to the two passes of an assembler: Pass 1 assigns addresses to all external symbols, and Pass 2 performs the actual loading, relocation, and linking.

The main data structure needed for our linking loader is an external symbol table ESTAB. This table, which is analogous to SYMTAB in our assembler algorithm, is used to store the name and address of each external symbol in the set of control sections being loaded. The table also often indicates in which control section the symbol is defined. A hashed organization is typically used for this table. Two other important variables are PROGADDR (program load address) and CSADDR (control section address). PROGADDR is the beginning address in memory where the linked program is to be loaded. Its value is supplied to the loader by the operating system. (In Chapter 6 we discuss how PROGADDR might be generated within the operating system.) CSADDR contains the starting address assigned to the control section currently being scanned by the loader. This value is added to all relative addresses within the control section to convert them to actual addresses.

The algorithm itself is presented in Fig. 3.11. As we discuss this algorithm, you may find it useful to refer to the example of loading and linking in the preceding section (Figs. 3.9 and 3.10).

During the first pass (Fig. 3.11a), the loader is concerned only with Header and Define record types in the control sections. The beginning

FIGURE 3.11(a) Algorithm for Pass 1 of a linking loader.

Pass 1:

```
begin
get PROGADDR from operating system
set CSADDR to PROGADDR {for first control section}
while not end of input do
    begin
        read next input record {Header record for control section}
        set CSLTH to control section length
        search ESTAB for control section name
        if found then
            set error flag (duplicate external symbol)
        else
            enter control section name into ESTAB with value CSADDR
        while record type <> 'E' do
            begin
                read next input record
                if record type = 'D' then
                    for each symbol in the record do
                        begin
                            search ESTAB for symbol name
                            if found then
                                set error flag (duplicate external symbol)
                            else
                                enter symbol into ESTAB with value
                                    (CSADDR + indicated address)
                        end {for}
            end {while <> 'E'}
        add CSLTH to CSADDR {starting address for next control section}
    end {while not EOF}
end {Pass 1}
```

Pass 2:

```
begin
set CSADDR to PROGADDR
set EXECADDR to PROGADDR
while not end of input do
    begin
        read next input record   {Header record}
        set CSLTH to control section length
        while record type <> 'E' do
            begin
                read next input record
                if record type = 'T' then
                    begin
                        {if object code is in character form, convert
                            into internal representation}
                        move object code from record to location
                            (CSADDR + specified address)
                    end {if 'T'}
                else if record type = 'M' then
                    begin
                        search ESTAB for modifying symbol name
                        if found then
                            add or subtract symbol value at location
                                (CSADDR + specified address)
                        else
                            set error flag (undefined external symbol)
                    end  {if 'M'}
            end  {while <> 'E'}
        if an address is specified {in End record} then
            set EXECADDR to (CSADDR + specified address)
        add CSLTH to CSADDR
    end  {while not EOF}
jump to location given by EXECADDR {to start execution of loaded program}
end  {Pass 2}
```

FIGURE 3.11(b) Algorithm for Pass 2 of a linking loader.

load address for the linked program (PROGADDR) is obtained from the operating system. This becomes the starting address (CSADDR) for the first control section in the input sequence. The control section name from the Header record is entered into ESTAB, with value given by CSADDR. All external symbols appearing in the Define record for the control section are also entered into ESTAB. Their addresses are obtained by adding the value specified in the Define record to CSADDR. When the End record is read, the control section length CSLTH (which was saved from the Header record) is added to CSADDR. This calculation gives the starting address for the next control section in sequence.

At the end of Pass 1, ESTAB contains all external symbols defined in the set of control sections together with the address assigned to each. Many loaders include as an option the ability to print a *load map* that shows these symbols and their addresses. This information is often useful in program debugging. For the example in Figs. 3.9 and 3.10, such a load map might look like the following:

Control section	Symbol name	Address	Length
PROGA		4000	0063
	LISTA	4040	
	ENDA	4054	
PROGB		4063	007F
	LISTB	40C3	
	ENDB	40D3	
PROGC		40E2	0051
	LISTC	4112	
	ENDC	4124	

This is essentially the same information contained in ESTAB at the end of Pass 1.

Pass 2 of our loader (Fig. 3.11b) performs the actual loading, relocation, and linking of the program. CSADDR is used in the same way it was in Pass 1: it always contains the actual starting address of the control section currently being loaded. As each Text record is read, the object code is moved to the specified address (plus the current value of CSADDR). When a Modification record is encountered, the symbol whose value is to be used for modification is looked up in ESTAB. This value is then added to or subtracted from the indicated location in memory.

The last step performed by the loader is usually the transferring of control to the loaded program to begin execution. (On some systems, the address where execution is to begin is simply passed back to the operating system. The user must then enter a separate Execute command.) The End record for each control section may contain the address of the first instruction in that control section to be executed. Our loader takes this as the transfer point to begin execution. If more than one control section specifies a transfer address, the loader arbitrarily uses the last one encountered. If no control section contains a transfer address, the loader uses the beginning of the linked program (i.e., PROGADDR) as the transfer point. This convention is typical of those found in most linking loaders. Normally, a transfer address would be placed in the End record for a main program, but not for a subroutine. Thus the correct execution address would be specified regardless of the order in which the control sections were presented for loading. (See Fig. 2.17 for an example of this.)

You should apply this algorithm (by hand) to load and link the object programs in Fig. 3.9. If PROGADDR is taken to be 4000, the result should be the same as that shown in Fig. 3.10.

This algorithm can be made more efficient if a slight change is made in the object program format. This modification involves assigning a *reference number* to each external symbol referred to in a control section. This reference number is used (instead of the symbol name) in Modification records.

Suppose we always assign the reference number 01 to the control section name. The other external reference symbols may be assigned numbers as part of the Refer record for the control section. Figure 3.12 shows the object programs from Fig. 3.9 with this change. The reference numbers are underlined in the Refer and Modification records for easier reading. The common use of a technique such as this is one reason we included Refer records in our object programs. You may have noticed that these records were not used in the algorithm of Fig. 3.11.

The main advantage of this reference-number mechanism is that it avoids multiple searches of ESTAB for the same symbol during the loading of a control section. An external reference symbol can be

FIGURE 3.12 Object programs corresponding to Fig. 3.8 using reference numbers for code modification. (Reference numbers are underlined for easier reading.)

```
H^PROGA ^000000^000063
D^LISTA ^000040^ENDA  ^000054
R^02LISTB ^03ENDB  ^04LISTC ^05ENDC
•
•
T^000020^0A^03201D^77100004^050014
•
•
T^000054^0F^000014^FFFFF6^00003F^000014^FFFFC0
M^000024^05^+02
M^000054^06^+04
M^000057^06^+05
M^000057^06^-04
M^00005A^06^+05
M^00005A^06^-04
M^00005A^06^+01
M^00005D^06^-03
M^00005D^06^+02
M^000060^06^+02
M^000060^06^-01
E^000020
```

```
H^PROGB ^000000^00007F
D^LISTB ^000060^ENDB  ^000070
R^02LISTA ^03ENDA  ^04LISTC ^05ENDC
•
•
T^000036^0B^03100000^772027^05100000
•
•
T^000070^0F^000000^FFFFF6^FFFFFF^FFFFF0^000060
M^000037^05^+02
M^00003E^05^+03
M^00003E^05^-02
M^000070^06^+03
M^000070^06^-02
M^000070^06^+04
M^000073^06^+05
M^000073^06^-04
M^000076^06^+05
M^000076^06^-04
M^000076^06^+02
M^000079^06^+03
M^000079^06^-02
M^00007C^06^+01
M^00007C^06^-02
E
```

```
H^PROGC ^000000^000051
D^LISTC ^000030^ENDC  ^000042
R^02LISTA ^03ENDA  ^04LISTB ^05ENDB
•
•
T^000018^0C^03100000^77100004^05100000
•
•
T^000042^0F^000030^000008^000011^000000^000000
M^000019^05^+02
M^00001D^05^+04
M^000021^05^+03
M^000021^05^-02
M^000042^06^+03
M^000042^06^-02
M^000042^06^+01
M^000048^06^+02
M^00004B^06^+03
M^00004B^06^-02
M^00004B^06^-05
M^00004B^06^+04
M^00004E^06^+04
M^00004E^06^-02
E
```

FIGURE 3.12 *(cont'd)*

looked up in ESTAB once for each control section that uses it. The values for code modification can then be obtained by simply indexing into an array of these values. You are encouraged to develop an algorithm that includes this technique, together with any additional data structures you may require.

3.3 MACHINE-INDEPENDENT LOADER FEATURES

In this section we discuss a number of loader features that are not directly related to machine structure and design. Loading and linking are often thought of as operating system service functions. The programmer's connection with such services is not as direct as it is with, for example, the assembler during program development. Therefore, most loaders include fewer different features (and less varied capabilities) than are found in a typical assembler.

Section 3.3.1 discusses the use of an automatic library search process for handling external references. This feature allows a programmer to use standard subroutines without explicitly including them in the program to be loaded. The routines are automatically retrieved from a library as they are needed during linking.

Section 3.3.2 presents some common options that can be selected at the time of loading and linking. These include such capabilities as specifying alternative sources of input, changing or deleting external references, and controlling the automatic processing of external references.

In Section 3.3.3 we expand on the concept of an overlay program briefly introduced in Section 2.4.1. We discuss methods of specifying such an overlay structure and consider how such overlays might be managed during program execution.

3.3.1 Automatic Library Search

Many linking loaders can automatically incorporate routines from a subprogram library into the program being loaded. In most cases there is a standard system library that is used in this way. Other libraries may be specified by control statements or by parameters to the loader. This feature allows the programmer to use subroutines from one or more libraries (for example, mathematical or statistical routines) almost as if they were a part of the programming language. The subroutines called by the program being loaded are automatically fetched from the library, linked with the main program, and loaded. The programmer does not need to take any action beyond mentioning the subroutine names as external references in the source program. On some systems, this feature is referred to as *automatic library call*. We use the term *library search* to avoid confusion with the call feature found in most programming languages.

Linking loaders that support automatic library search must keep track of external symbols that are referred to, but not defined, in the primary input to the loader. One easy way to do this is to enter symbols from each Refer record into the symbol table (ESTAB) unless these symbols are already present. These entries are marked to indicate that the symbol has not yet been defined. When the definition is encountered, the address assigned to the symbol is filled in to complete the entry. At the end of Pass 1, the symbols in ESTAB that remain undefined represent *unresolved* external references. The loader searches the library or libraries specified for routines that contain the definitions of these symbols, and processes the subroutines found by this search exactly as if they had been part of the primary input stream.

Note that the subroutines fetched from a library in this way may themselves contain external references. It is therefore necessary to repeat the library search process until all references are resolved (or until no further resolution can be made). If unresolved external references remain after the library search is completed, these must be treated as errors.

The process just described allows the programmer to override the standard subroutines in the library by supplying his or her own routines. For example, suppose that the main program refers to a standard subroutine named SQRT. Ordinarily the subroutine with this name would automatically be included via the library search function. A programmer who for some reason wanted to use a different version of SQRT could do so simply by including it as input to the loader. By the end of Pass 1 of the loader, SQRT would already be defined, so it would not be included in any library search that might be necessary.

The libraries to be searched by the loader ordinarily contain assembled or compiled versions of the subroutines (that is, object programs). It is possible to search these libraries by scanning the Define records for all of the object programs on the library, but this might be quite inefficient. In most cases a special file structure is used for the libraries. This structure contains a *directory* that gives the name of each routine and a pointer to its address within the file. If a subroutine is to be callable by more than one name (using different entry points), both names are entered into the directory. The object program itself, of course, is only stored once. Both directory entries point to the same copy of the routine. Thus the library search itself really involves a search of the directory, followed by reading the object programs indicated by this search. Some operating systems can keep the directory for commonly used libraries permanently in memory. This can expedite the search process if a large number of external references are to be resolved.

The process of library search has been discussed as the resolution of a call to a subroutine. Obviously the same technique applies equally well to the resolution of any other type of external reference.

3.3.2 Loader Options

Many loaders allow the user to specify options that modify the standard processing described in Section 3.2. In this section we discuss some typical loader options and give examples of their use. Many loaders have a special command language that is used to specify options. Sometimes there is a separate input file to the loader that contains such control statements. Sometimes these same statements can also be embedded in the primary input stream between object programs. On a few systems the programmer can even include loader control statements in the source program, and the assembler or compiler retains these commands as a part of the object program.

We discuss loader options in this section as though they were specified using a command language, but there are other possibilities. On some systems options are specified as a part of the job control language that is processed by the operating system. When this approach is used, the operating system incorporates the options specified into a control block that is made available to the loader when it is invoked. The implementation of such options is, of course, the same regardless of the means used to select them.

One typical loader option allows the selection of alternative sources of input. For example, the command

```
INCLUDE     program-name(library-name)
```

might direct the loader to read the designated object program from a library and treat it as if it were part of the primary loader input.

Other commands allow the user to delete external symbols or entire control sections. It may also be possible to change external references within the programs being loaded and linked. For example, the command

```
DELETE      csect-name
```

might instruct the loader to delete the named control section(s) from the set of programs being loaded. The command

```
CHANGE      name1,name2
```

might cause the external symbol name1 to be changed to name2 wherever it appears in the object programs. An illustration of the use of such commands is given in the following example.

Consider the source program in Fig. 2.15 and the corresponding object program in Fig. 2.17. There is a main program (COPY) that uses two subprograms (RDREC and WRREC); each of these is a separate control section. If RDREC and WRREC are designed only for use with COPY, it is likely that the three control sections will be assembled at the same time. This means that the three control sections of the object program will appear in the same file (or as part of the same library member).

Suppose now that a set of utility subroutines is made available on the computer system. Two of these, READ and WRITE, are designed to perform the same functions as RDREC and WRREC. It would probably be desirable to change the source program of COPY to use these utility routines. As a temporary measure, however, a sequence of loader commands could be used to make this change without reassembling the program. This might be done, for example, to test the utility routines before the final conversion is made.

Suppose that a file containing the object programs in Fig. 2.17 is the primary loader input with the loader commands

```
INCLUDE    READ(UTLIB)
INCLUDE    WRITE(UTLIB)
DELETE     RDREC,WRREC
CHANGE     RDREC,READ
CHANGE     WRREC,WRITE
```

These commands would direct the loader to include control sections READ and WRITE from the library UTLIB, and to delete the control sections RDREC and WRREC from the load. The first CHANGE command would cause all external references to symbol RDREC to be changed to refer to symbol READ. Similarly, references to WRREC would be changed to WRITE. The result would be exactly the same as if the source program for COPY had been changed to use READ and WRITE. You are encouraged to think for yourself about how the loader might handle such commands to perform the specified processing.

Another common loader option involves the automatic inclusion of library routines to satisfy external references (as described in the preceding section). Most loaders allow the user to specify alternative libraries to be searched, using a statement such as

```
LIBRARY    MYLIB
```

Such user-specified libraries are normally searched before the standard system libraries. This allows the user to use special versions of these standard routines.

Loaders that perform automatic library search to satisfy external references often allow the user to specify that some references not be resolved in this way. Suppose, for example, that a certain program has as its main function the gathering and storing of data. However, the program can also perform an analysis of the data using the routines STDDEV, PLOT, and CORREL from a statistical library. The user may request this analysis at execution time. Since the program contains external references to these three routines, they would ordinarily be loaded and linked with the program. If it is known that the statistical analysis is not to be performed in a particular execution of this program, the user could include a command such as

```
NOCALL      STDDEV,PLOT,CORREL
```

to instruct the loader that these external references are to remain unresolved. This avoids the overhead of loading and linking the unneeded routines, and saves the memory space that would otherwise be required.

It is also possible to specify that *no* external references be resolved by library search. Of course, this means an error will result if the program attempts to make such an external reference during execution. This option is more useful when programs are to be linked but not executed immediately. It is often desirable to postpone the resolution of external references in such a case. In Section 3.4.1 we discuss linkage editors that perform this sort of function.

Another common option involves output from the loader. In Section 3.2.3 we gave an example of a load map that might be generated during the loading process. Through control statements the user can often specify whether or not such a map is to be printed at all. If a map is desired, the level of detail can be selected. For example, the map may include control section names and addresses only. It may also include external symbol addresses or even a cross-reference table that shows references to each external symbol.

Loaders often include a variety of other options. One such option is the ability to specify the location at which execution is to begin (overriding any information given in the object programs). Another is the ability to control whether or not the loader should attempt to execute the program if errors are detected during the load (for example, unresolved external references).

3.3.3 Overlay Programs

In Section 2.4.1 we discussed the implementation of a two-pass assembler using a simple overlay structure. This method of implementation reduced the total amount of memory required for the assembly by allowing Pass 1 and Pass 2 to occupy the same locations in memory. In this section we examine a more complex example of an overlay program and discuss how the overlay process itself is managed. As an introduction to the topic, you may want to review Figs. 2.18 and 2.19 together with the discussion in Section 2.4.1.

Figure 3.13(a) shows the overlay structure for our example program. The letters represent control section names, and the lines represent transfers of control between control sections. Thus the *root* control section (corresponding to the driver routine in Fig. 2.18) is named A. Control section A can call B, C, or D/E; B can call F/G or H, etc. The

FIGURE 3.13 Example of an overlay program.

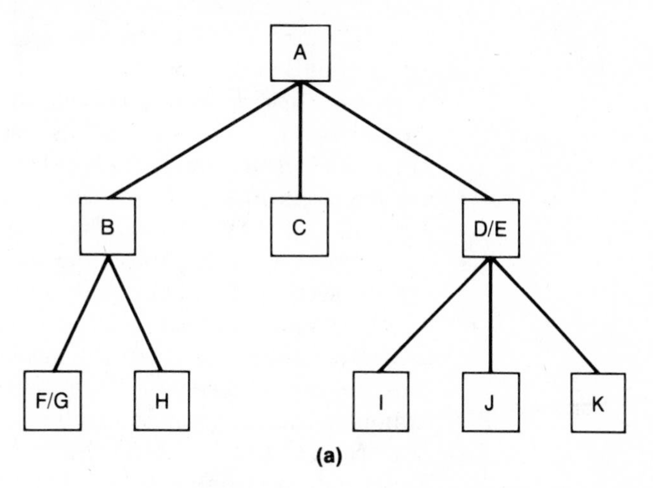

(a)

Control section	Length (bytes)
A	1000
B	1800
C	4000
D	2800
E	800
F	1000
G	400
H	800
I	1000
J	800
K	2000

(b)

notation D/E represents two control sections (D and E) that are closely related and are always used together. Although these control sections are separately handled by the assembler and the loader, they will be treated as a unit during the overlay process. For example, D might be a control section containing executable instructions, and E might contain the associated data items, or E might be a subroutine frequently called by D. The length (in hexadecimal) of each of the control sections is shown in Fig. 3.13(b).

Most systems that support overlay programs require that the overlay be a tree structure such as that shown in Fig. 3.13(a); the nodes in this structure are called *segments*. The root segment is loaded when execution of the program is begun, and it remains in memory until the program terminates. The other segments are loaded as they are called. Suppose, for example, that control section A calls B. The segment containing B will automatically be loaded because of this call (unless it is already in memory). If at some later time A calls D, the segment containing D will be loaded. The ordering of segments that are at the same level in the overlay structure is not significant. It implies nothing about the order in which the segments are used. Thus, for example, A might call C first, then B, then D/E, then C again, and so on.

Suppose that at some time during the running of the program control section H is being executed. H must have been called by B, which was called by A; thus, these three segments are *active,* and must be present in memory. On the other hand, the other segments in the program cannot be active since there is no path from them to H. If, for example, the segment containing K was called previously, it must have returned to D/E (and then to A) before B could be called by A. This line of reasoning leads to the following rule, which is found in most systems that allow overlay: If a segment S is present in memory, all of the other segments that lie on the path from S to the root must also be present. This simplifies the handling of the overlay process. If a segment calls another that is lower in the tree structure (i.e., farther from the root), that segment may need to be loaded. On the other hand, if a segment refers to another segment that lies between it and the root, that segment must already be in memory.

Because segments at the same level (for example, B, C, and D/E) can be called only from the level above, they cannot be required at the same time. Thus they can be assigned to the same locations in memory. If a segment is loaded because of a transfer of control, it *overlays* any segments at the same level (and their subordinate segments) that may be in memory. Thus the entire program can be executed in a smaller total amount of memory, which is the main reason for the use of overlay structures. In the remainder of this section we examine in more detail how the overlay process is handled by the loader.

The structure of an overlay program is defined to the loader using commands similar to those discussed in the preceding section. Figure 3.14 shows a typical set of commands for defining the overlay structure in Fig. 3.13(a). The statement

```
SEGMENT     seg-name(control-section . . .)
```

defines a segment (i.e., a node in the tree structure), gives it a name, and lists the control sections to be included in it. The first segment defined is the root. Two consecutive SEGMENT statements specify a parent–child relationship between the segments being defined. Thus the first three SEGMENT statements in Fig. 3.14 define the leftmost path (from the root to F/G) in the structure shown in Fig. 3.13(a).

The other type of statement in Fig. 3.14 allows the definition of multiple segments having the same parent. The statement

```
PARENT      seg-name
```

identifies the (already existing) segment that is to be the parent of the next segment defined. Thus the first PARENT statement in Fig. 3.14 causes SEG2 to be the parent of SEG4 in the tree structure. If this PARENT statement were not present, the parent of SEG4 would be SEG3.

FIGURE 3.14 Overlay definition using control statements.

```
SEGMENT     SEG1(A)
SEGMENT     SEG2(B)
SEGMENT     SEG3(F,G)
PARENT      SEG2
SEGMENT     SEG4(H)
PARENT      SEG1
SEGMENT     SEG5(C)
PARENT      SEG1
SEGMENT     SEG6(D,E)
SEGMENT     SEG7(I)
PARENT      SEG6
SEGMENT     SEG8(J)
PARENT      SEG6
SEGMENT     SEG9(K)
```

You should carefully examine the remainder of the definition in Fig. 3.14 to verify that it corresponds to the tree structure shown in Fig. 3.13(a).

Once the overlay structure has been defined, it is easy to find the starting addresses for the segments because each segment begins immediately after the end of its parent. The rule governing which segments are in memory assures that all currently loaded segments will be contiguous at the beginning of the area of memory assigned to the job. Figure 3.15(a) shows the length and relative starting address of each segment in our example program. It also shows the actual starting address (assuming that the beginning load address for the program is 8000). During the execution of the program there are many different sets of segments that may be in memory together. Figure 3.15(b) shows three of these possibilities; you are encouraged to work out the others for yourself. As we discuss later, the loader may add one or more special control sections containing overlay control information. The length of these added sections must, of course, be taken into account in assigning addresses.

As we have seen, the loader can assign an actual starting address to every segment in the overlay program once the initial load address is supplied. Thus the addresses for all external symbols are known. This means that all relocation and linking operations can be performed in the usual way with one exception: transfers of control to a segment from its parent must allow for the possibility that the segment being called is not in memory. The root segment can be loaded directly into memory; the other segments (with all relocation and linking operations performed) are written onto a special working file SEGFILE that is created by the loader.

The overlay process itself—that is, the loading of a segment when control is transferred to it—can be handled in several different ways. We describe a simple mechanism that does not require operating system intervention. Techniques similar to those we describe in connection with dynamic linking (see Section 3.4.2) could also be applied.

The actual loading of segments during program execution is handled by an *overlay manager*. In our implementation, this is a special control section named OVLMGR, which is automatically included in the root segment of the overlay program by the loader. To manage the overlay process properly, OVLMGR must have information about the overlay program structure. This information is stored in a *segment table* SEGTAB, which is created by the loader and included as a separate control section in the root segment. SEGTAB describes the tree structure, specifying the level at which each segment is placed. It also contains the address at which the segment is to be loaded, the address

Segment	Starting address: Relative	Starting address: Actual	Length
1	0000	8000	1000
2	1000	9000	1800
3	2800	A800	1400
4	2800	A800	800
5	1000	9000	4000
6	1000	9000	3000
7	4000	C000	1000
8	4000	C000	800
9	4000	C000	2000

(a)

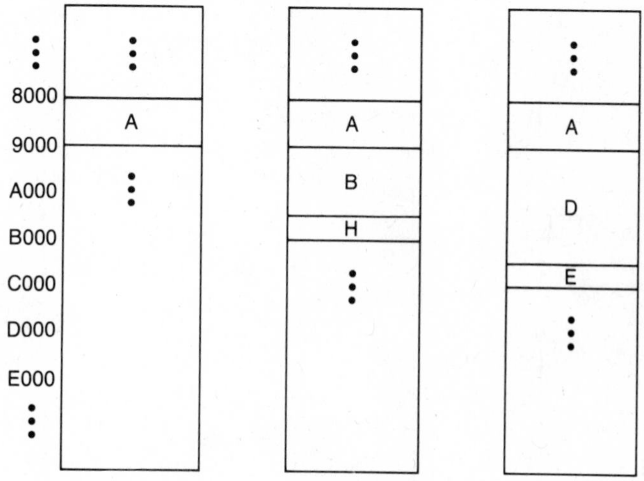

(b)

FIGURE 3.15 Memory assignment for an overlay program.

of its entry point, and the location of the segment on SEGFILE. (In this implementation, we assume that a segment can be called at only one entry point.)

SEGTAB also includes a special *transfer area* for each segment except the root. This transfer area contains instructions that are used in passing control. If a segment is currently loaded in memory, the transfer area contains a jump to the entry point of that segment. If the segment is not currently loaded, the transfer area contains instructions that invoke OVLMGR and pass to it information concerning the segment to be loaded. The instructions in the transfer areas are changed as necessary by the overlay manager to reflect the current status of each segment. A transfer of control from a segment to another segment lower in the tree structure (that is, a transfer that may cause overlay to occur) is converted by the loader into a jump to the transfer area for that segment. Because SEGTAB is placed in the root segment, this transfer area is always in memory. The instructions in the transfer area then pass control either directly to the target segment or to OVLMGR. In the latter case, OVLMGR will pass control to the target segment after loading it and updating SEGTAB.

This process is illustrated in Fig. 3.16. Figure 3.16(a) represents

FIGURE 3.16 Example of overlay management.

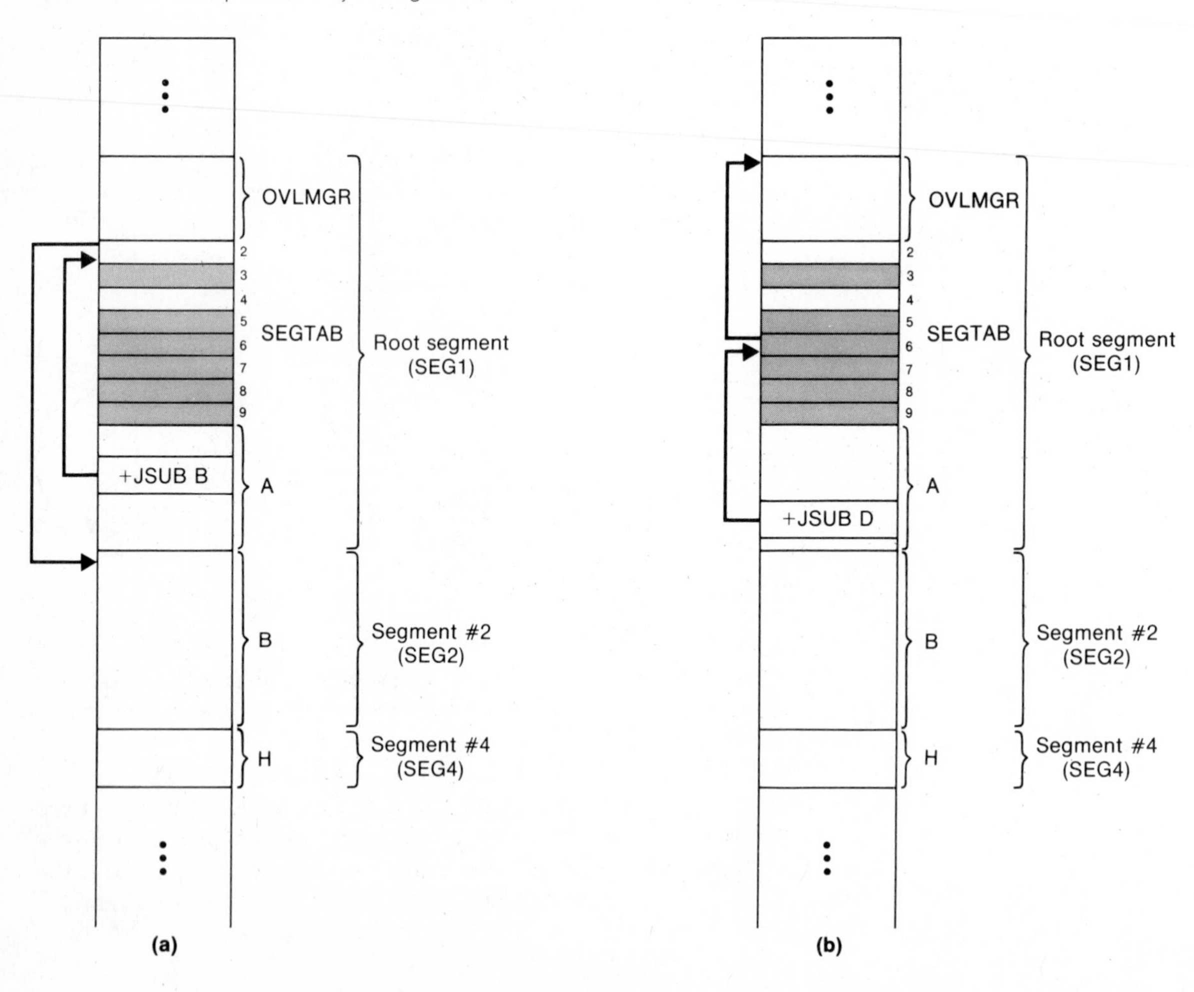

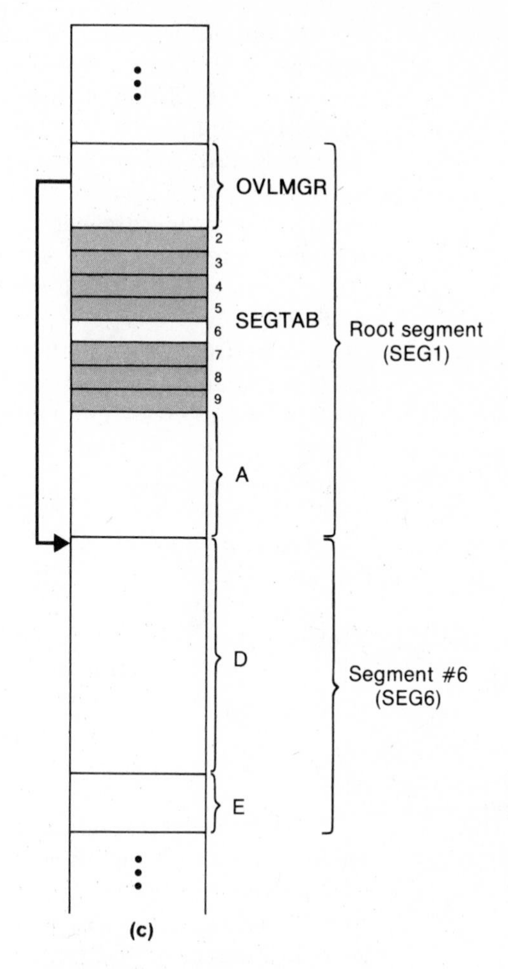

FIGURE 3.16 *(cont'd)*

the contents of memory at some time during execution. Segments 1, 2, and 4 are currently loaded, and instructions from control section A are being executed. Segments 2 and 4 were loaded in response to previous transfers of control. Control has now returned to segment 1; however, segments 2 and 4 will remain in memory until overlaid by other segments. The transfer areas for segments 2 and 4 contain jumps to control sections B and H. The other transfer areas contain instructions that invoke OVLMGR (indicated by the shaded areas in SEGTAB). If control section A now executes an instruction that appeared in the source program as "+JSUB B," the result is a jump to the transfer area for segment 2. This causes a direct transfer of control to B, as shown in the diagram.

On the other hand, suppose A executes a call to D. The transfer area for segment 6 contains instructions that invoke OVLMGR, as shown in Fig. 3.16(b). The overlay manager loads segment 6 from SEGFILE and places it at the proper address in memory. It then updates SEGTAB to indicate that segment 6 is currently loaded, and that segments 2 and 4 are not. Segment 4 must be removed (even if its memory locations are not needed for the new segment) because its parent segment has been removed. Control is then passed to the entry point for segment 6, completing the transfer of control to D. (See Fig. 3.16c.)

The return of control from a called segment (for example, from D back to A) can be accomplished in the usual way; the overlay manager need not be involved. For example, the JSUB instruction illustrated in Fig. 3.16(b) would place the return address in register L. This value would be preserved by the overlay manager. Control section D could then return to A by executing an RSUB instruction. This is exactly the same mechanism that would be used if the program were loaded without overlays.

On some systems, the functions of OVLMGR are built into a part of the operating system called the *overlay supervisor*. In this case, SEGTAB might contain statements to call the appropriate operating system service function when a segment must be loaded.

3.4 LOADER DESIGN OPTIONS

In this section we discuss some common alternatives for organizing the loading functions, including relocation and linking. Linking loaders, as described in Section 3.2.3, perform all linking and relocation at load time. We discuss two alternatives to this: linkage editors, which perform linking prior to load time, and dynamic linking, in which the linking function is performed at execution time.

Section 3.4.1 discusses linkage editors, which are found on many computing systems instead of or in addition to the linking loader. A linkage editor performs linking and some relocation; however, the linked program is written on a file or library instead of being immediately loaded into memory. This approach reduces the overhead when the program is executed. All that is required at load time is a very simple form of relocation.

Section 3.4.2 introduces the topic of dynamic linking, which uses facilities of the operating system to load subprograms at the time they are first called. By delaying the linking process in this way, additional flexibility can be achieved. However, this approach usually involves more overhead than does a linking loader.

In Section 3.4.3 we discuss bootstrap loaders. Such loaders can be used to run stand-alone programs independent of the operating system or the system loader. They can also be used to load the operating system or the loader itself into memory.

3.4.1 Linkage Editors

The essential difference between a linkage editor and a linking loader is illustrated in Fig. 3.17. The source program is first assembled or compiled, producing an object program (which may contain several different control sections). A linking loader performs all linking and relocation operations, including automatic library search if specified, and loads the linked program directly into memory for execution. A *linkage editor,* on the other hand, produces a linked version of the program (often called a *load module* or an *executable image*), which is written onto a file or library for later execution.

When the user is ready to run the linked program, a simple relocating loader can be used to load the program into memory. The only

FIGURE 3.17 Processing of an object program using (a) linking loader and (b) linkage editor.

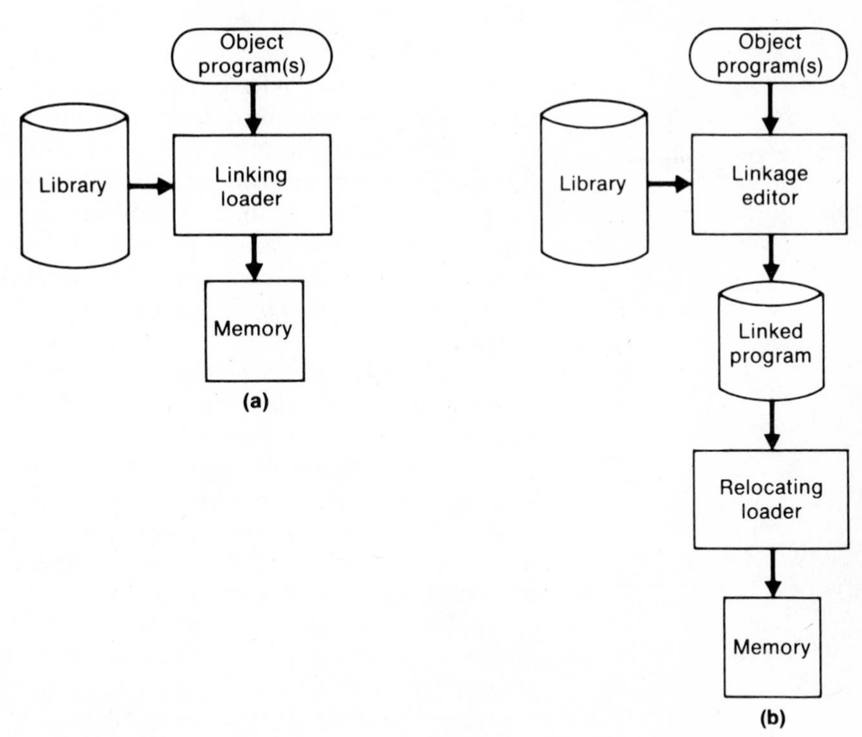

object code modification necessary is the addition of an actual load address to relative values within the program. The linkage editor performs relocation of all control sections relative to the start of the linked program. Thus, all items that need to be modified at load time have values that are relative to the start of the linked program. This means that the loading can be accomplished in one pass with no external symbol table required. This involves much less overhead than using a linking loader.

If a program is to be executed many times without being reassembled, the use of a linkage editor substantially reduces the overhead required. Resolution of external references and library searching are only performed once (when the program is link edited). In contrast, a linking loader searches libraries and resolves external references every time the program is executed.

Sometimes, however, a program is reassembled for nearly every execution. This situation might occur in a program development and testing environment (for example, student programs). It also occurs when a program is used so infrequently that it is not worthwhile to store the assembled version in a library. In such cases it is more efficient to use a linking loader, which avoids the steps of writing and reading the linked program.

The linked program produced by the linkage editor is generally in a form that is suitable for processing by a relocating loader. All external references are resolved, and relocation is indicated by some mechanism such as Modification records or a bit mask. Even though all linking has been performed, information concerning external references is often retained in the linked program. This allows subsequent relinking of the program to replace control sections, modify external references, etc. If this information is not retained, the linked program cannot be reprocessed by the linkage editor; it can only be loaded and executed.

If the actual address at which the program will be loaded is known in advance, the linkage editor can perform all of the needed relocation. The result is a linked program that is an exact image of the way the program will appear in memory during execution. The content and processing of such an image are the same as for an absolute object program. Normally, however, the added flexibility of being able to load the program at any location is easily worth the slight additional overhead for performing relocation at load time.

Linkage editors can perform many useful functions besides simply preparing an object program for execution. Consider, for example, a program (PLANNER) that uses a large number of subroutines. Suppose that one subroutine (PROJECT) used by the program is changed to correct an error or to improve efficiency. After the new version of

PROJECT is assembled or compiled, the linkage editor can be used to replace this subroutine in the linked version of PLANNER. It is not necessary to go back to the original (separate) versions of all of the other subroutines. The following is a typical sequence of linkage editor commands used to accomplish this. The command language is similar to that discussed in Section 3.3.2:

```
INCLUDE   PLANNER(PROGLIB)
DELETE    PROJECT             {DELETE from existing PLANNER}
INCLUDE   PROJECT(NEWLIB)     {INCLUDE new version}
REPLACE   PLANNER(PROGLIB)
```

Linkage editors can also be used to build packages of subroutines or other control sections that are generally used together. This can be useful when dealing with subroutine libraries that support high-level programming languages. In a typical FORTRAN implementation, for example, there are a large number of subroutines that are used to handle formatted input and output. These include routines to read and write data blocks, to block and deblock records, and to encode and decode data items according to FORMAT specifications. There are a large number of cross-references between these subprograms because of their closely related functions. However, it is desirable that they remain as separate control sections for reasons of program modularity and maintainability.

If a program using formatted I/O were linked in the usual way, all of the cross-references between these library subroutines would have to be processed individually. Exactly the same set of cross-references would need to be processed for almost every FORTRAN program linked. This represents a substantial amount of overhead. The linkage editor could be used to combine the appropriate subroutines into a package with a command sequence like the following:

```
INCLUDE   READR(FTNLIB)
INCLUDE   WRITER(FTNLIB)
INCLUDE   BLOCK(FTNLIB)
INCLUDE   DEBLOCK(FTNLIB)
INCLUDE   ENCODE(FTNLIB)
INCLUDE   DECODE(FTNLIB)
   .
   .
   .
SAVE      FTNIO(SUBLIB)
```

The linked module named FTNIO could be indexed in the directory of SUBLIB under the same names as the original subroutines. Thus a search of SUBLIB before FTNLIB would retrieve FTNIO instead of the separate routines. Since FTNIO already has all of the cross-references between subroutines resolved, these linkages would not be reprocessed when each user's program is linked. The result would be a much more efficient linkage editing operation for each program and a considerable overall savings for the system.

Linkage editors often allow the user to specify that external references are not to be resolved by automatic library search. Suppose, for example, that 100 FORTRAN programs using the I/O routines described above were to be stored on a library. If all external references were resolved, this would mean that a total of 100 copies of FTNIO would be stored. If library space were an important resource, this might be highly undesirable. Using commands like those discussed in Section 3.3.2, the user could specify that no library search be performed during linkage editing. Thus only the external references between user-written routines would be resolved. A linking loader could then be used to combine the linked user routines with FTNIO at execution time. Because this process involves two separate linking operations, it would require slightly more overhead; however, it would result in a large savings in library space.

Linkage editors often include a variety of other options and commands like those discussed for linking loaders. Compared to linking loaders, linkage editors in general tend to offer more flexibility and control, with a corresponding increase in complexity and overhead.

3.4.2 Dynamic Linking

Linkage editors perform linking operations before the program is loaded for execution. Linking loaders perform these same operations at load time. In this section we discuss a scheme that postpones the linking function until execution time: a subroutine is loaded and linked to the rest of the program when it is first called. This type of function is usually called *dynamic linking, dynamic loading,* or *load on call.*

Dynamic linking offers some advantages over the other types of linking we have discussed. Suppose, for example, that a program contains subroutines that correct or clearly diagnose errors in the input data during execution. If such errors are rare, the correction and diagnostic routines may not be used at all during most executions of the program. However, if the program were completely linked before execution, these subroutines would need to be loaded and linked every time the program is run. Dynamic linking provides the ability to load the routines only when (and if) they are needed. If the subroutines involved are large, or have many external references, this can result in substantial savings of time and memory space.

Similarly, suppose that in any one execution a program uses only a few out of a large number of possible subroutines, but the exact routines needed cannot be predicted until the program examines its input. This situation could occur, for example, with a program that allows its user to interactively call any of the subroutines of a large mathematical and statistical library. Input data could be supplied by the user from a timesharing terminal, and results could be displayed at the terminal. In this case, all of the library subroutines could potentially be needed, but only a few will actually be used in any one terminal session. Dynamic linking avoids the necessity of loading the entire library for each execution. As a matter of fact, dynamic linking may make it unnecessary for the program even to know the possible set of subroutines that might be used. The subroutine name might simply be treated as another input item.

There are a number of different mechanisms that can be used to accomplish the actual loading and linking of a called subroutine. We shall discuss a method in which routines that are to be dynamically loaded must be called via an operating system service request. This method could also be thought of as a request to a part of the loader that is kept in memory during execution of the program.

Instead of executing a JSUB instruction that refers to an external symbol, the program makes a load-and-call service request to the operating system. The parameter of this request is the symbolic name of the routine to be called. (See Fig. 3.18a.) The operating system examines its internal tables to determine whether or not the routine is already loaded. If necessary, the routine is loaded from the specified user or system libraries as shown in Fig. 3.18(b). Control is then passed from the operating system to the routine being called (Fig. 3.18c).

When the called subroutine completes its processing, it returns to its caller (that is, to the operating system routine that handles the load-and-call service request). The operating system then returns control to the program that issued the request. This process is illustrated in Fig. 3.18(d). It is important that control be returned in this way so that the operating system knows when the called routine has completed its execution. After the subroutine is completed, the memory that was allocated to load it may be released and used for other purposes. However, this is not always done immediately. Sometimes it is desirable to retain the routine in memory for later use as long as the storage space is not needed for other processing. If a subroutine is still in memory, a second call to it may not require another load operation.

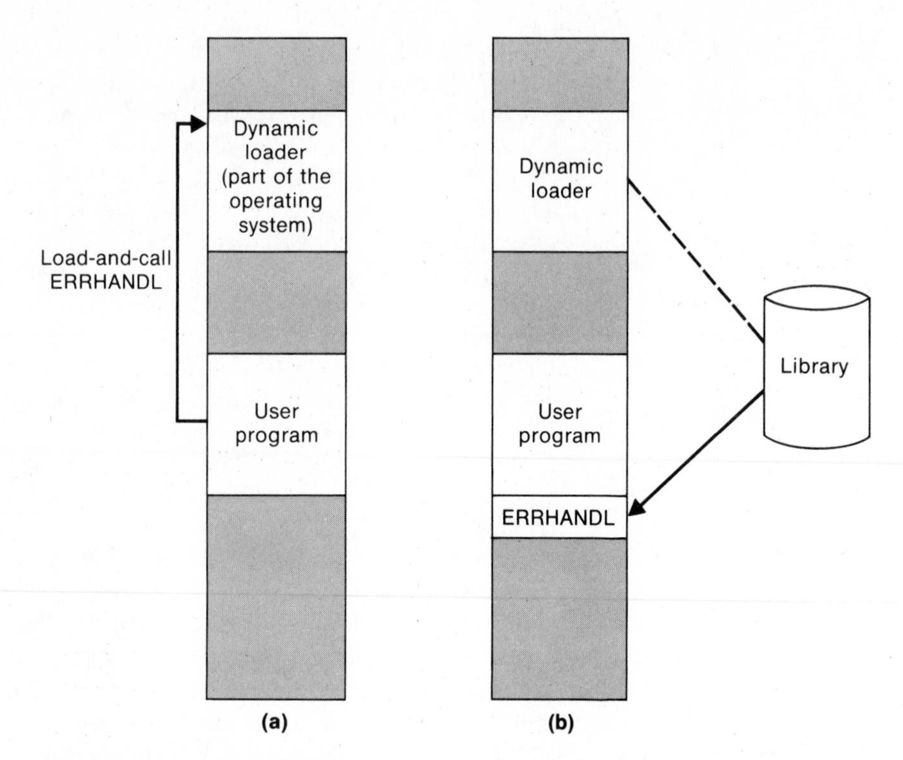

FIGURE 3.18 Loading and calling of a subroutine using dynamic linking.

Control may simply be passed from the dynamic loader to the called routine, as shown in Fig. 3.18(e).

When dynamic linking is used, the association of an actual address with the symbolic name of the called routine is not made until the call statement is executed. Another way of describing this is to say that the *binding* of the name to an actual address is delayed from load time until execution time. This delayed binding results in greater flexibility, as we have discussed. It also requires more overhead since the operating system must intervene in the calling process. In later chapters we see other examples of delayed binding. In those examples, too, delayed binding gives more capabilities at a higher cost.

3.4.3 Bootstrap Loaders

In our discussions of loaders we have neglected to answer one important question: How is the loader itself loaded into memory? Of course, we could say that the operating system loads the loader; however, we are then left with the same question with respect to the operating

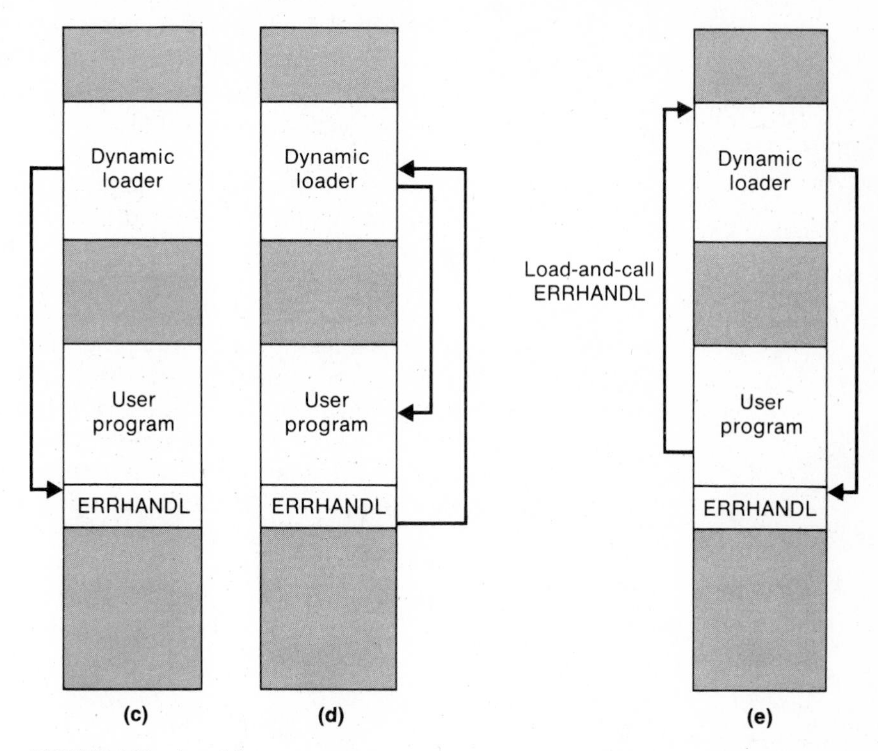

FIGURE 3.18 *(cont'd)*

system. More generally, the question is this: Given an idle computer with no program in memory, how do we get things started?

In this situation, with the machine empty and idle, there is no need for program relocation. We can simply specify the absolute address for whatever program is first loaded. Most often, this program will be the operating system, which occupies a predefined location in memory. This means that we need some means of accomplishing the functions of an absolute loader. One option is to have the operator enter into memory the object code for an absolute loader, using switches on the computer console. Some computers have required that the operator do just that. However, this process is much too inconvenient and error-prone to be a really good solution to the problem.

Another possibility is to have the absolute loader program permanently resident in a read-only memory (ROM). When some hardware signal occurs (for example, the operator pressing a "system start" button), the machine begins to execute this ROM program. On some computers, the program is executed directly in the ROM; on others, the program is copied from ROM to main memory and executed there. However, some machines do not have such read-only storage. In addition, it can be inconvenient to change a ROM program if modifications in the absolute loader are required.

An intermediate solution is to have a built-in hardware function (or a very short ROM program) that reads a fixed-length record from some device into memory at a fixed location. The particular device to be used can often be selected via console switches. After the read operation is complete, control is automatically transferred to the address in memory where the record was stored. This record contains machine instructions that load the absolute program that follows. If the loading process requires more instructions than can be read in a single record, this first record causes the reading of others, and these in turn can cause the reading of still more records—hence the term *bootstrap*. The first record (or records) is generally referred to as a *bootstrap loader*. (A simple example of such a bootstrap loader was given in Section 3.1.2.) Such a loader is added to the beginning of all object programs that are to be loaded into an empty and idle system. This includes, for example, the operating system itself and all stand-alone programs that are to be run without an operating system.

3.5 IMPLEMENTATION EXAMPLES

In this section we briefly examine linkers and loaders for actual computers. Because of the relationships between language translation and loading, we use as examples the same three machines we considered in Section 2.5. As you read this section, you may want to refer to that discussion and to the machine descriptions in Chapter 1.

As in our previous discussions, we make no attempt to give a full description of the linkers and loaders used as examples. Instead we concentrate on any particularly interesting or unusual features, and on differences between these implementations and the more general model discussed earlier in this chapter. We also point out areas in which the linker or loader design is related to the assembler design or to the structure and characteristics of the machine.

3.5.1 System/370 Linkage Editor

The format of the object program handled by the System/370 linkage editor is very similar to the one we discussed for SIC/XE. The reference-number technique discussed in Section 3.2.3 is used to improve efficiency. The output program from the linkage editor is called a *load module*. Load modules can be loaded into memory for execution. They also (normally) contain enough information to allow reprocessing by the linkage editor. It is possible for a user to specify that a load module is to be "not editable," in which case much of the control information can be omitted to produce a smaller load module.

The System/370 linkage editor can perform all the standard functions we have discussed. Control sections can be deleted, replaced, or rearranged. Symbols used in external references can be changed or deleted. To facilitate editing, the linkage editor provides for automatic replacement of control sections. If two or more control sections being processed have the same name, only the first is included in the load module. The others are deleted without being regarded as errors. The linkage editor will automatically search system or user-specified libraries to resolve external references. The user can, however, suppress this search for some or all external references.

The linkage editor also stores a variety of other information with the load module. This includes the language translator used for each control section and the date of assembly or compilation (taken from the object program). The linkage editor also supplies dates of editing and modification to maintain a processing history. When a load module is placed into a library, the linkage editor makes an entry in the directory for that library. This directory entry specifies whether the module can be reprocessed by the linkage editor, whether it is an overlay program, whether it is reusable or shareable, and a variety of other attributes. Some of these attributes are specified by the user; others are generated by the linkage editor from information gathered during processing.

The 370 linkage editor supports overlay programs of the type discussed in Section 3.3.3, with a number of extended features. Multiple entry points are allowed in each segment. To support this feature, *entry tables* are automatically created by the linkage editor in all segments that contain a call that might cause overlay. The entry table relates each symbol used as an external reference to the segment in which it is located. There is also a segment table similar to that discussed in Section 3.3.3. An overlay program is permitted, under certain conditions, to make *exclusive references*—that is, references that cause overlay of the calling segment. This would happen, for example, if there were a call from B to C in Fig. 3.13(a). However, the use of this feature is recommended only under unusual circumstances. Unless a special option is specified, such references are treated by the linkage editor as errors.

The overlay process occurs automatically when a segment is called, as discussed in Section 3.3.3. It is also possible for the program to explicitly request that a particular segment be loaded (via a call to the operating system). Some versions of the operating system allow the requesting segment to continue execution while the specified segment is being loaded, thus overlapping the overlay process with program execution. Of course, this type of parallel operation must be carefully planned when the overlay program is being designed.

Overlay programs on the 370 may be divided into multiple *regions* to improve storage utilization. Each region contains a tree-structured

FIGURE 3.19 Multiple-region overlay program.

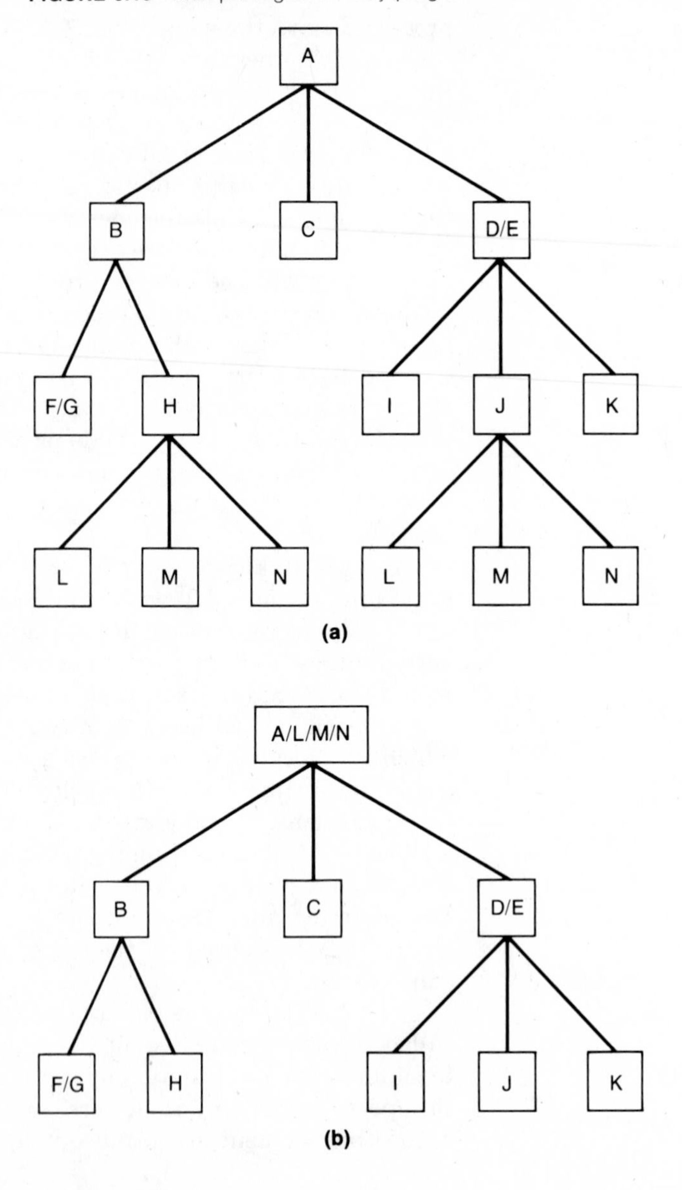

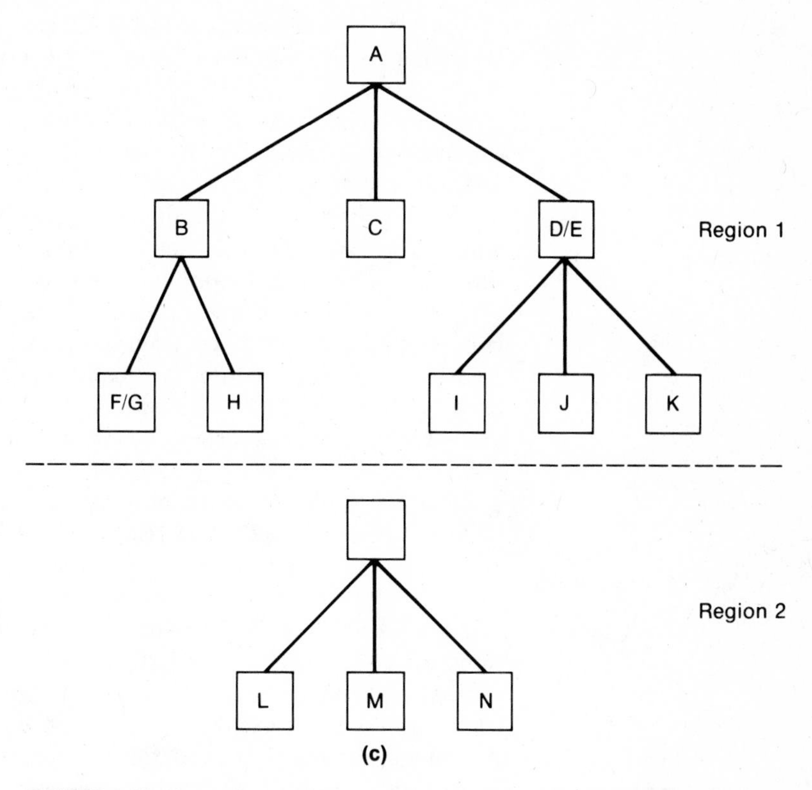

FIGURE 3.19 *(cont'd)*

overlay; within a single region, the usual rules for references between segments apply. However, the regions are independent of each other. A segment in one region is free to call any segment in a different region.

Consider, for example, the structure shown in Fig. 3.19(a). This is the same as the overlay structure in Fig. 3.13(a), with the addition of control sections L, M, and N, each of which may be called by either H or J. Figure 3.19(a) does not represent a legal overlay structure because L, M, and N are placed at two different locations in the tree—this would result in duplicate definitions of external symbols. To create a legal overlay program in a single region, we would need to place L, M, and N in the root segment so that they are accessible by both H and J, as shown in Fig. 3.19(b). However, this would mean that all three of these control sections would remain loaded throughout the execution of the program. If they were not in the root segment, the three control sections could overlay each other. Thus a potential benefit of the overlay structure would be lost. If L, M, and N are large segments, the memory requirements could be increased substantially.

In Fig. 3.19(c), this problem is solved with a multiple-region overlay structure. Region 1 contains an overlay structure that is the same as in Fig. 3.13(a). Region 2 contains segments L, M, and N; these segments are shown connected to a dummy root to emphasize that they can overlay one another. This solves the storage problem just discussed and allows both H and J to call any of the segments in Region 2.

Most 370 systems include a linking loader as well as the linkage editor. The loader provides fewer options and capabilities than does the linkage editor. For example, the loader does not support overlay programs or produce load modules for storing in a program library. However, because of its reduced complexity, and because it avoids the creation of a load module, the loader can reduce editing and loading time by about one-half. If the specialized functions of the linkage editor are not required, it is recommended that the loader be used instead for efficiency.

Further information about a typical System/370 linkage editor can be found in IBM (1978) and IBM (1972a).

3.5.2 VAX Linker

The VAX linker is a linkage editor that performs the same basic functions discussed earlier in this chapter. The object program format is somewhat more complex than the one we discussed for SIC/XE. The linked version of the program consists of one or more *image sections*. An image section is a collection of program sections (PSECTs) that have similar values of attributes such as writability, executability, and shareability.

The action of the linker in creating the image sections is controlled by the assembler or compiler through a sequence of commands that are a part of the object program. (The Modification record that was a part of our SIC/XE object programs is a simple example of a command of this type.) The linker uses an internal stack as working storage. Commands in the object program may specify stacking values from a variety of sources, storing values from the stack into the image being created, and performing operations on values from the stack. The command language offers a large variety of capabilities: there are more than 50 different possible command codes.

A simple example of this approach to relocation and linking is given in Fig. 3.20. Figure 3.20(a) shows three PSECTs that are assembled together. These PSECTs are essentially the same as the example programs in Fig. 3.8; however, we have shown only the information that is needed for processing REF4 in PROGA. Figure 3.20(b) shows a slightly simplified version of the commands in the object program that would generate the value for REF4.

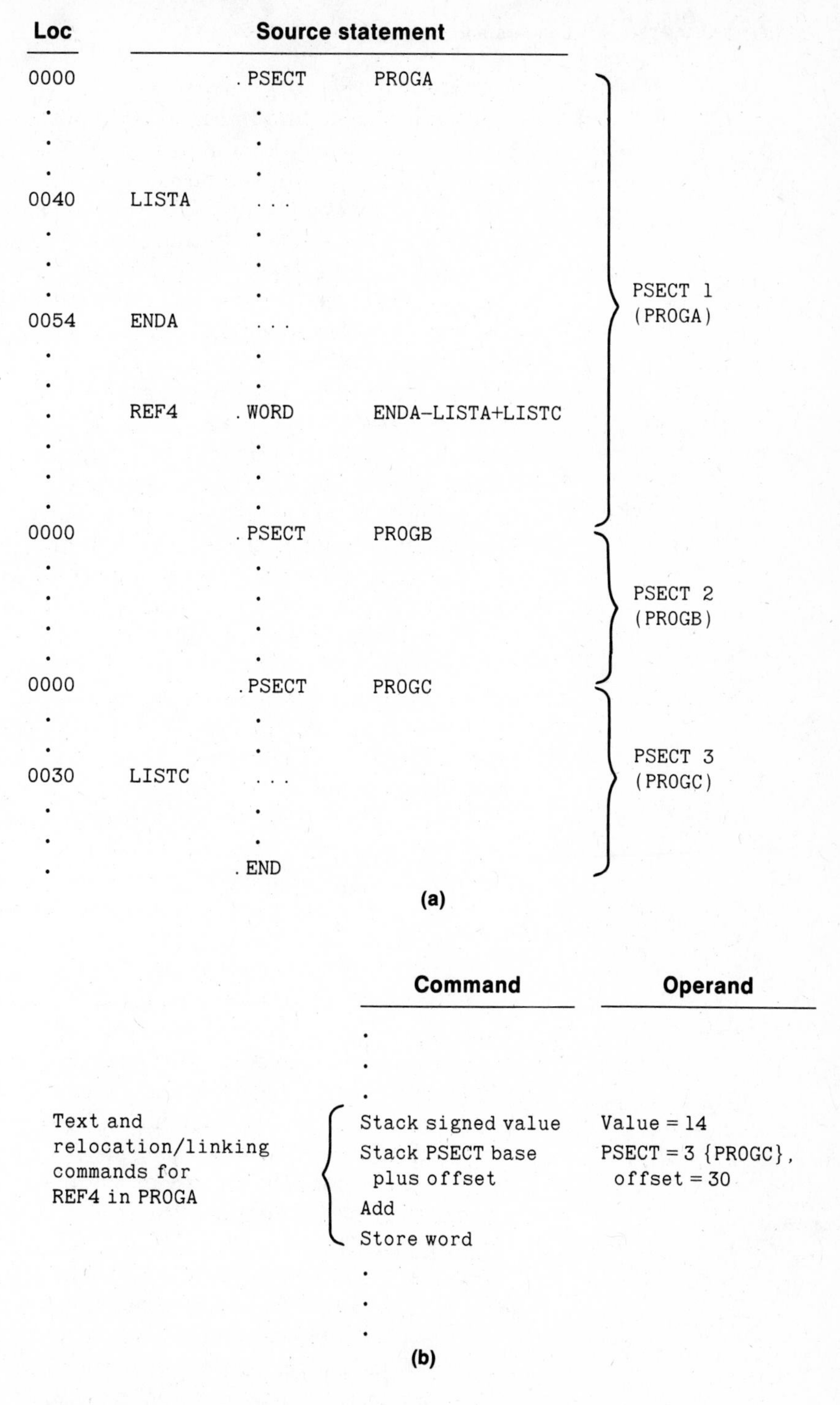

FIGURE 3.20 Example of VAX relocation and linking mechanism.

The first command in Fig. 3.20(b) pushes the value (hexadecimal) 14 onto the top of the linker's stack; this represents the value of (ENDA − LISTA). The second command pushes onto the stack the base address for the PSECT PROGC (which is assigned by the linker) plus the offset (hexadecimal) 30; this sum is the value of LISTC. The third command adds the top two values on the stack, and leaves the result on the top of the stack. The last command stores this sum as the next word of the image section being generated, which is the value of the word designated by REF4.

There are three types of images that can be produced by the VAX linker. An *executable image* is one that is suitable for loading and execution. However, this type of image cannot be reprocessed by the linker. A *shareable image* is not executable, but can be reprocessed by the linker. Such an image might be used, for example, as an intermediate stage in linking a very large program. Shareable images also make it possible for different programs to share the same copy of certain instructions or data areas. Only one copy of a shareable image is maintained on disk, or loaded into memory. This copy is shared by all application programs that require it. Thus the use of shareable images can result in savings in memory, disk space, and paging I/O.

The third type of image that can be created by the linker is a *system image*. A system image is one that is intended to run directly on the VAX machine without the services of the operating system. (The VAX operating system itself is a system image.) This type of image is used only in special circumstances. The structure and content of a system image are simpler than for either of the other types.

The VAX linker performs the usual linking and relocation functions. In addition, it does part of the work that, on some other systems, is accomplished by the assembler or compiler. For example, the VAX assembler does not gather together all of the parts of one PSECT in the object program; this rearrangement of PSECTs is done by the linker. Also, references between different PSECTs that are assembled together are resolved by the linker under control of the command sequence generated by the assembler. The symbols involved do not need to be declared as external symbols, as they would on many other systems. (This is one illustration of the power of the command-driven approach to linking taken on the VAX.)

References between PSECTs that are not assembled together are handled as described in Section 3.2. The symbols involved must have been declared as *global* (i.e., external) symbols. The definitions of these global symbols appear as part of the object program and are used by the linker to resolve external references between PSECTs. External references may be either *strong* or *weak*. Strong references are processed in the ordinary way. Weak references are resolved only if the program that defines the external symbol is part of the primary input to the linker; they are not resolved with the library search mechanism. Weak references that remain unresolved are not treated as errors. Instead the symbol involved is assumed to have a value of zero. One use of such weak references occurs in modular program testing when it might be desirable to test part of a program before all of the subroutines have been written. By identifying calls to the missing routines as weak references, the program can be linked and executed without an error indication. Of course, an execution error will result if the program attempts to invoke a subroutine that has not been linked. (The System/370 linkage editor and the CYBER loader allow similar types of weak references.)

The definitions of global symbols are not normally present in an executable image since this type of image is not intended for relinking. However, when certain debugging tools are in use, the global symbols are retained in a table in the image to aid in providing diagnostic output. In a shareable image, the definitions of certain global symbols are retained to allow the linking of the image with other programs. These symbols, called *universal* symbols, are identified by linker control statements. Global symbols that are not identified as universal are not retained in the shareable image.

The VAX linker does not support overlay programs. This is in part because of the large virtual memory available on the VAX. The system designers felt that the size of this virtual memory, together with the memory management algorithms, made the use of overlays unnecessary. (Recall, however, that the System/370 linkage editor still supports overlays, even on virtual memory systems.)

Further information about the VAX linker can be found in DEC (1982) and DEC (1978).

3.5.3 CYBER Loader

The object program format used by the CYBER loader is somewhat more complex than the one we described for SIC/XE; however, the same basic information is present. Linking is specified using a mechanism similar to our Modification record. A closely related technique is available for describing relocation. However, it is also possible to specify relocation using a method similar to the bit masks discussed in Section 3.2.1. The bit mask technique is particularly useful on the CYBER because relative addressing is not available. This means that CYBER programs usually contain many more relocatable values than do VAX or 370 programs.

A CYBER word may contain more than one instruction, so it is not possible to use a single relocation bit per word. Because only 30-bit instructions on the CYBER may contain memory addresses, there are five possibilities for relocatable values within a word:

1. no relocation
2. relocatable value in upper half of word only
3. relocatable value in lower half of word only
4. relocatable values in both upper and lower halves of word
5. relocatable value in middle 30 bits of word.

When the bit mask technique is used, there is a 4-bit field associated with each word of object code. These four bits are used to encode the possibilities listed above, and also to specify whether the program base address is to be added to or subtracted from each relocatable value.

The CYBER loader supports overlay programs of a more restricted type than we have described. An overlay structure is limited to a maximum of three levels. Each segment is identified by an ordered pair of integers, and a segment may have only one entry point. Each segment except the root must be loaded by an explicit request. There is no provision for automatic loading of segments. The application program may request the loading of a segment via a service call to the operating system. As an alternative, a small resident loader may be included in the root segment to handle the overlay process.

There is also a more powerful alternative to overlays, called *segmentation*. A segmented program also uses a tree structure; however, there can be more than three levels, and each segment can have multiple entry points. It is also possible to have multiple regions like those described for the System/370 linkage editor (although different terminology is used). The programmer may use a command language to assign programs to segments in the tree structure explicitly. If this is not done, the loader will assign programs to segments by analyzing the external references between them.

Segments are loaded automatically under the control of a resident loader (SEGRES). If a segment is not currently loaded, all references to entry points in that segment are replaced with calls to SEGRES. Execution of such a call causes SEGRES to load the needed segment and restore the original reference.

In addition to the standard loading functions, the CYBER loader provides the ability to create *capsules*. A capsule consists of one or more object programs bound together in a special format to allow fast loading. Application programs can call a Fast Dynamic Loader (FDL) to load such a capsule into memory. The FDL facility is more efficient than other methods of loading and uses less memory; however, it is more restrictive and more difficult to use.

Further information about the CYBER loader can be found in CDC (1982b).

EXERCISES

Section 3.1

1. Define a binary object program format for SIC and write an absolute loader (in SIC assembler language) to load programs in this format.
2. Describe a method for performing the packing required when loading an object program such as that in Fig. 3.1(a), which uses character representation of assembled code. How could you implement this method in SIC assembler language?
3. What would be the advantages and disadvantages of writing a loader using a high-level language such as Pascal? What problems would you encounter, and how might these be solved?

Section 3.2

1. Modify the algorithm given in Fig. 3.10 to use the bit-mask approach to relocation. Linking will still be performed using Modification records.
2. Suppose that a computer primarily uses direct addressing, but has several different instruction formats. What problems does this create for the relocation-bit approach to program relocation? How might these problems be solved?
3. Apply the algorithm described in Fig. 3.11 to link and load the object programs in Fig. 3.9. Compare your results with those shown in Fig. 3.10.
4. Assume that PROGA, PROGB, and PROGC are the same as in Fig. 3.8. Show how the object programs would change (including Text and Modification records) if the following statements were added to each program:

```
REF9     WORD    LISTC
REF10    WORD    LISTB-3
REF11    WORD    LISTA+LISTB
REF12    WORD    ENDC-LISTC-100
REF13    WORD    LISTA-LISTB-ENDA+ENDB
```

5. Apply the algorithm described in Fig. 3.11 to link and load the revised object programs you generated in Exercise 4.
6. Using the methods outlined in Chapter 8, develop a modular design for a relocating and linking loader.
7. Extend the algorithm in Fig. 3.11 to include the detection of improper external reference expressions as suggested in the text. (See Section 2.3.5 for the set of rules to be applied.) What problems arise in performing this kind of error checking?

8. Modify the algorithm in Fig. 3.11 to use the reference-number technique for code modification that is described in Section 3.2.3.

9. Suppose that you are implementing an assembler and loader and wish to allow *absolute-valued* external symbols. For example, one control section might contain the statements

```
          EXTDEF MAXLEN
            .
            .
            .
MAXLEN EQU    4096
```

and other control sections could refer to the value of MAXLEN as an external symbol. Describe a way of implementing this new feature, including any needed changes in the loader logic and object program format.

10. Suppose that you have been given the task of writing an "unloader"—that is, a piece of software that can take the image of a program that has been loaded and write out an object program that could later be loaded and executed. The computer system uses a relocating loader, so the object program you produce must be capable of being loaded at a location in memory that is different from where your unloader took it. What problems do you see that would prevent you from accomplishing this task?

11. Suppose that you are given two images of a program as it would appear after loading at two *different* locations in memory. Assume that the images represent the program *after* it is loaded and relocated, but *before* any of the program's instructions are actually executed. Describe how this information could be used to accomplish the "unloading" task mentioned in Exercise 10.

12. Some loaders have used an indirect linking scheme. To use such a technique with SIC/XE, the assembler would generate a list of pointer words from the EXTREF directive (one pointer word for each external reference symbol). Modification records would direct the loader to insert the external symbol addresses into the corresponding words in the pointer list. External references would then be accomplished with indirect addressing using these pointers. Thus, for example, an instruction like

```
LDA      XYZ
```

(where XYZ is an external reference) would be assembled as if it were

```
LDA    @PXYZ,
```

where PXYZ is the pointer word containing the address of XYZ. What would be the advantages and disadvantages of using such a method?

13. Suggest a design for a *one-pass* linking loader. What restrictions (if any) would be required? What would be the advantages and disadvantages of such a one-pass loader?

14. Some programming languages allow data items to be placed in common areas. There may be more than one common area (with different names) in a source program. We may think of each common area as being a separate control section in the object program.

When object programs are linked and loaded, all of the common areas with the same name are assigned the same starting address in memory. (These common areas may be of different lengths in the different programs declaring them.) This assignment of memory establishes an equivalence between the variables that were declared in common by the different programs. Any data value stored into a common area by one program is thus available to the others.

How might the loader handle such common areas? (Suggest modifications to the algorithm of Fig. 3.11 that will perform the necessary processing.)

Section 3.3

1. Modify the algorithm in Fig. 3.11 to include automatic library search to resolve external references. You may assume that the details of library access are handled by operating system service routines.

2. Modify the algorithm in Fig. 3.11 to implement CHANGE, DELETE, and INCLUDE directives as described in Section 3.3.2. If you need to place any restrictions on the use of these commands, be sure to state what they are.

3. Suppose that the loader is to produce a listing that shows not only the addresses assigned to external symbols, but also the cross-references between control sections in the program being loaded. What information might be useful in such a listing? Briefly describe how you might implement this feature and include a description of any data structures needed.

4. Consider the calling structure shown in Fig. 3.13(a). Suppose that routine H could be called by C as well as by B. How could this be handled in an overlay program? (Assume that the rules for overlays given in Section 3.3.3 are not to be changed.)

5. Describe an algorithm to assign starting addresses to segments in an overlay structure, taking into account the memory required by SEGTAB and OVLMGR.

6. During the execution of an overlay program, there are many different sets of segments that can be loaded in memory at various times. Figure 3.15(b) shows three of these possibilities for the program structure in Fig. 3.13(a). Diagram the other possibilities using a similar format.

7. Briefly describe an algorithm for the overlay management routine we have called OVLMGR. Include descriptions of any data structures required.

8. Suppose that we wanted to allow more than one entry point in a segment of an overlay program. For the program of Fig. 3.13(a), for example, we might want to be able to call either D or E directly from A. The entire segment D/E would be loaded, regardless of which entry point was called.

How might this feature be implemented by the loader and/or the overlay manager?

9. As we have described the overlay process, only transfers of control to another segment can cause overlay. How could overlays be implemented so that data references can also cause segments to be loaded? Be sure to describe any restrictions you feel are necessary.

Section 3.4

1. Define a module format suitable for representing linked programs produced by a linkage editor. Assume that the linked program is not to be reprocessed by the linkage editor. Describe an algorithm for a relocating loader that would be suitable for the loading of linked programs in this format.

2. Define a module format suitable for representing linked programs produced by a linkage editor. This format should allow for the loading of the linked program by a one-pass relocating loader, as in Exercise 1. However, it should also allow for the linked program to be reprocessed by the linkage editor. Describe how your format allows for both one-pass loading and relinking.

3. Consider the following possibilities for the storage, linking, and execution of a user's program.

    a) Store the source program only; reassemble the program and use a linking loader each time it is to be executed.

    b) Store the source and object versions of the program; use a linking loader each time the program is to be executed.

    c) Store the source program and the linked version with external references to library subroutines left unresolved. Use a linking loader each time the program is to be executed.

    d) Store the source program and the linked version with all external references resolved. Use a relocating loader each time the program is to be executed.

    e) Store the source program and a linked version that has all external references resolved and all relocation performed. Use an absolute loader each time the program is to be executed.

    Under what conditions might each of these approaches be appropriate? Assume that no changes are required in the source program from one execution to the next.

4. Dynamic linking and overlays are two features that have a number of similarities. Compare and contrast these two features, considering efficiency, ease of use, and any other factors you think are important. When might the use of each of these capabilities be preferred over the other?

5. Dynamic linking, as described in Section 3.4.2, works for transfers of control only. How could the implementation be extended so that data references could also cause dynamic loading to occur?

6. Suppose that routines that are brought into memory by dynamic loading need not be removed until the termination of the main program. Suggest a way to improve the efficiency of dynamic linking by making it unnecessary for the operating system to be involved in the transfer of control after the routine is loaded.

7. Suppose that it may be necessary to remove from memory routines that were dynamically loaded (to reuse the space). Will the method that you suggested in Exercise 6 still work? What problems arise, and how might they be solved?

8. What kinds of errors might occur during bootstrap loading? What action should the bootstrap loader take for such errors? Modify the SIC/XE bootstrap loader shown in Fig. 3.3 to include such error checking.

CHAPTER **4**

MACRO PROCESSORS

In this chapter we study the design and implementation of macro processors. A *macro instruction* (often abbreviated to *macro*) is simply a notational convenience for the programmer. A macro represents a commonly used group of statements in the source programming language. The macro processor replaces each macro instruction with the corresponding group of source language statements. This is called *expanding* the macros. Thus macro instructions allow the programmer to write a shorthand version of a program, and leave the mechanical details to be handled by the macro processor.

The functions of a macro processor essentially involve the substitution of one group of characters or lines for another. Except in a few specialized cases, the macro processor performs no analysis of the text it handles. The design and capabilities of a macro processor may be influenced by the *form* of the programming language statements involved. However, the *meaning* of these statements, and their translation into machine language, are of no concern whatsoever during macro expansion. This means that the design of a macro processor is *not* directly related to the structure of the computer on which it is to run.

The most common use of macro processors is in assembler language programming. We use SIC assembler language examples to illustrate most of the concepts being discussed. However, macro processors can also be used with high-level programming languages, operating system command languages, etc. In addition, there are general-purpose macro processors that are not tied to any particular language. In the later sections of this chapter, we briefly discuss these more general uses of macros.

Section 4.1 introduces the basic concepts of macro processing, including macro definition and expansion. We also present an algorithm for a simple macro processor. Section 4.2 discusses extended features that are commonly found in macro processors. These features include the generation of unique labels within macro expansions, conditional macro expansion, and the use of keyword parameters in macros. All these features are machine-independent. Because the macro processor is not directly related to machine structure, this chapter contains no section on machine-dependent features.

Section 4.3 describes some macro processor design options. One of these options (recursive macro expansion) involves the internal structure of the macro processor itself. The other options are concerned with how the macro processor is related to other pieces of system software such as assemblers or compilers.

Finally, Section 4.4 briefly presents three examples of actual macro processors. Two of these are contrasting macro processors designed for use by assembler language programmers; the third is a general-purpose macro processor. Additional examples may be found in the references cited throughout this chapter.

4.1 BASIC MACRO PROCESSOR FUNCTIONS

In this section we examine the fundamental functions that are common to all macro processors. Section 4.1.1 discusses the processes of macro definition, invocation, and expansion with substitution of parameters. These functions are illustrated with examples using the SIC/XE assembler language. Section 4.1.2 presents a one-pass algorithm for a simple macro processor together with a description of the data structures needed for macro processing. Later sections in this chapter discuss extensions to the basic capabilities introduced in this section.

4.1.1 Macro Definition and Expansion

Figure 4.1 shows an example of a SIC/XE program using macro instructions. This program has the same functions and logic as the sample program in Fig. 2.5; however, the numbering scheme used for the source statements has been changed.

This program defines and uses two macro instructions, RDBUFF and WRBUFF. The functions and logic of the RDBUFF macro are similar to those of the RDREC subroutine in Fig. 2.5; likewise, the WRBUFF macro is similar to the WRREC subroutine. The definitions of these macro instructions appear in the source program following the START statement.

Two new assembler directives (MACRO and MEND) are used in macro definitions. The first MACRO statement (line 10) identifies the beginning of a macro definition. The symbol in the label field (RDBUFF) is the name of the macro, and the entries in the operand field identify the *parameters* of the macro instruction. In our macro language, each parameter begins with the character &, which facilitates the substitution of parameters during macro expansion. The macro name and parameters define a pattern or *prototype* for the macro instructions used by the programmer. Following the MACRO directive are the statements that make up the *body* of the macro definition (lines 15 through 90). These are the statements that will be generated as the expansion of the macro. The MEND assembler directive (line 95) marks the end of the macro definition. The definition of the WRBUFF macro (lines 100 through 160) follows a similar pattern.

The main program itself begins on line 180. The statement on line 190 is a *macro invocation* statement that gives the name of the macro instruction being invoked and the *arguments* to be used in expanding the macro. (A macro invocation statement is often referred to as a *macro call.* To avoid confusion with the call statements used for procedures and subroutines, we prefer to use the term *invocation.* As we shall see, the processes of macro invocation and subroutine call are quite different.) You should compare the logic of the main program in Fig. 4.1 with that of the main program in Fig. 2.5, remembering the similarities in function between RDBUFF and RDREC and between WRBUFF and WRREC.

The program in Fig. 4.1 could be supplied as input to a macro processor. Figure 4.2 shows the output that would be generated. The macro instruction definitions have been deleted since they are no longer needed after the macros are expanded. Each macro invocation statement has been expanded into the statements that form the body of the macro, with the arguments from the macro invocation substituted

```
  5     COPY     START    0                   COPY FILE FROM INPUT TO OUTPUT
 10     RDBUFF   MACRO    &INDEV,&BUFADR,&RECLTH
 15     .
 20     .   MACRO TO READ RECORD INTO BUFFER
 25     .
 30              CLEAR    X                   CLEAR LOOP COUNTER
 35              CLEAR    A
 40              CLEAR    S
 45             +LDT     #4096                SET MAXIMUM RECORD LENGTH
 50              TD      =X'&INDEV'           TEST INPUT DEVICE
 55              JEQ      *-3                 LOOP UNTIL READY
 60              RD      =X'&INDEV'           READ CHARACTER INTO REGISTER A
 65              COMPR    A,S                 TEST FOR END OF RECORD
 70              JEQ      *+11                EXIT LOOP IF EOR
 75              STCH     &BUFADR,X           STORE CHARACTER IN BUFFER
 80              TIXR     T                   LOOP UNLESS MAXIMUM LENGTH
 85              JLT      *-19                   HAS BEEN REACHED
 90              STX      &RECLTH             SAVE RECORD LENGTH
 95              MEND
100     WRBUFF   MACRO    &OUTDEV,&BUFADR,&RECLTH
105     .
110     .   MACRO TO WRITE RECORD FROM BUFFER
115     .
120              CLEAR    X                   CLEAR LOOP COUNTER
125              LDT      &RECLTH
130              LDCH     &BUFADR,X           GET CHARACTER FROM BUFFER
135              TD      =X'&OUTDEV'          TEST OUTPUT DEVICE
140              JEQ      *-3                 LOOP UNTIL READY
145              WD      =X'&OUTDEV'          WRITE CHARACTER
150              TIXR     T                   LOOP UNTIL ALL CHARACTERS
155              JLT      *-14                  HAVE BEEN WRITTEN
160              MEND
165     .
170     .   MAIN PROGRAM
175     .
180     FIRST    STL      RETADR                      SAVE RETURN ADDRESS
190     CLOOP    RDBUFF   F1,BUFFER,LENGTH            READ RECORD INTO BUFFER
195              LDA      LENGTH                      TEST FOR END OF FILE
200              COMP    #0
205              JEQ      ENDFIL                      EXIT IF EOF FOUND
210              WRBUFF   05,BUFFER,LENGTH            WRITE OUTPUT RECORD
215              J        CLOOP                       LOOP
220     ENDFIL   WRBUFF   05,EOF,THREE                INSERT EOF MARKER
225              J       @RETADR
230     EOF      BYTE     C'EOF'
235     THREE    WORD     3
240     RETADR   RESW     1
245     LENGTH   RESW     1                           LENGTH OF RECORD
250     BUFFER   RESB     4096                        4096-BYTE BUFFER AREA
255              END      FIRST
```

FIGURE 4.1 Use of macros in a SIC/XE program.

```
  5    COPY     START    0                   COPY FILE FROM INPUT TO OUTPUT
180    FIRST    STL      RETADR                SAVE RETURN ADDRESS
190    .CLOOP    RDBUFF  F1,BUFFER,LENGTH      READ RECORD INTO BUFFER
190a   CLOOP    CLEAR    X                   CLEAR LOOP COUNTER
190b            CLEAR    A
190c            CLEAR    S
190d           +LDT      #4096               SET MAXIMUM RECORD LENGTH
190d            TD       =X'F1'           TEST INPUT DEVICE
190e            JEQ      *-3                 LOOP UNTIL READY
190f            RD       =X'F1'           READ CHARACTER INTO REGISTER A
190g            COMPR    A,S                 TEST FOR END OF RECORD
190h            JEQ      *+11                EXIT LOOP IF EOR
190i            STCH     BUFFER,X          STORE CHARACTER IN BUFFER
190j            TIXR     T                   LOOP UNLESS MAXIMUM LENGTH
190k            JLT      *-19                  HAS BEEN REACHED
190l            STX      LENGTH            SAVE RECORD LENGTH
195             LDA      LENGTH                TEST FOR END OF FILE
200             COMP     #0
205             JEQ      ENDFIL                EXIT IF EOF FOUND
210              WRBUFF  05,BUFFER,LENGTH      WRITE OUTPUT RECORD
210a            CLEAR    X                   CLEAR LOOP COUNTER
210b            LDT      LENGTH
210c            LDCH     BUFFER,X          GET CHARACTER FROM BUFFER
210d            TD       =X'05'         TEST OUTPUT DEVICE
210e            JEQ      *-3                 LOOP UNTIL READY
210f            WD       =X'05'         WRITE CHARACTER
210g            TIXR     T                   LOOP UNTIL ALL CHARACTERS
210h            JLT      *-14                  HAVE BEEN WRITTEN
215             J        CLOOP                 LOOP
220    .ENDFIL  WRBUFF   05,EOF,THREE          INSERT EOF MARKER
220a   ENDFIL   CLEAR    X                   CLEAR LOOP COUNTER
220b            LDT      THREE
220c            LDCH     EOF,X            GET CHARACTER FROM BUFFER
220d            TD       =X'05'         TEST OUTPUT DEVICE
220e            JEQ      *-3                 LOOP UNTIL READY
220f            WD       =X'05'         WRITE CHARACTER
220g            TIXR     T                   LOOP UNTIL ALL CHARACTERS
220h            JLT      *-14                  HAVE BEEN WRITTEN
225             J        @RETADR
230    EOF      BYTE     C'EOF'
235    THREE    WORD     3
240    RETADR   RESW     1
245    LENGTH   RESW     1                     LENGTH OF RECORD
250    BUFFER   RESB     4096                  4096-BYTE BUFFER AREA
255             END      FIRST
```

FIGURE 4.2 Program from Fig. 4.1 with macros expanded.

for the parameters in the macro prototype. The arguments and parameters are associated with one another according to their positions. The first argument in the macro invocation corresponds to the first parameter in the macro prototype, and so on. In expanding the macro invocation on line 190, for example, the argument F1 is substituted for the parameter &INDEV wherever it occurs in the body of the macro. Similarly, BUFFER is substituted for &BUFADR, and LENGTH is substituted for &RECLTH.

Lines 190a through 190l show the complete expansion of the macro invocation on line 190. The comment lines within the macro body have been deleted, but comments on individual statements have been retained. Note that the macro invocation statement itself has been included as a comment line. This serves as documentation of the statement written by the programmer. The label on the macro invocation statement (CLOOP) has been retained as a label on the first statement generated in the macro expansion. This allows the programmer to use a macro instruction in exactly the same way as an assembler language mnemonic. The macro invocations on lines 210 and 220 are expanded in the same way. Note that the two invocations of WRBUFF specify different arguments, so they produce different expansions.

After macro processing, the expanded file (Fig. 4.2) can be used as input to the assembler. The macro invocation statements will be treated as comments, and the statements generated from the macro expansions will be assembled exactly as though they had been written directly by the programmer.

A comparison of the expanded program in Fig. 4.2 with the program in Fig. 2.5 shows the most significant differences between macro invocation and subroutine call. In Fig. 4.2, the statements from the body of the macro WRBUFF are generated twice: lines 210a through 210h and lines 220a through 220h. In the program of Fig. 2.5, the corresponding statements appear only once: in the subroutine WRREC (lines 210 through 240). In general, the statements that form the expansion of a macro are generated (and assembled) each time the macro is invoked. Statements in a subroutine appear only once, regardless of how many times the subroutine is called.

Note also that our macro instructions have been written so that the body of the macro contains no labels. In Fig. 4.1, for example, line 140 contains the statement "JEQ *–3" and line 155 contains "JLT *–14." The corresponding statements in the WRREC subroutine (Fig. 2.5) are "JEQ WLOOP" and "JLT WLOOP," where WLOOP is a label on the TD instruction that tests the output device. If such a label appeared on line 135 of the macro body, it would be generated twice—on lines 210d and 220d of Fig. 4.2. This would result in an error (a duplicate label definition) when the program is assembled. To avoid duplication of symbols, we have eliminated labels from the body of our macro definitions.

The use of statements like "JLT *–14" is generally considered to be a poor programming practice. It is somewhat less objectionable within a macro definition; however, it is still an inconvenient and error-prone method. In Section 4.2.2 we discuss ways of avoiding this problem.

4.1.2 Macro Processor Tables and Logic

It is easy to design a two-pass macro processor in which all macro definitions are processed during the first pass, and all macro invocation statements are expanded during the second pass. However, such a two-pass macro processor would not allow the body of one macro instruction to contain definitions of other macros (because all macros would have to be defined during the first pass before any macro invocations were expanded).

Such definitions of macros by other macros can be useful in certain cases. Consider, for example, the two macro instruction definitions in Fig. 4.3. The body of the first macro (MACROS) contains statements that define RDBUFF, WRBUFF, and other macro instructions for a SIC system (standard version). The body of the second macro instruction (MACROX) defines these same macros for a SIC/XE system. A program that is to be run on a standard SIC system could invoke MACROS to define the other utility macro instructions. A program for a SIC/XE system could invoke MACROX to define these same macros in their XE versions. In this way, the same program could run on either a standard SIC machine or a SIC/XE machine (taking advantage of the extended features). The only change required would be the invocation of either MACROS or MACROX. It is important to understand that *defining* MACROS or MACROX does not define RDBUFF and the other macro instructions. These definitions are processed only when an invocation of MACROS or MACROX is *expanded.*

A one-pass macro processor that can alternate between macro definition and macro expansion is able to handle macros like those in Fig. 4.3. In this section we present an algorithm and a set of data structures for such a macro processor. Because of the one-pass structure, the definition of a macro must appear in the source program before any statements that invoke that macro. This restriction does not create any real inconvenience for the programmer. In fact, a macro invocation statement that preceded the definition of the macro would be confusing for anyone reading the program.

There are three main data structures involved in our macro processor. The macro definitions themselves are stored in a definition table (DEFTAB), which contains the macro prototype and the statements that make up the macro body (with a few modifications). Comment lines from the macro definition are not entered into DEFTAB because they will not be part of the macro expansion. References to the macro instruction parameters are converted to a positional notation for effi-

```
1    MACROS   MACRO   {Defines SIC standard version macros}
2    RDBUFF   MACRO   &INDEV,&BUFADR,&RECLTH
                .
                .     {SIC standard version}
                .
3             MEND    {End of RDBUFF}
4    WRBUFF   MACRO   &OUTDEV,&BUFADR,&RECLTH
                .
                .     {SIC standard version}
                .
5             MEND    {End of WRBUFF}
                .
                .
                .
6             MEND    {End of MACROS}
```

(a)

```
1    MACROX   MACRO   {Defines SIC/XE macros}
2    RDBUFF   MACRO   &INDEV,&BUFADR,&RECLTH
                .
                .     {SIC/XE version}
                .
3             MEND    {End of RDBUFF}
4    WRBUFF   MACRO   &OUTDEV,&BUFADR,&RECLTH
                .
                .     {SIC/XE version}
                .
5             MEND    {End of WRBUFF}
                .
                .
                .
6             MEND    {End of MACROX}
```

(b)

FIGURE 4.3 Example of the definition of macros within a macro body.

ciency in substituting arguments. The macro names are also entered into NAMTAB, which serves as an index to DEFTAB. For each macro instruction defined, NAMTAB contains pointers to the beginning and end of the definition in DEFTAB.

The third data structure is an argument table (ARGTAB), which is used during the expansion of macro invocations. When a macro invocation statement is recognized, the arguments are stored in ARGTAB according to their position in the argument list. As the macro is expanded, arguments from ARGTAB are substituted for the corresponding parameters in the macro body.

Figure 4.4 shows portions of the contents of these tables during the processing of the program in Fig. 4.1. Figure 4.4(a) shows the defini-

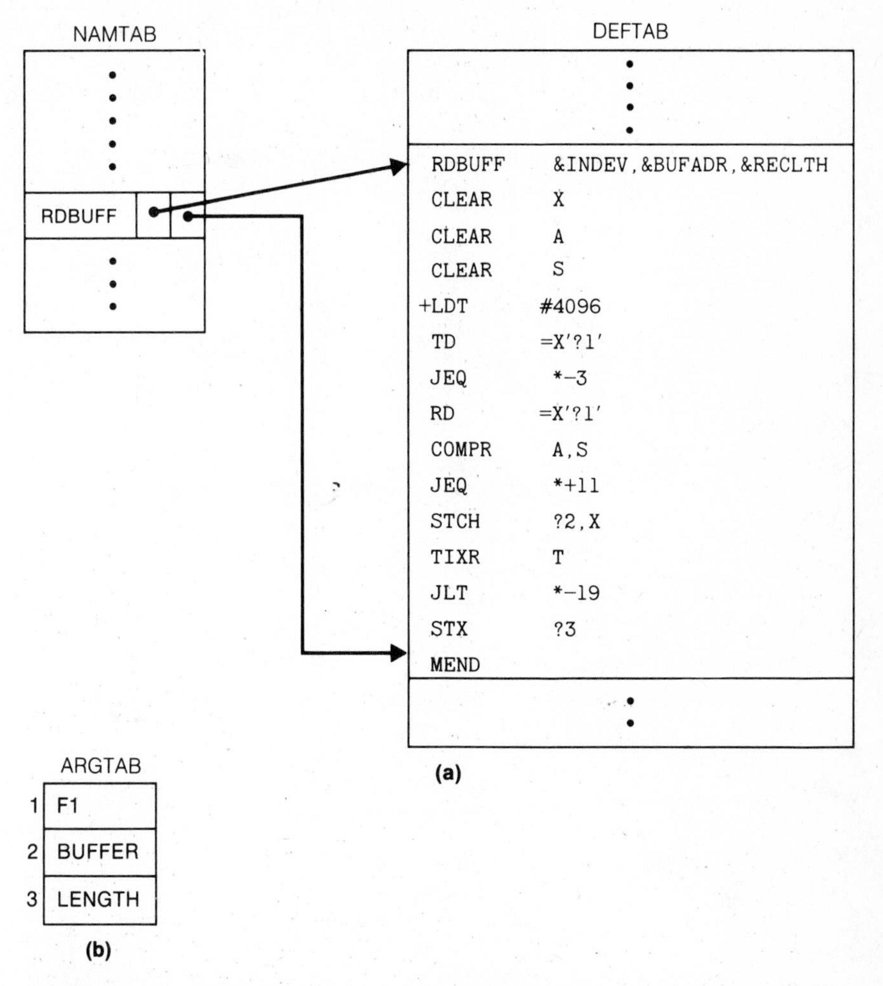

FIGURE 4.4 Contents of macro processor tables for the program in Fig. 4.1: (a) entries in NAMTAB and DEFTAB defining macro RDBUFF, (b) entries in ARGTAB for invocation of RDBUFF on line 190.

tion of RDBUFF stored in DEFTAB, with an entry in NAMTAB identifying the beginning and end of the definition. Note the positional notation that has been used for the parameters: the parameter &INDEV has been converted to *?1* (indicating the first parameter in the prototype), &BUFADR has been converted to *?2,* and so on. Figure 4.4(b) shows ARGTAB as it would appear during expansion of the RDBUFF statement on line 190. For this invocation, the first argument is F1, the second is BUFFER, etc. This scheme makes substitution of macro arguments much more efficient. When the *?n* notation is recognized in a line from DEFTAB, a simple indexing operation supplies the proper argument from ARGTAB.

The macro processor algorithm itself is presented in Fig. 4.5. The procedure DEFINE, which is called when the beginning of a macro definition is recognized, makes the appropriate entries in DEFTAB and NAMTAB. EXPAND is called to set up the argument values in

FIGURE 4.5 Algorithm for a one-pass macro processor.

```
begin {macro processor}
   EXPANDING := FALSE
   while OPCODE <> 'END' do
      begin
         GETLINE
         PROCESSLINE
      end {while}
end {macro processor}


procedure PROCESSLINE
   begin
      search NAMTAB for OPCODE
      if found then
         EXPAND
      else if OPCODE = 'MACRO' then
         DEFINE
      else write source line to expanded file
   end {PROCESSLINE}


procedure DEFINE
   begin
      enter macro name into NAMTAB
      enter macro prototype into DEFTAB
      LEVEL := 1
      while LEVEL > 0 do
         begin
            GETLINE
            if this is not a comment line then
               begin
                  substitute positional notation for parameters
                  enter line into DEFTAB
                  if OPCODE = 'MACRO' then
                     LEVEL := LEVEL + 1
                  else if OPCODE = 'MEND' then
                     LEVEL := LEVEL - 1
               end {if not comment}
         end {while}
      store in NAMTAB pointers to beginning and end of definition
   end {DEFINE}
```

```
procedure EXPAND
   begin
      EXPANDING := TRUE
      get first line of macro definition {prototype} from DEFTAB
      set up arguments from macro invocation in ARGTAB
      write macro invocation to expanded file as a comment
      while not end of macro definition do
         begin
            GETLINE
            PROCESSLINE
         end {while}
      EXPANDING := FALSE
   end {EXPAND}


procedure GETLINE
   begin
      if EXPANDING then
         begin
            get next line of macro definition from DEFTAB
            substitute arguments from ARGTAB for positional notation
         end {if}
      else
         read next line from input file
   end {GETLINE}
```

FIGURE 4.5 *(cont'd)*

ARGTAB and expand a macro invocation statement. The procedure GETLINE, which is called at several points in the algorithm, gets the next line to be processed. This line may come from DEFTAB (the next line of a macro being expanded), or from the input file, depending upon whether the Boolean variable EXPANDING is set to TRUE or FALSE.

One aspect of this algorithm deserves further comment: the handling of macro definitions within macros (as illustrated in Fig. 4.3). When a macro definition is being entered into DEFTAB, the normal approach would be to continue until an MEND directive is reached. This would not work for the example in Fig. 4.3, however. The MEND on line 3 (which actually marks the end of the definition of RDBUFF) would be taken as the end of the definition of MACROS. To solve this problem, our DEFINE procedure maintains a counter named LEVEL. Each time a MACRO directive is read, the value of LEVEL is increased by 1; each time an MEND directive is read, the value of LEVEL is decreased by 1. When LEVEL reaches 0, the MEND that corresponds to the original MACRO directive has been found. This process is very much like matching left and right parentheses when scanning an arithmetic expression.

You may want to apply this algorithm by hand to the program in Fig. 4.1 to be sure you understand its operation. The result should be the same as shown in Fig. 4.2.

Most macro processors allow the definitions of commonly used macro instructions to appear in a standard system library, rather than in the source program. This makes the use of such macros much more convenient. Definitions are retrieved from this library as they are needed during macro processing. The extension of the algorithm in Fig. 4.5 to include this sort of processing appears as an exercise at the end of this chapter.

4.2 MACHINE-INDEPENDENT MACRO PROCESSOR FEATURES

In this section we discuss several extensions to the basic macro processor functions presented in Section 4.1. As we have mentioned before, these extended features are not directly related to the architecture of the computer for which the macro processor is written. Section 4.2.1 describes a method for concatenating macro instruction parameters with other character strings. Section 4.2.2 discusses one method for generating unique labels within macro expansions, which avoids the need for extensive use of relative addressing at the source statement level. Section 4.2.3 introduces the important topic of conditional macro expansion and illustrates the concepts involved with several examples. This ability to alter the expansion of a macro by using control statements makes macro instructions a much more powerful and useful tool for the programmer. Section 4.2.4 describes the definition and use of keyword parameters in macro instructions.

4.2.1 Concatenation of Macro Parameters

Most macro processors allow parameters to be concatenated with other character strings. Suppose, for example, that a program contains one series of variables named by the symbols XA1, XA2, XA3, . . . , another series named by XB1, XB2, XB3, . . . , etc. If similar processing is to be performed on each series of variables, the programmer might want to incorporate this processing into a macro instruction. The parameter to such a macro instruction could specify the series of variables to be operated on (A, B, etc.). The macro processor would use this parameter to construct the symbols required in the macro expansion (XA1, XB1, etc.).

Suppose that the parameter to such a macro instruction is named &ID. The body of the macro definition might contain a statement like

```
LDA     X&ID1
```

in which the parameter &ID is concatenated after the character string X and before the character string 1. Closer examination, however, reveals a problem with such a statement. The beginning of the macro parameter is identified by the starting symbol &; however, the end of the parameter is not marked. Thus the operand in the foregoing statement could equally well represent the character string X followed by the parameter &ID1. In this particular case, the macro processor could potentially deduce the meaning that was intended. However, if the macro definition contained both &ID and &ID1 as parameters, the situation would be unavoidably ambiguous.

Most macro processors deal with this problem by providing a special *concatenation operator*. In the SIC macro language, this operator is the character →. Thus the previous statement would be written as

```
LDA       X&ID→1
```

so that the end of the parameter &ID is clearly identified. The macro processor deletes all occurrences of the concatenation operator immediately after performing parameter substitution, so the character → will not appear in the macro expansion.

Figure 4.6(a) shows a macro definition that uses the concatenation operator as previously described. Figures 4.6(b) and (c) show macro invocation statements and the corresponding macro expansions. You should work through the generation of these macro expansions for yourself to be sure you understand how the concatenation operators are handled. You are also encouraged to think about how the concatenation operator would be handled in a macro processing algorithm like the one given in Fig. 4.5.

4.2.2 Generation of Unique Labels

As we discussed in Section 4.1, it is in general not possible for the body of a macro instruction to contain labels of the usual kind. This leads to the use of relative addressing at the source statement level. Consider, for example, the definition of WRBUFF in Fig. 4.1. If a label were placed on the TD instruction on line 135, this label would be defined twice—once for each invocation of WRBUFF. This duplicate definition would prevent correct assembly of the resulting expanded program.

Because it was not possible to place a label on line 135 of this macro definition, the Jump instructions on lines 140 and 155 were written using the relative operands *–3 and *–14. This sort of relative addressing in a source statement may be acceptable for short jumps such as "JEQ *–3." However, for longer jumps spanning several instructions, such notation is very inconvenient, error-prone, and diffi-

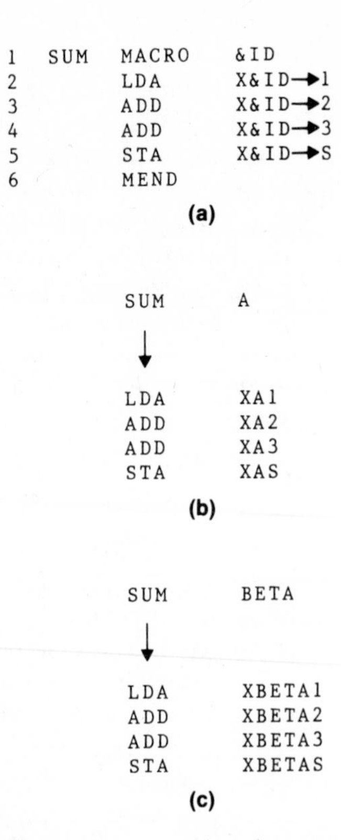

FIGURE 4.6 Concatenation of macro parameters.

cult to read. Many macro processors avoid these problems by allowing the creation of special types of labels within macro instructions.

Figure 4.7 illustrates one technique for generating unique labels within a macro expansion. A definition of the RDBUFF macro is shown in Fig. 4.7(a). Labels used within the macro body begin with the special character *$*. Figure 4.7(b) shows a macro invocation statement and the resulting macro expansion. Each symbol beginning with *$* has been modified by replacing *$* with $AA. More generally, the character *$* will be replaced by *$xx,* where *xx* is a two-character alphanumeric counter of the number of macro instructions expanded. For the first macro expansion in a program, *xx* will have the value AA. For succeeding macro expansions, *xx* will be set to AB, AC, etc. (If only alphabetic and numeric characters are allowed in *xx,* such a two-character counter provides for as many as 1296 macro expansions in a single program.) This results in the generation of unique labels for each expansion of a macro instruction. For further examples, see Figs. 4.8 and 4.10.

The SIC assembler language allows the use of the character *$* in

```
25      RDBUFF   MACRO   &INDEV,&BUFADR,&RECLTH
30               CLEAR   X                  CLEAR LOOP COUNTER
35               CLEAR   A
40               CLEAR   S
45              +LDT     #4096              SET MAXIMUM RECORD LENGTH
50      $LOOP    TD      =X'&INDEV'         TEST INPUT DEVICE
55               JEQ     $LOOP              LOOP UNTIL READY
60               RD      =X'&INDEV'         READ CHARACTER INTO REGISTER A
65               COMPR   A,S                TEST FOR END OF RECORD
70               JEQ     $EXIT              EXIT LOOP IF EOR
75               STCH    &BUFADR,X          STORE CHARACTER IN BUFFER
80               TIXR    T                  LOOP UNLESS MAXIMUM LENGTH
85               JLT     $LOOP                HAS BEEN REACHED
90      $EXIT    STX     &RECLTH            SAVE RECORD LENGTH
95               MEND
```

(a)

```
        .        RDBUFF  F1,BUFFER,LENGTH

30               CLEAR   X                  CLEAR LOOP COUNTER
35               CLEAR   A
40               CLEAR   S
45              +LDT     #4096              SET MAXIMUM RECORD LENGTH
50      $AALOOP  TD      =X'F1'             TEST INPUT DEVICE
55               JEQ     $AALOOP            LOOP UNTIL READY
60               RD      =X'F1'             READ CHARACTER INTO REGISTER A
65               COMPR   A,S                TEST FOR END OF RECORD
70               JEQ     $AAEXIT            EXIT LOOP IF EOR
75               STCH    BUFFER,X           STORE CHARACTER IN BUFFER
80               TIXR    T                  LOOP UNLESS MAXIMUM LENGTH
85               JLT     $AALOOP              HAS BEEN REACHED
90      $AAEXIT  STX     LENGTH             SAVE RECORD LENGTH
```

(b)

FIGURE 4.7 Generation of unique labels within macro expansion.

symbols; however, programmers are instructed not to use this character in their source programs. This avoids any possibility of conflict between programmer-generated symbols and those created by the macro processor.

4.2.3 Conditional Macro Expansion

In all of our previous examples of macro instructions, each invocation of a particular macro was expanded into the same sequence of statements. These statements could be varied by the substitution of parameters, but the form of the statements, and the order in which they appeared, were unchanged. A simple macro facility such as this can be a useful tool. Most macro processors, however, can also modify the sequence of statements generated for a macro expansion, depending on the arguments supplied in the macro invocation. Such a capability adds greatly to the power and flexibility of a macro language. In this section we present a typical set of conditional macro expansion statements. Other examples are found in the macro processor descriptions in Section 4.4.

The term *conditional assembly* is commonly used to describe features such as those discussed in this section. However, there are appli-

FIGURE 4.8 Use of macro-time conditional statements.

```
25    RDBUFF   MACRO   &INDEV,&BUFADR,&RECLTH,&EOR,&MAXLTH
26             IF      (&EOR NE '')
27    &EORCK   SET     1
28             ENDIF
30             CLEAR   X                  CLEAR LOOP COUNTER
35             CLEAR   A
38             IF      (&EORCK EQ 1)
40             LDCH    =X'&EOR'           SET EOR CHARACTER
42             RMO     A,S
43             ENDIF
44             IF      (&MAXLTH EQ '')
45            +LDT     #4096              SET MAX LENGTH = 4096
46             ELSE
47            +LDT     #&MAXLTH           SET MAXIMUM RECORD LENGTH
48             ENDIF
50    $LOOP    TD      =X'&INDEV'         TEST INPUT DEVICE
55             JEQ     $LOOP              LOOP UNTIL READY
60             RD      =X'&INDEV'         READ CHARACTER INTO REGISTER A
63             IF      (&EORCK EQ 1)
65             COMPR   A,S                TEST FOR END OF RECORD
70             JEQ     $EXIT              EXIT LOOP IF EOR
73             ENDIF
75             STCH    &BUFADR,X          STORE CHARACTER IN BUFFER
80             TIXR    T                  LOOP UNLESS MAXIMUM LENGTH
85             JLT     $LOOP                 HAS BEEN REACHED
90    $EXIT    STX     &RECLTH            SAVE RECORD LENGTH
95             MEND
```

(a)

```
      .        RDBUFF  F3,BUF,RECL,04,2048

30             CLEAR   X                  CLEAR LOOP COUNTER
35             CLEAR   A
40             LDCH    =X'04'           SET EOR CHARACTER
42             RMO     A,S
47            +LDT     #2048              SET MAXIMUM RECORD LENGTH
50    $AALOOP  TD      =X'F3'         TEST INPUT DEVICE
55             JEQ     $AALOOP            LOOP UNTIL READY
60             RD      =X'F3'         READ CHARACTER INTO REGISTER A
65             COMPR   A,S                TEST FOR END OF RECORD
70             JEQ     $AAEXIT            EXIT LOOP IF EOR
75             STCH    BUF,X          STORE CHARACTER IN BUFFER
80             TIXR    T                  LOOP UNLESS MAXIMUM LENGTH
85             JLT     $AALOOP               HAS BEEN REACHED
90    $AAEXIT  STX     RECL             SAVE RECORD LENGTH
```

(b)

```
        .        RDBUFF   0E,BUFFER,LENGTH,,80

30                CLEAR    X                 CLEAR LOOP COUNTER
35                CLEAR    A
47               +LDT      #80            SET MAXIMUM RECORD LENGTH
50      $ABLOOP   TD       =X'0E'          TEST INPUT DEVICE
55                JEQ      $ABLOOP           LOOP UNTIL READY
60                RD       =X'0E'          READ CHARACTER INTO REGISTER A
75                STCH     BUFFER,X           STORE CHARACTER IN BUFFER
80                TIXR     T                  LOOP UNLESS MAXIMUM LENGTH
87                JLT      $ABLOOP               HAS BEEN REACHED
90      $ABEXIT   STX      LENGTH             SAVE RECORD LENGTH
```

(c)

```
        .        RDBUFF   F1,BUFF,RLENG,04

30                CLEAR    X                 CLEAR LOOP COUNTER
35                CLEAR    A
40                LDCH     =X'04'          SET EOR CHARACTER
42                RMO      A,S
45               +LDT      #4096             SET MAX LENGTH = 4096
50      $ACLOOP   TD       =X'F1'          TEST INPUT DEVICE
55                JEQ      $ACLOOP           LOOP UNTIL READY
60                RD       =X'F1'          READ CHARACTER INTO REGISTER A
65                COMPR    A,S               TEST FOR END OF RECORD
70                JEQ      $ACEXIT           EXIT LOOP IF EOR
75                STCH     BUFF,X           STORE CHARACTER IN BUFFER
80                TIXR     T                 LOOP UNLESS MAXIMUM LENGTH
85                JLT      $ACLOOP              HAS BEEN REACHED
90      $ACEXIT   STX      RLENG            SAVE RECORD LENGTH
```

(d)

FIGURE 4.8 *(cont'd)*

cations of macro processors that are not related to assemblers or assembler language programming. For this reason, we prefer to use the term *conditional macro expansion.*

The use of one type of conditional macro expansion statement is illustrated in Fig. 4.8. Figure 4.8(a) shows a definition of a macro RDBUFF, the logic and functions of which are similar to those previously discussed. However, this definition of RDBUFF has two additional parameters: &EOR, which specifies a hexadecimal character code that marks the end of a record, and &MAXLTH, which specifies the maximum length record that can be read. (As we shall see, it is possible for either or both of these parameters to be omitted in an invocation of RDBUFF.)

The statements on lines 44 through 48 of this definition illustrate a simple macro-time conditional structure. The IF statement evaluates a Boolean expression that is its operand. If the value of this expression is TRUE, the statements following the IF are generated until an ELSE is encountered. Otherwise, these statements are skipped, and the statements following the ELSE are generated. The ENDIF statement terminates the conditional expression that was begun by the IF statement. (As usual, the ELSE clause can be omitted entirely.) Thus if the parameter &MAXLTH is equal to the null string (that is, if the corresponding argument was omitted in the macro invocation statement), the statement on line 45 is generated. Otherwise, the statement on line 47 is generated.

A similar structure appears on lines 26 through 28. In this case, however, the statement controlled by the IF is not a line to be generated into the macro expansion. Instead, it is another macro processor directive (SET). This SET statement assigns the value 1 to &EORCK. The symbol &EORCK is a *macro-time variable* (also often called a *set symbol*), which can be used to store working values during the macro expansion. Any symbol that begins with the character & and that is not a macro instruction parameter is assumed to be a macro-time variable. All such variables are initialized to a value of 0. Thus if there is an argument corresponding to &EOR (that is, if &EOR is not null), the variable &EORCK is set to 1. Otherwise, it retains its default value of 0. The value of this macro-time variable is used in the conditional structures on lines 38 through 43 and 63 through 73.

In the previous example the value of the macro-time variable &EORCK was used to store the result of the comparison involving &EOR (line 26). The IF statements that use this value (lines 38 and 63) could, of course, simply have repeated the original test. However, the use of a macro-time variable makes it clear that the same logical condition is involved in both IF statements. Examining the value of the variable may also be faster than repeating the original test, especially if the test involves a complicated Boolean expression rather than just a single comparison.

Figure 4.8(b–d) show the expansion of three different macro invocation statements that illustrate the operation of the IF statements in Fig. 4.8(a). You should carefully work through these examples to be sure you understand how the given macro expansion was obtained from the macro definition and the macro invocation statement.

The implementation of the conditional macro expansion features just described is relatively simple. The macro processor must maintain a symbol table that contains the values of all macro-time variables used. Entries in this table are made or modified when SET statements are processed. The table is used to look up the current value of a macro-time variable whenever it is required.

When an IF statement is encountered during the expansion of a macro, the specified Boolean expression is evaluated. If the value of this expression is TRUE, the macro processor continues to process lines from DEFTAB until it encounters the next ELSE or ENDIF statement. If an ELSE is found, the macro processor then skips lines in DEFTAB until the next ENDIF. Upon reaching the ENDIF, it resumes expanding the macro in the usual way. If the value of the specified Boolean expression is FALSE, the macro processor skips ahead in DEFTAB until it finds the next ELSE or ENDIF statement. The macro processor then resumes normal macro expansion.

The implementation outlined above does not allow for nested IF structures. You are encouraged to think about how this technique could be modified to handle such nested structures (see Exercise 4.2.8).

It is extremely important to understand that the testing of Boolean expressions in IF statements occurs at the time macros are *expanded*. By the time the program is assembled, all such decisions have been made. There is only one sequence of source statements (for example, the statements in Fig. 4.8c), and the conditional macro expansion directives have been removed. Thus macro-time IF statements correspond to options that might have been selected by the programmer in writing the source code. They are fundamentally different from statements such as COMPR (or IF statements in a high-level programming language), which test data values during program *execution*. The same applies to the assignment of values to macro-time variables, and to the other conditional macro expansion directives we discuss.

The macro-time IF–ELSE–ENDIF structure provides a mechanism for either generating (once) or skipping selected statements in the macro body. A different type of conditional macro expansion statement is illustrated in Fig. 4.9. Figure 4.9(a) shows another definition of RDBUFF. The purpose and function of the macro are the same as before. With this definition, however, the programmer can specify a list of end-of-record characters. In the macro invocation statement in Fig. 4.9(b), for example, there is a list (00,03,04) corresponding to the parameter &EOR. Any one of these characters is to be interpreted as marking the end of a record. To simplify the macro definition, the parameter &MAXLTH has been deleted; the maximum record length will always be 4096.

The definition in Fig. 4.9(a) uses a macro-time looping statement WHILE. The WHILE statement specifies that the following lines, until the next ENDW statement, are to be generated repeatedly as long as a particular condition is true. As before, the testing of this condition, and the looping, are done while the macro is being expanded. The condi-

```
 25    RDBUFF   MACRO   &INDEV,&BUFADR,&RECLTH,&EOR
 27    &EORCT   SET     %NITEMS(&EOR)
 30             CLEAR   X                      CLEAR LOOP COUNTER
 35             CLEAR   A
 45            +LDT     #4096                  SET MAX LENGTH = 4096
 50    $LOOP    TD      =X'&INDEV'             TEST INPUT DEVICE
 55             JEQ      $LOOP                 LOOP UNTIL READY
 60             RD      =X'&INDEV'             READ CHARACTER INTO REGISTER A
 63    &CTR     SET     1
 64             WHILE   (&CTR LE &EORCT)
 65             COMP    =X'0000&EOR[&CTR]'
 70             JEQ     $EXIT
 71    &CTR     SET     &CTR+1
 73             ENDW
 75             STCH    &BUFADR,X              STORE CHARACTER IN BUFFER
 80             TIXR    T                      LOOP UNLESS MAXIMUM LENGTH
 85             JLT     $LOOP                    HAS BEEN REACHED
 90    $EXIT    STX     &RECLTH                SAVE RECORD LENGTH
100             MEND
```

(a)

```
                RDBUFF   F2,BUFFER,LENGTH,(00,03,04)

 30              CLEAR   X                     CLEAR LOOP COUNTER
 35              CLEAR   A
 45             +LDT     #4096                 SET MAX LENGTH = 4096
 50    $AALOOP   TD      =X'F2'                TEST INPUT DEVICE
 55              JEQ      $AALOOP              LOOP UNTIL READY
 60              RD      =X'F2'                READ CHARACTER INTO REGISTER A
 65              COMP    =X'000000'
 70              JEQ      $AAEXIT
 65              COMP    =X'000003'
 70              JEQ      $AAEXIT
 65              COMP    =X'000004'
 70              JEQ      $AAEXIT
 75              STCH     BUFFER,X             STORE CHARACTER IN BUFFER
 80              TIXR     T                    LOOP UNLESS MAXIMUM LENGTH
 85              JLT      $AALOOP                HAS BEEN REACHED
 90    $AAEXIT   STX      LENGTH               SAVE RECORD LENGTH
```

(b)

FIGURE 4.9 Use of macro-time looping statements.

tions to be tested involve macro-time variables and arguments, not run-time data values.

The use of the WHILE–ENDW structure is illustrated on lines 63 through 73 of Fig. 4.9(a). The macro-time variable &EORCT has previously been set (line 27) to the value %NITEMS(&EOR). %NITEMS is a macro processor function that returns as its value the number of members in an argument list. For example, if the argument corresponding to &EOR is (00,03,04), then %NITEMS(&EOR) has the value 3.

The macro-time variable &CTR is used to count the number of times the lines following the WHILE statement have been generated. The value of &CTR is initialized to 1 (line 63), and incremented by 1 each time through the loop (line 71). The WHILE statement itself specifies that the macro-time loop will continue to be executed as long as the value of &CTR is less than or equal to the value of &EORCT. This means that the statements on lines 65 and 70 will be generated once for each member of the list corresponding to the parameter &EOR. The value of &CTR is used as a subscript to select the proper member of the list for each iteration of the loop. Thus on the first iteration the expression &EOR[&CTR] on line 65 has the value 00; on the second iteration it has the value 03, and so on.

Figure 4.9(b) shows the expansion of a macro invocation statement using the definition in Fig. 4.9(a). You should examine this example carefully to be sure you understand how the WHILE statements are handled.

The implementation of a macro-time looping statement such as WHILE is also relatively simple. When a WHILE statement is encountered during macro expansion, the specified Boolean expression is evaluated. If the value of this expression is FALSE, the macro processor skips ahead in DEFTAB until it finds the next ENDW statement, and then resumes normal macro expansion. If the value of the Boolean expression is TRUE, the macro processor continues to process lines from DEFTAB in the usual way until the next ENDW statement. When the ENDW is encountered, the macro processor returns to the preceding WHILE, re-evaluates the Boolean expression, and takes action based on the new value of this expression as previously described.

This method of implementation does not allow for nested WHILE structures. You are encouraged to think about how such nested structures might be supported (see Exercise 4.2.12).

4.2.4 Keyword Macro Parameters

All the macro instruction definitions we have seen thus far used *positional parameters*. That is, parameters and arguments were associated with each other according to their positions in the macro prototype and the macro invocation statement. With positional parameters, the programmer must be careful to specify the arguments in the proper order. If an argument is to be omitted, the macro invocation statement must contain a null argument (two consecutive commas) to maintain the correct argument positions. (See, for example, the macro invocation statement in Fig. 4.8c.)

Positional parameters are quite suitable for most macro instructions. However, if a macro has a large number of parameters, and only a few of these are given values in a typical invocation, a different form of parameter specification is more useful. (Such a macro may occur in a situation in which a large and complex sequence of statements—perhaps even an entire operating system—is to be generated from a macro invocation. In such cases, most of the parameters may have acceptable default values; the macro invocation specifies only the changes from the default set of values.)

For example, suppose that a certain macro instruction GENER has 10 possible parameters, but in a particular invocation of the macro, only the third and ninth parameters are to be specified. If positional parameters were used, the macro invocation statement might look like

```
GENER    ,,DIRECT,,,,,,3.
```

Using a different form of parameter specification, called *keyword parameters,* each argument value is written with a keyword that names the corresponding parameter. Arguments may appear in any order. If the third parameter in the previous example is named &TYPE and the ninth parameter is named &CHANNEL, the macro invocation statement would be

```
GENER    TYPE=DIRECT,CHANNEL=3.
```

This statement is obviously much easier to read, and much less error-prone, than the positional version.

Figure 4.10(a) shows a version of the RDBUFF macro definition using keyword parameters. Except for the method of specification, the parameters are the same as those in Fig. 4.8(a). In the macro prototype, each parameter name is followed by an equal sign, which identifies a keyword parameter. After the equal sign, a default value is specified for some of the parameters. The parameter is assumed to have this default value if its name does not appear in the macro invocation statement. Thus the default value for the parameter &INDEV is F1. There is no default value for the parameter &BUFADR.

Default values can simplify the macro definition in many cases. For example, the macro definitions in both Fig. 4.10(a) and Fig. 4.8(a) provide for setting the maximum record length to 4096 unless a different value is specified by the user. The default value established in Fig. 4.10(a) takes care of this automatically. In Fig. 4.8(a), an IF–ELSE–ENDIF structure is required to accomplish the same thing.

The other parts of Fig. 4.10 contain examples of the expansion of keyword macro invocation statements. In Fig. 4.10(b), all the default values are accepted. In Fig. 4.10(c), the value of &INDEV is specified as F3, and the value of &EOR is specified as null. These values override the corresponding defaults. Note that the arguments may appear

```
25    RDBUFF   MACRO   &INDEV=F1,&BUFADR=,&RECLTH=,&EOR=04,&MAXLTH=4096
26             IF      (&EOR NE '')
27    &EORCK   SET     1
28             ENDIF
30             CLEAR   X               CLEAR LOOP COUNTER
35             CLEAR   A
38             IF      (&EORCK EQ 1)
40             LDCH    =X'&EOR'        SET EOR CHARACTER
42             RMO     A,S
43             ENDIF
47            +LDT     #&MAXLTH        SET MAXIMUM RECORD LENGTH
50    $LOOP    TD      =X'&INDEV'      TEST INPUT DEVICE
55             JEQ     $LOOP           LOOP UNTIL READY
60             RD      =X'&INDEV'      READ CHARACTER INTO REGISTER A
63             IF      (&EORCK EQ 1)
65             COMPR   A,S             TEST FOR END OF RECORD
70             JEQ     $EXIT           EXIT LOOP IF EOR
73             ENDIF
75             STCH    &BUFADR,X       STORE CHARACTER IN BUFFER
80             TIXR    T               LOOP UNLESS MAXIMUM LENGTH
85             JLT     $LOOP              HAS BEEN REACHED
90    $EXIT    STX     &RECLTH         SAVE RECORD LENGTH
95             MEND
```

(a)

```
      .        RDBUFF  BUFADR=BUFFER,RECLTH=LENGTH

30             CLEAR   X               CLEAR LOOP COUNTER
35             CLEAR   A
40             LDCH    =X'04'          SET EOR CHARACTER
42             RMO     A,S
47            +LDT     #4096           SET MAXIMUM RECORD LENGTH
50    $AALOOP  TD      =X'F1'          TEST INPUT DEVICE
55             JEQ     $AALOOP         LOOP UNTIL READY
60             RD      =X'F1'          READ CHARACTER INTO REGISTER A
65             COMPR   A,S             TEST FOR END OF RECORD
70             JEQ     $AAEXIT         EXIT LOOP IF EOR
75             STCH    BUFFER,X        STORE CHARACTER IN BUFFER
80             TIXR    T               LOOP UNLESS MAXIMUM LENGTH
85             JLT     $AALOOP            HAS BEEN REACHED
90    $AAEXIT  STX     LENGTH          SAVE RECORD LENGTH
```

(b)

```
      .        RDBUFF  RECLTH=LENGTH,BUFADR=BUFFER,EOR=,INDEV=F3

30             CLEAR   X               CLEAR LOOP COUNTER
35             CLEAR   A
47            +LDT     #4096           SET MAXIMUM RECORD LENGTH
50    $ABLOOP  TD      =X'F3'          TEST INPUT DEVICE
55             JEQ     $ABLOOP         LOOP UNTIL READY
60             RD      =X'F3'          READ CHARACTER INTO REGISTER A
75             STCH    BUFFER,X        STORE CHARACTER IN BUFFER
80             TIXR    T               LOOP UNLESS MAXIMUM LENGTH
85             JLT     $ABLOOP            HAS BEEN REACHED
90    $ABEXIT  STX     LENGTH          SAVE RECORD LENGTH
```

(c)

FIGURE 4.10 Use of keyword parameters in macro instructions.

in any order in the macro invocation statement. You may want to work through these macro expansions for yourself, concentrating on how the default values are handled.

4.3 MACRO PROCESSOR DESIGN OPTIONS

In this section we discuss some major design options for a macro processor. The algorithm presented in Fig. 4.5 does not work properly if a macro invocation statement appears within the body of a macro instruction. However, it is often desirable to allow macros to be used in this way. Section 4.3.1 examines the problems created by such macro invocation statements, and suggests some possibilities for the solution of these problems.

Although the most common use of macro instructions is in connection with assembler language programming, there are other possibilities. Section 4.3.2 discusses general-purpose macro processors that are not tied to any particular language. An example of such a macro processor can be found in Section 4.4. Section 4.3.3 examines the other side of this issue: the integration of a macro processor with a particular assembler or compiler. We discuss the possibilities for cooperation between the macro processor and the language translator, and briefly indicate some of the potential benefits and problems of such integration.

4.3.1 Recursive Macro Expansion

In Fig. 4.3 we presented an example of the *definition* of one macro instruction by another. We have not, however, dealt with the *invocation* of one macro by another. Figure 4.11 shows an example of such a use of macros. The definition of RDBUFF in Fig. 4.11(a) is essentially the same as the one in Fig. 4.1. The order of the parameters has been changed to make the point of the example clearer. In this case, however, we have assumed that a related macro instruction (RDCHAR) already exists. The purpose of RDCHAR is to read one character from a specified device into register A, taking care of the necessary test-and-wait loop. The definition of this macro appears in Fig. 4.11(b). It is convenient to use a macro like RDCHAR in the definition of RDBUFF so that the programmer who is defining RDBUFF need not worry about the details of device access and control. (RDCHAR might be written at a different time, or even by a different programmer.) The advantages of using RDCHAR in this way would be even greater on a more complex machine, where the code to read a single character might be longer and more complicated than our simple three-line version.

Unfortunately, the macro processor design we have discussed pre-

```
10        RDBUFF   MACRO    &BUFADR,&RECLTH,&INDEV
15        .
20        .   MACRO TO READ RECORD INTO BUFFER
25        .
30                 CLEAR    X                   CLEAR LOOP COUNTER
35                 CLEAR    A
40                 CLEAR    S
45                +LDT      #4096               SET MAXIMUM RECORD LENGTH
50        $LOOP    RDCHAR   &INDEV              READ CHARACTER INTO REGISTER A
65                 COMPR    A,S                 TEST FOR END OF RECORD
70                 JEQ      $EXIT               EXIT LOOP IF EOR
75                 STCH     &BUFADR,X           STORE CHARACTER IN BUFFER
80                 TIXR     T                   LOOP UNLESS MAXIMUM LENGTH
85                 JLT      $LOOP                  HAS BEEN REACHED
90        $EXIT    STX      &RECLTH             SAVE RECORD LENGTH
95                 MEND
```

(a)

```
 5        RDCHAR   MACRO    &IN
10        .
15        .   MACRO TO READ CHARACTER INTO REGISTER A
20        .
25                 TD       =X'&IN'             TEST INPUT DEVICE
30                 JEQ       *-3                LOOP UNTIL READY
35                 RD       =X'&IN'             READ CHARACTER
40                 MEND
```

(b)

```
                   RDBUFF   BUFFER,LENGTH,F1
```

(c)

FIGURE 4.11 Example of nested macro invocation.

viously cannot handle such invocations of macros within macros. For example, suppose that the algorithm of Fig. 4.5 were applied to the macro invocation statement in Fig. 4.11(c). The procedure EXPAND would be called when the macro is recognized. The arguments from the macro invocation would be entered into ARGTAB as follows:

ARGTAB:	Parameter	Value
	1	BUFFER
	2	LENGTH
	3	F1
	4	(unused)
	.	.
	.	.

The Boolean variable EXPANDING would be set to TRUE, and expansion of the macro invocation statement would begin. The processing would proceed normally until line 50, which contains a statement invoking RDCHAR. At that point, PROCESSLINE would call EXPAND again. This time, ARGTAB would look like

ARGTAB:	**Parameter**	**Value**
	1	F1
	2	(unused)
	.	.

The expansion of RDCHAR would also proceed normally. At the end of this expansion, however, a problem would appear. When the end of the definition of RDCHAR is recognized, EXPANDING would be set to FALSE. Thus the macro processor would "forget" that it had been in the middle of expanding a macro when it encountered the RDCHAR statement. In addition, the arguments from the original macro invocation (RDBUFF) would be lost because the values in ARGTAB were overwritten with the arguments from the invocation of RDCHAR.

The cause of these difficulties is the recursive call of the procedure EXPAND. When the RDBUFF macro invocation is encountered, EXPAND is called. Later, it calls PROCESSLINE for line 50, which results in another call to EXPAND before a return is made from the original call. A similar problem would occur with PROCESSLINE since this procedure too would be called recursively. For example, there might be confusion about whether the return from PROCESSLINE should be made to the main (outermost) loop of the macro processor logic, or to the loop within EXPAND.

These problems are not difficult to solve if the macro processor is being written in a programming language (such as Pascal or PL/I) that allows recursive calls. The compiler would be sure that previous values of any variables declared within a procedure are saved when that procedure is called recursively. It would also take care of other details involving return from the procedure. (In Chapter 5 we consider in detail how such recursive calls are handled by a compiler.)

If a programming language that supports recursion is not available, the programmer must take care of handling such items as return addresses and values of local variables. In such a case, PROCESSLINE and EXPAND would probably not be procedures at all. Instead, the same logic would be incorporated into a looping structure, with data values being saved on a stack. The concepts involved are the same as those that we discuss when we consider recursion in Chapter 5. An example of such an implementation can be found in Donovan (1972).

4.3.2 General-Purpose Macro Processors

The most common use of macro processors is as an aid to assembler language programming. Often such macro processors are combined with, or closely related to, the assembler. Macro processors have also been developed for some high-level programming languages. (See, for example, Kernighan, 1976.) These special-purpose macro processors are similar in general function and approach; however, the details differ from language to language. In this section we discuss general-purpose macro processors. These are not dependent on any particular programming language, but can be used with a variety of different languages.

The advantages of such a general-purpose approach to macro processing are obvious. The programmer does not need to learn about a different macro facility for each compiler or assembler language, so much of the time and expense involved in training are eliminated. The costs involved in producing a general-purpose macro processor are somewhat greater than those for developing a language-specific processor. However, this expense does not need to be repeated for each language; the result is a substantial overall saving in software development cost. Similar savings in software maintenance effort should also be realized. Over a period of years, these maintenance costs may be even more significant than the original cost for software development.

In spite of the advantages noted, there are still relatively few general-purpose macro processors. One of the reasons for this situation is the large number of details that must be dealt with in a real programming language. A special-purpose macro processor can have these details built into its logic and structure. A general-purpose facility, on the other hand, must provide some way for a user to define the specific set of rules to be followed.

In a typical programming language, there are several situations in which normal macro parameter substitution should not occur. For example, comments should usually be ignored by a macro processor (at least in scanning for parameters). However, each programming language has its own methods for identifying comments. They may be introduced by a keyword (as in ALGOL), or delimited by predefined start and end characters (as in Pascal). Some languages use special symbols to flag an entire line as a comment line (as in FORTRAN). In most assembler languages, any characters on a line following the end of the instruction operand field are automatically taken as comments.

Some languages provide a special escape character. Anything between this character and the end of the line is considered to be a comment.

Another difference between programming languages is related to their facilities for grouping together terms, expressions, or statements. Most languages use parentheses for such functions. A general-purpose macro processor may need to take these groupings into account in scanning the source statements. However, some languages use characters such as [and] instead of parentheses. Other languages use keywords such as **begin** and **end** for some such functions.

A more general problem involves the *tokens* of the programming language—for example, identifiers, constants, operators, and keywords. Languages differ substantially in their restrictions on the length of identifiers and the rules for the formation of constants. Sometimes the rules for such tokens are different in certain parts of the program (for example, within a FORMAT statement in FORTRAN or a DATA DIVISION in COBOL). In some languages, there are multiple-character operators such as ** in FORTRAN and := in Pascal. Problems may arise if these are treated by a macro processor as two separate characters rather than as a single operator. Even the arrangement of the source statements in the input file may create difficulties. The macro processor must be concerned with whether or not blanks are significant, with the way statements are continued from one line to another, and with special statement formatting conventions such as those found in FORTRAN and COBOL.

Another potential problem with general-purpose macro processors involves the syntax used for macro definitions and macro invocation statements. With most special-purpose macro processors, macro invocations are very similar in form to statements in the source programming language. (For example, the invocation of RDBUFF in Fig. 4.1 has the same form as a SIC assembler language statement.) This similarity of form tends to make the source program easier to write and read. However, it is difficult to achieve with a general-purpose macro processor that is to be used with programming languages having different basic statement forms.

In Section 4.4.3 we briefly describe a general-purpose macro processor that deals with many of the above issues. Discussions of general-purpose macro processors and macro processors for high-level languages can also be found in Cole (1981), Brown (1974), and Campbell–Kelley (1973).

4.3.3 Macro Processing within Language Translators

The macro processors that we have discussed so far might be termed *pre-processors*. That is, they process macro definitions and expand macro invocations, producing an expanded version of the source program. This expanded program is then used as input to an assembler or compiler. In this section we discuss an alternative: combining the macro processing functions with the language translator itself.

The simplest method of achieving this sort of combination is a *line-by-line* macro processor. Using this approach, the macro processor reads the source program statements and performs all of its functions as previously described. However, the output lines are passed to the language translator as they are generated (one at a time), instead of being written to an expanded source file. Thus the macro processor operates as a sort of input routine for the assembler or compiler.

This line-by-line approach has several advantages. It avoids making an extra pass over the source program (writing and then reading the expanded source file), so it can be more efficient than using a macro pre-processor. Some of the data structures required by the macro processor and the language translator can be combined. For example, OPTAB in an assembler and NAMTAB in the macro processor could be implemented in the same table. In addition, many utility subroutines and functions can be used by both the language translator and the macro processor. These include such operations as scanning input lines, searching tables, and converting numeric values from external to internal representations. A line-by-line macro processor also makes it easier to give diagnostic messages that are related to the source statement containing the error (i.e., the macro invocation statement). With a macro pre-processor, such an error might be detected only in relation to some statement in the macro expansion. The programmer would then need to backtrack to discover the original source of trouble.

Although a line-by-line macro processor may use some of the same utility routines as the language translator, the functions of macro processing and program translation are still relatively independent. The main form of communication between the two functions is the passing of source statements from one to the other. It is possible to have even closer cooperation between the macro processor and the assembler or compiler. Such a scheme can be thought of as a language translator with an *integrated* macro processor.

An integrated macro processor can potentially make use of any information about the source program that is extracted by the language translator. The actual degree of integration varies considerably from one system to another. At a relatively simple level of cooperation, the macro processor may use the results of such translator operations as scanning for symbols, constants, etc. Such operations must be performed by the assembler or compiler in any case; the macro processor can simply use the results without being involved in such details as multiple-character operators, continuation lines, and the rules for token formation. This is particularly useful when the rules for such details vary from one part of the program to another (for example, within FORMAT statements and character string constants in FORTRAN).

The sort of token scan just mentioned is conceptually quite simple. However, many real programming languages have certain characteristics that create unpleasant difficulties. One classic example is the FORTRAN statement

```
DO 100 I =1,20
```

This is a DO statement: DO is recognized as a keyword, 100 as a statement number, I as a variable name, etc. However, blanks are not significant in FORTRAN statements (except within character string constants). Thus the similar statement

```
DO 100 I = 1
```

has a quite different meaning. This is an assignment statement that gives the value 1 to the variable DO100I. Thus the proper interpretation of the characters DO, 100, etc., cannot be decided until the rest of the statement is examined. Such interpretations would be very important if, for example, a macro involved substituting for the variable name I. A FORTRAN compiler must be able to recognize and handle situations such as this. However, it would be very difficult for an ordinary macro processor (not integrated with a compiler) to do so. Such a macro processor would be concerned only with character strings, not with the interpretation of source statements.

With an even closer degree of cooperation, an integrated macro processor can support macro instructions that depend upon the context in which they occur. For example, a macro could specify a substitution to be applied only to variables or constants of a certain type, or only to variables appearing as loop indices in DO statements. The expansion of a macro could also depend upon a variety of characteristics of its arguments. (For an example of this, see the description of the System/370 macro processor in Section 4.5.1.)

There are, of course, disadvantages to integrated and line-by-line macro processors. They must be specially designed and written to work with a particular implementation of an assembler or compiler (not just with a particular programming language). The costs of macro processor development must therefore be added to the cost of the language translator, which results in a more expensive piece of software. In addition, the assembler or compiler will be considerably larger and more complex than it would be if a macro pre-processor were used. The size may be a problem if the translator is to run on a computer with limited memory. In any case, the additional complexity will add to the overhead of language translation. (For example, some assemblers with integrated macro processors consume more time per line of source code than do some compilers on the same computing system.) Decisions about what type of macro processor to use should be based on considerations such as the frequency and complexity of macro processing that is anticipated, and other characteristics of the computing environment.

4.4 IMPLEMENTATION EXAMPLES

In this section we briefly present three examples of actual macro processors. As before, we do not attempt to cover all the characteristics of each system. Instead, we focus on the more interesting or unusual features. The first two examples discussed are macro processors that are integrated with assemblers (for the System/370 and the VAX). The third example is a general-purpose macro pre-processor that is intended for use with a variety of different programming languages.

4.4.1 System/370 Macro Processor

The System/370 macro processor discussed in this section is closely integrated with the 370 assembler. It supports all the main macro processor functions that we have discussed, including the definition and invocation of macro instructions within macros. Macro instructions may use positional or keyword parameters, or a mixture of these two parameter types. Control statements allow the user to specify whether or not the expansions of macro instructions appear in the assembly listing, and to select a variety of other options. Comments in the body of a macro instruction may or may not appear in the assembly listing, depending upon the notation used in writing them.

One important difference between the System/370 macro processor and the one we discussed for SIC lies in the nature of the conditional macro expansion statements. On the 370, these are called *conditional assembly* statements. Although the main use of such statements is in connection with macro instructions, they can also appear outside of macros (in *open code*).

The 370 conditional assembly language includes *set symbols* that are similar to our macro-time variables. Set symbols may be assigned arithmetic, binary, or character values. The three different types of set symbols correspond to these three value types. Set symbols may be declared to be *local* or *global* in scope. Local set symbols may be referred to only within the macro (or the open code) that contains their definition. If the same local symbol is defined in more than one macro, each definition is considered to represent a different macro-time variable. Global set symbols may be referred to anywhere in the program. Thus, for example, a global symbol can be assigned a value in one macro expansion, and this value can be tested in another macro expansion.

The conditional assembly statements that control the generation of source statements are quite different from those we discussed for SIC. The basic control structure used is an *assembly-time* conditional "go to" statement. When executed, such a statement directs the assembler (or macro processor) to skip to another point in the source program. These skips may be either forward or backward in the input stream.

Figure 4.12 shows an example of macro definition and expansion using the 370 macro language. The macro being defined is intended to compute and store the sum of two data items. Different instruction sequences can be generated, according to whether the arguments are fullword (integer) or floating-point variables. An error message will be produced if the arguments are of some different data type, or if the two arguments are not of the same type. The form of the macro definition in Fig. 4.12(a) is generally the same as in our earlier examples. The definition is introduced with the MACRO statement and terminated by MEND. The macro prototype appears immediately following the MACRO statement (line 2). All macro parameters and set symbols begin with the character &. In this example, all the parameters are positional.

The LCLC statement on line 3 defines a set symbol &TYPE, which is to contain a character value. This symbol is declared to be local to the macro instruction containing its definition. The default initial value given to this type of symbol is the null character string. The statement on line 4 (AIF) is a conditional assembly-time "go to" statement. The assembler evaluates the Boolean expression in parentheses, and jumps to another point in the source program if the resulting value is TRUE. Otherwise, assembly proceeds with the next statement in sequence. In this case, the target of the assembly-time "go to" is identified with the *sequence symbol* .MIXTYP; this symbol appears on line 15. The Boolean expression in the AIF statement involves the *type* of each argument. The type of &OP1 (denoted by T′&OP1) is the character F if the corresponding argument is a fullword integer, and E if it is a floating-point data item. Thus the AIF statement on line 4 causes the assembler to skip to line 15 if the types of the two arguments are different. If the types are the same, assembly continues with line 5. The other AIF statements follow a similar pattern.

The AGO on line 7 is an unconditional assembly-time "go to." The ANOP statements (lines 8 and 10) are assembly-time "no operation"

```
 1                MACRO
 2      &NAME     ADD        &OP1,&OP2,&SUM
 3                LCLC       &TYPE
 4                AIF        (T'&OP1  NE  T'&OP2) .MIXTYP
 5                AIF        (T'&OP1  EQ  'F') .INTGR
 6                AIF        (T'&OP1  EQ  'E') .FLOAT
 7                AGO        .TYPERR
 8      .FLOAT    ANOP
 9      &TYPE     SETC       'E'
10      .INTGR    ANOP
11      &NAME     L&TYPE     2,&OP1
12                A&TYPE     2,&OP2
13                ST&TYPE    2,&SUM
14                MEXIT
15      .MIXTYP   MNOTE      'MIXED OPERAND TYPES'
16                MEXIT
17      .TYPERR   MNOTE      'ILLEGAL OPERAND TYPE'
18                MEND
```

(a)

```
LAB      ADD      I,J,K

          ↓

LAB      L        2,I
         A        2,J
         ST       2,K
```

(b)

```
         ADD      X,Y,Z

          ↓

         LE       2,X
         AE       2,Y
         STE      2,Z
```

(c)

```
         ADD      I,Y,Z

          ↓

    **** MIXED OPERAND TYPES
```

(d)

FIGURE 4.12 Examples of System/370 macro definition and expansion.

directives. These directives simply provide a place to attach the sequence symbols .FLOAT and .INTGR. Lines 14 through 17 contain two new types of macro processing directives. The MNOTE statement causes the generation of an error message that will be printed in the assembly listing. This message will be recognized by the assembler as constituting an assembly error. The MEXIT statement instructs the macro processor to terminate the expansion of the macro even though the MEND statement has not yet been reached.

Parts (b) through (d) of Fig. 4.12 show macro invocation statements and the corresponding macro expansions. It is assumed that variables I, J, and K are fullword integers, and that X, Y, and Z are floating-point variables. You may want to follow through the macro expansion process, using the definition in Fig. 4.12(a), to be sure you understand how the AIF and AGO statements work.

The example just discussed illustrates the use of AIF and AGO to accomplish the equivalent of an IF–ELSE–ENDIF structure. Because looping structures such as WHILE can also be programmed using AIF and AGO, these statements provide a powerful conditional assembly capability. However, AIF and AGO statements are harder to implement (and generally less convenient to use) than the IF and WHILE statements we discussed for SIC. The SIC conditional macro expansion statements may be compared to the control structures in a structured language like Pascal. The 370 conditional assembly statements more closely resemble an unstructured language like FORTRAN IV.

In the example in Fig. 4.12, AIF statements were used to test the type of the macro arguments. There are many other characteristics of symbols and arguments that can be tested in this way. These characteristics are called *attributes*. It is possible for a conditional assembly statement to refer to the attributes of symbols whose definitions do not appear until later in the source program. Therefore, the 370 assembler must make a pre-pass through the entire source program, noting the attributes of symbols for use in conditional assembly statements. This pre-pass is made before any macro instructions are processed, so it is not possible to refer to the attributes of a symbol that is defined within a macro expansion.

The System/370 macro processor also provides a number of *system variable symbols,* which are like predefined set symbols that can be used during macro expansion. Among the values made available through such symbols are the date and time of assembly, and the name of the current control section. One system variable symbol, named &SYSNDX, is intended for use in generating unique labels within macro expansions. The value of &SYSNDX is a four-digit number that is initialized at 0001 and incremented by 1 for each macro invocation statement processed. By using this symbol as part of a label, the programmer can avoid duplicating symbols. For example, L&SYSNDX might become L0001, L0002, etc., during successive expansions of a macro.

Further information about a typical System/370 macro processor can be found in IBM (1979) and IBM (1974).

4.4.2 VAX Macro Processor

The VAX macro processor is also closely integrated with the assembler—so closely, in fact, that the assembler language itself is named VAX-11 MACRO. The general approach to macro instruction definition and invocation is similar to that we have discussed for SIC; however, there are some interesting differences.

VAX macro instruction parameters do not begin with &, or any other special character. This means that the process of scanning for occurrences of parameters is more complex. Concatenation operators must also be used more frequently. Consider, for example, the macro definition in Fig. 4.13(a). In the statement

```
3       TSTL    R'NUM
```

the operand is intended to be the character R concatenated with the value of the parameter NUM. If this operand were written simply as RNUM, the string NUM would not be recognized as a macro parameter. Thus a concatenation operator is necessary. The VAX macro language uses the apostrophe to denote concatenation.

Compare this situation with the example in Fig. 4.12. On line 13 of that example, no concatenation operator is required in ST&TYPE because the ampersand identifies &TYPE as a parameter. It would be possible for the macro processor to recognize the substring NUM in Fig. 4.13 without any special concatenation operator. However, we would then need some way of *preventing* parameter substitution where it is not desired (for example, we might not want NUMBER to be converted into 5BER). Of course, if a macro parameter is not concatenated with any other characters, no apostrophe is required.

The VAX macro processor also provides a facility for creating local labels within macro expansions. The programmer specifies a local label by including it as a parameter, with a question mark (?) placed before the parameter name. The macro invocation statement may specify a value for such a label. If the value is not specified, the assembler creates a new local label. Labels created by the assembler begin at 30000$, and can range up to 65535$.

This process is illustrated by the macro definition in Fig. 4.13(a) and the macro invocations and expansions in Fig. 4.13(b–c). In the first

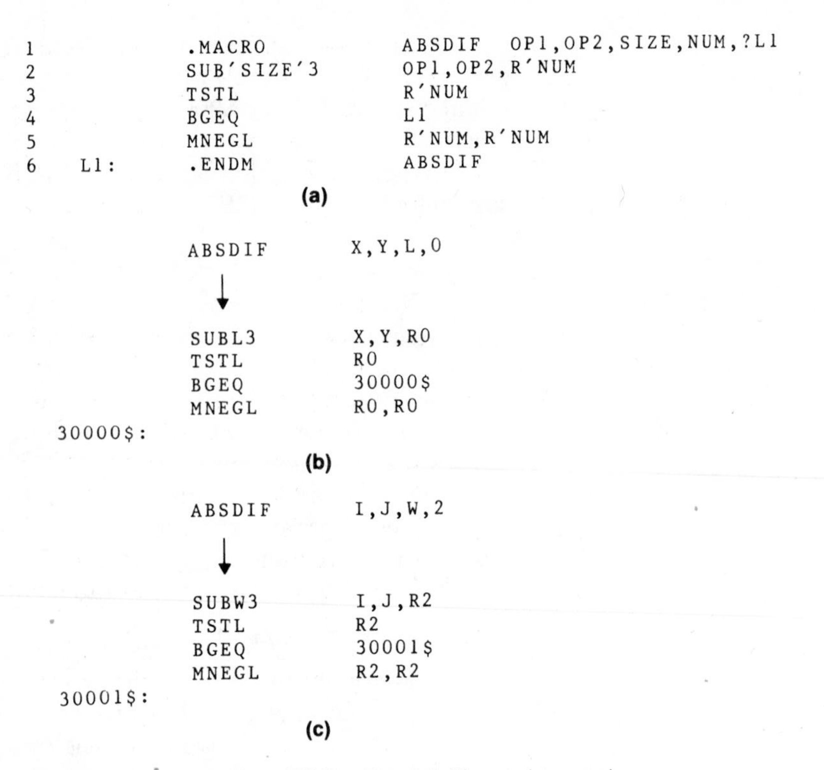

```
1          .MACRO        ABSDIF  OP1,OP2,SIZE,NUM,?L1
2          SUB'SIZE'3    OP1,OP2,R'NUM
3          TSTL          R'NUM
4          BGEQ          L1
5          MNEGL         R'NUM,R'NUM
6   L1:    .ENDM         ABSDIF
```

(a)

```
           ABSDIF        X,Y,L,0
             ↓
           SUBL3         X,Y,R0
           TSTL          R0
           BGEQ          30000$
           MNEGL         R0,R0
  30000$:
```

(b)

```
           ABSDIF        I,J,W,2
             ↓
           SUBW3         I,J,R2
           TSTL          R2
           BGEQ          30001$
           MNEGL         R2,R2
  30001$:
```

(c)

FIGURE 4.13 Examples of VAX macro definition and expansion.

macro expansion, L1 is replaced by 30000$; in the second, it is replaced by 30001$. It is possible for the programmer to define local labels outside of macro instructions. However, to prevent conflict with macro-generated labels, the assembler documentation specifies that no user-defined label should be in the range 30000$ through 65535$. You may want to work through the macro expansions shown in Fig. 4.13 for yourself, concentrating on how the concatenation operators and local labels are handled.

Further information about the VAX macro processor can be found in DEC (1982) and DEC (1979).

4.4.3 The PM Macro Processor*

In this section we give a brief description of a general-purpose macro processor called PM (for Pattern Matching). A more complete description of this system may be found in Sassa (1979).

* Adapted from "A Pattern Matching Macro Processor," by Masataka Sassa. *Software: Practice and Experience,* copyright 1979 by John Wiley & Sons, Ltd. Reprinted by permission.

PM macro definition and invocation statements are quite different from those we have discussed previously. Macro *patterns* (which correspond in function to our macro definitions) are specified using a notation that allows for a variety of different possibilities. Some portions of the pattern may be omitted entirely in a particular macro invocation; some may contain a number of different alternative structures; some may be repeated as often as necessary. Macro invocations are analyzed with a pattern matching process. A macro invocation statement (or argument) may occupy several lines of source code. Macro bodies are written using an ALGOL-like notation.

PM allows for user-defined local and global macro-time variables. There are also system variables similar to those available in the System/370 macro processor. One of these system variables contains a count of the number of macro invocation statements processed. The value of this variable can be used to generate unique labels. PM also provides conditional macro expansion statements that are similar in approach (although not in notation) to those discussed for SIC.

One important characteristic of PM is that the user can declare a number of language-dependent constructs. As we discussed in Section 4.3.2, such language dependencies can be a problem for a general-purpose macro processor. A *skip* in PM, which is a generalization from the idea of a comment, specifies the portion of the input text that should be skipped or deleted. PM allows the declaration of the syntax of skips. It is also possible to specify replacement of skips by some other character string. A *copy* in PM, which is a generalization from the idea of a character string constant, specifies the portion of the input text that is to be copied directly into the output text without being analyzed by the macro processor. PM allows the declaration of the syntax for copies. The characters or tokens that mark the beginning and end of the copy can be changed in the output text. In addition to skips and copies, PM provides for the specification of matching parentheses, multiple-character tokens, rules for continuation statements, and the handling of blanks and linefeeds.

Figure 4.14 shows an example of macro definition and invocation using the PM language. The macro being defined in Fig. 4.14(a) is designed for use with FORTRAN. The macro invocation is an ALGOL-like **for** statement; the output text is in FORTRAN format. Line 1 gives the general pattern for the macro invocation statement: the keyword **for,** followed by a parameter *id,* followed by the keyword **from,** and so on. The parenthesized expression on this line indicates an *alternative.* The first possibility in this alternative is the **by** clause; the second possibility is empty (indicated by /). Thus the pattern indicates that a macro invocation statement may optionally contain a **by** clause. The *noneg* specification attached to the parameter *body* specifies that

```
 1   macro 'for' id 'from' f ('by' b | /) 'to' t 'do' body:noneg 'od'

 2   begin
 3     <%id = %f
 4      %snum IF (%id .GT. %t) GO TO %(snum+1)
 5      %body
 6      %id = %id + >;
 7          if b ≠ '' then <(%b)> else <1> fi;
 8     <%/GO TO %snum
 9      %(snum+1) CONTINUE %/>;
10          snum := snum + 2;
11   end
```

(a)

```
for I from 0 to n-1
   do
      S := S + A(I)
   od
```

(b)

```
       I = 0
9000   IF (I .GT. N-1) GO TO 9001
       S = S + A(I)
       I = I + 1
       GO TO 9000
9001   CONTINUE
```

(c)

FIGURE 4.14 Examples of PM macro definition and expansion.

blanks and linefeeds are to be considered significant (i.e., not neglected) in reading the actual macro argument.

The macro body is defined, using an ALGOL-like notation, on lines 2 through 11. Text that is to be written to the expanded output is enclosed between $<$ and $>$. Lines 3 through 6 contain such an output string, which consists of portions of four lines of code. Within an output text specification, the character % is used as a flag to indicate parameters and macro-time variables. Thus line 3 specifies that the value of the parameter *id* is to be written to the output, followed by an equal sign and the value of the parameter *f*. The specification then continues on the next line, indicating that a linefeed should be inserted in the output. It is assumed that the macro-time variable *snum* has been declared elsewhere with an initial value of 9000. This variable is used in the generation of FORTRAN statement numbers. The **if** statement on line 7 will output either the value of parameter *b* or the value 1, depending upon whether or not a value was specified for *b*. This value will be on the same output line started on line 6. The %/ that begins the next output specification (on line 8) indicates a linefeed, so that the GO TO statement will begin on a separate output line. The macro-time assignment statement on line 10 increments the value of *snum*, so that new FORTRAN statement numbers will be used in the next macro expansion.

Figure 4.14(b) shows a source statement that matches the pattern previously defined. Figure 4.14(c) shows the output generated by expanding this macro invocation. (The details of specifying the FORTRAN statement format for the output lines are not shown.) It is assumed that another macro has been defined that converts := into = in the assignment statement "S := S + A(I)." You are encouraged to follow through this macro expansion to better understand the PM macro invocation and expansion process.

EXERCISES

Section 4.1

1. Apply the algorithm in Fig. 4.5 to process the source program in Fig. 4.1; the results should be the same as shown in Fig. 4.2.
2. Using the methods outlined in Chapter 8, develop a modular design for a one-pass macro processor.
3. Macro invocation statements are a part of the source program. In many cases, the programmer may not be concerned with the statements in the macro expansion. How could the macro processor and assembler cooperate to list only the macro invocation, and not the expanded version?
4. Suppose we want macro definitions to appear as a part of the assembly listing. How could the macro processor and the assembler accomplish this?
5. In most cases, character strings that occur in comments should not be replaced by macro arguments, even if they happen to match a macro parameter. How could parameter substitution in comments be prevented?
6. How should a programmer decide whether to use a macro or a subroutine to accomplish a given logical function?
7. Suppose that a certain logical task must be performed at ten different places in an assembler language program. This task could be implemented either as a macro or as a subroutine. Describe a situation where using a macro would take *less* central memory than using a subroutine.
8. Some macros simply expand into instructions that call a subroutine. What are the advantages of this approach, as compared to using either a "pure" macro or a "pure" subroutine?
9. Write an algorithm for a two-pass macro processor in which all macro definitions are processed in the first pass, and all macro invocations are expanded in the second pass. You do not need to allow for macro definitions or invocations within macros.

10. Modify the algorithm in Fig.4.5 to allow macro definitions to be retrieved from a library if they are not specified by the programmer.
11. Suggest appropriate ways of organizing and accessing the tables DEFTAB and NAMTAB.
12. Suppose that the occurrences of macro parameters in DEFTAB were not replaced by the positional notation *?n*. What changes would be required in the macro processor algorithm of Fig. 4.5?

Section 4.2

1. The macro definitions in Fig. 4.1 contain several statements in which macro parameters are concatenated with other characters (for example, lines 50 and 75). Why was it not necessary to use concatenation operators in these statements?
2. Modify the algorithm in Fig. 4.5 to include the handling of concatenation operators.
3. Modify the algorithm in Fig. 4.5 to include the generation of unique labels within macro expansions.
4. Suppose that we wished to allow labels within macro expansions without requiring them to have any special form (such as beginning with *$*). Each such label would be considered to be defined only within the macro expansion in which it occurs; this would eliminate the problem caused by duplicate labels. How could the macro processor and the assembler work together to allow this?
5. What is the most important difference between the following two sequences of statements?

a)

```
        LDA     ALPHA
        COMP    #0
        JEQ     SKIP
        LDA     #3
        STA     BETA
SKIP    . . .
```

b)

```
        IF      (&ALPHA NE 0)
&BETA   SET     3
        ENDIF
```

6. Expand the following macro invocation statements, using the macro definition in Fig. 4.8(a):

a)
```
        RDBUFF  F1,BUFFER,LENGTH,00,1024
```

b)
```
LOOP    RDBUFF  F2,BUFFER,LTH
```

7. Modify the algorithm in Fig. 4.5 to include SET statements and the IF–ELSE–ENDIF structure. You do not need to allow for nested IFs.
8. Modify your answer to Exercise 7 to allow nested IFs.

9. What is the most important difference between the following two control structures?

a)

```
        LDT     #8
        CLEAR   X
LOOP    .
        .
        .
        TIXR    T
        JLT     LOOP
```

b)

```
&CTR    SET     0
        WHILE   (&CTR LT 8)
        .
        .
        .
&CTR    SET     &CTR+1
        ENDW
```

10. Using the definition in Fig. 4.9(a), expand the following macro invocation statements:

a)
```
        RDBUFF  F1,BUFFER,LENGTH,(04,12)
```
b)
```
LABEL   RDBUFF  F1,BUFFER,LENGTH,00
```
c)
```
        RDBUFF  F1,BUFFER,LENGTH
```

What value should the function %NITEMS(&EOR) return in the last two cases?

11. Modify your answer to Exercise 7 to include WHILE statements. You do not need to allow for nested WHILEs.

12. Modify your answer to Exercise 11 to allow nested WHILEs.

13. The values of macro-time variables are usually considered to be local to a macro definition. That is, a value assigned to a macro-time variable can be used only within the same macro definition. Sometimes, however, it might be useful to be able to communicate the value of a macro-time variable between two related macros. How could this be accomplished?

14. Modify the algorithm in Fig. 4.5 to include keyword parameters.

15. Some macro processors allow macro instructions in which some of the parameters are keyword parameters and some are positional parameters. How could a macro processor handle such mixed-mode macro instructions?

16. How could default values be specified for positional parameters? What changes in the algorithm of Fig. 4.5 would be necessary to handle such defaults?

17. Refer to the definition of RDBUFF that appears in Fig. 4.8(a). Each of the following macro invocation statements contains an error. Which of these errors would be detected by the macro processor, and which would be detected by the assembler?

a) `RDBUFF   F3,BUF,RECL,ZZ`
{illegal value specified for `&EOR`}

b) `RDBUFF   F3,BUF,RECL,04,2048,01`
{too many arguments}

c) `RDBUFF   F3,,RECL,04`
{no value specified for `&BUFADR`}

d) `RDBUFF   F3,RECL,BUF`
{arguments specified in wrong order}

Section 4.3

1. Suppose that a macro processor with logic similar to that in Fig. 4.5 is to perform recursive macro expansion. The text points out the need to save values of EXPANDING and ARGTAB when making a recursive call to EXPAND. Depending upon how the algorithm is implemented, what other values might it also be necessary to save?
2. How could a recursive macro processor be implemented in assembler language?
3. How could a *nonrecursive* macro processor allow for the invocation of macros within macros? What would be the advantages and disadvantages of such an approach?
4. Select two different high-level programming languages with which you are familiar. What differences between these languages might be significant to a macro processor that is intended for use with the language?
5. Select one high-level language and one (real) assembler language with which you are familiar. What differences between these languages might be significant to a macro processor that is intended for use with the language?
6. Outline an algorithm for combining a line-by-line macro processor with an assembler.
7. List utility functions and routines that might be shared by an assembler and an integrated macro processor.
8. Using the methods outlined in Chapter 8, develop a modular design for a two-pass assembler with an integrated macro processor.

CHAPTER **5**

COMPILERS

In this chapter we discuss the design and functions of compilers for high-level programming languages. Many textbooks are entirely devoted to compiler construction, and we cannot hope to cover the subject thoroughly in a single chapter. Instead, we introduce the most important concepts and issues related to compilers, and illustrate them with examples. As each subject is discussed, we give references for those readers who want to explore the topic in more detail.

Section 5.1 presents the basic functions of a simple one-pass compiler. We illustrate the operation of such a compiler by following an example program through the entire translation process. This section contains somewhat more detail than the other parts of the chapter because of the fundamental importance of the material.

Section 5.2 discusses machine-dependent extensions to the basic scheme presented in Section 5.1. These extensions are mainly in the area of object code generation and optimization. Section 5.3 describes some machine-independent extensions to the basic scheme.

Section 5.4 describes some compiler design alternatives. These include multi-pass compilers, interpreters, P-code compilers, and compiler-compilers. Finally, Section 5.5 presents four examples of actual compilers and compiler-writing systems, and relates them to the concepts introduced in previous sections.

5.1 BASIC COMPILER FUNCTIONS

This section introduces the fundamental operations that are necessary in compiling a typical high-level language program. We use as an example the Pascal program in Fig. 5.1; however, the concepts and approaches that we discuss can also be applied to the compilation of programs in other languages.

For the purposes of compiler construction, a high-level programming language is usually described in terms of a *grammar*. This grammar specifies the form, or *syntax,* of legal statements in the language. For example, an assignment statement might be defined by the grammar as a variable name, followed by an assignment operator (:=), followed by an expression. The problem of compilation then becomes one of matching statements written by the programmer to structures defined by the grammar, and generating the appropriate object code for each statement.

It is convenient to regard a source program statement as a sequence of *tokens* rather than simply as a string of characters. Tokens may be thought of as the fundamental building blocks of the language. For example, a token might be a keyword, a variable name, an integer, an arithmetic operator, etc. The task of scanning the source statement, recognizing and classifying the various tokens, is known as *lexical*

FIGURE 5.1 Example of a Pascal program.

```
 1  PROGRAM STATS
 2  VAR
 3     SUM,SUMSQ,I,VALUE,MEAN,VARIANCE : INTEGER
 4  BEGIN
 5     SUM := 0;
 6     SUMSQ := 0;
 7     FOR I := 1 TO 100 DO
 8        BEGIN
 9        READ(VALUE);
10        SUM := SUM + VALUE;
11        SUMSQ := SUMSQ + VALUE * VALUE
12        END;
13     MEAN := SUM DIV 100;
14     VARIANCE := SUMSQ DIV 100 - MEAN * MEAN;
15     WRITE(MEAN,VARIANCE)
16  END.
```

analysis. The part of the compiler that performs this analytic function is commonly called the *scanner*.

After the token scan, each statement in the program must be recognized as some language construct, such as a declaration or an assignment statement, described by the grammar. This process, which is called *syntactic analysis*, or *parsing*, is performed by a part of the compiler that is usually called the *parser*. The last step in the basic translation process is the generation of object code. Most compilers create machine-language programs directly instead of producing a symbolic program for later translation by an assembler.

Although we have mentioned three steps in the compilation process—scanning, parsing, and code generation—it is important to realize that a compiler does not necessarily make three passes over the program being translated. For some languages, it is quite possible to compile a program in a single pass. Our discussions in this section describe how such a one-pass compiler might work. On the other hand, compilers for other languages and compilers that perform sophisticated code optimization or other analysis of the program generally make several passes. We discuss the division of a compiler into passes in Section 5.4. Section 5.5 gives several examples of the structure of actual compilers.

In the following sections we discuss the basic elements of a simple compilation process, illustrating their application to the example program in Fig. 5.1. Section 5.1.1 introduces some concepts and notation used in specifying grammars for programming languages. Sections 5.1.2 through 5.1.4 discuss, in turn, the functions of lexical analysis, syntactic analysis, and code generation.

5.1.1 GRAMMARS

A grammar for a programming language is a formal description of the *syntax*, or form, of programs and individual statements written in the language. The grammar does not describe the *semantics*, or meaning, of the various statements; such knowledge must be supplied in the code-generation routines. As an illustration of the difference between syntax and semantics, consider the two statements

```
I := J + K
```

and

```
I := X + Y
```

where X and Y are REAL variables and I, J, K are INTEGER variables. These two statements have identical syntax. Each is an assignment statement, with the value to be assigned given by an expression that consists of two variable names separated by the operator +. However, the semantics of the two statements are quite different. The first statement specifies that the variables in the expression are to be added using integer arithmetic operations, with the result being assigned to the variable I. The second statement specifies a floating-point addition, with the result being converted to an integer before being assigned to I. Obviously, the two statements would be compiled into very different sequences of machine instructions. However, they would be described in the same way by the grammar. The differences between the statements would be recognized during code generation.

A number of different notations can be used for writing grammars. The one we describe is called BNF (for Backus–Naur Form). BNF is not the most powerful syntax description tool available. In fact, it is not even totally adequate for the description of some real programming languages. It does, however, have the advantages of being simple and widely used, and it provides capabilities that are sufficient for most purposes. Figure 5.2 gives one possible BNF grammar for a highly restricted subset of the Pascal language. A complete BNF grammar for Pascal can be found in Jensen (1974). In the remainder of this section, we discuss this grammar and show how it relates to the example program in Fig. 5.1.

A BNF grammar consists of a set of *rules,* each of which defines the syntax of some construct in the programming language. Consider, for example, Rule 13 in Fig. 5.2:

```
⟨read⟩  ::= READ ( ⟨id-list⟩ )
```

FIGURE 5.2 Simplified Pascal grammar.

```
 1   ⟨prog⟩        ::=  PROGRAM ⟨prog-name⟩ VAR ⟨dec-list⟩ BEGIN ⟨stmt-list⟩ END.
 2   ⟨prog-name⟩   ::=  id
 3   ⟨dec-list⟩    ::=  ⟨dec⟩ | ⟨dec-list⟩ ; ⟨dec⟩
 4   ⟨dec⟩         ::=  ⟨id-list⟩ : ⟨type⟩
 5   ⟨type⟩        ::=  INTEGER
 6   ⟨id-list⟩     ::=  id | ⟨id-list⟩ , id
 7   ⟨stmt-list⟩   ::=  ⟨stmt⟩ | ⟨stmt-list⟩ ; ⟨stmt⟩
 8   ⟨stmt⟩        ::=  ⟨assign⟩ | ⟨read⟩ | ⟨write⟩ | ⟨for⟩
 9   ⟨assign⟩      ::=  id := ⟨exp⟩
10   ⟨exp⟩         ::=  ⟨term⟩ | ⟨exp⟩ + ⟨term⟩ | ⟨exp⟩ - ⟨term⟩
11   ⟨term⟩        ::=  ⟨factor⟩ | ⟨term⟩ * ⟨factor⟩ | ⟨term⟩ DIV ⟨factor⟩
12   ⟨factor⟩      ::=  id | int | ( ⟨exp⟩ )
13   ⟨read⟩        ::=  READ ( ⟨id-list⟩ )
14   ⟨write⟩       ::=  WRITE ( ⟨id-list⟩ )
15   ⟨for⟩         ::=  FOR ⟨index-exp⟩ DO ⟨body⟩
16   ⟨index-exp⟩   ::=  id := ⟨exp⟩ TO ⟨exp⟩
17   ⟨body⟩        ::=  ⟨stmt⟩ | BEGIN ⟨stmt-list⟩ END
```

This is a definition of the syntax of a Pascal READ statement that is denoted in the grammar as ⟨read⟩. The symbol ::= can be read "is defined to be." On the left of this symbol is the language construct being defined, ⟨read⟩, and on the right is a description of the syntax being defined for it. Character strings enclosed between the angle brackets ⟨ and ⟩ are called *nonterminal symbols* (i.e., the names of constructs defined in the grammar). Entries not enclosed in angle brackets are *terminal symbols* of the grammar (i.e., tokens). In this rule, the nonterminal symbols are ⟨read⟩ and ⟨id-list⟩, and the terminal symbols are the tokens READ, (, and). Thus this rule specifies that a ⟨read⟩ consists of the token READ, followed by the token "(", followed by a language construct ⟨id-list⟩, followed by the token ")". The blank spaces in the grammar rules are not significant. They have been included only to improve readability.

To recognize a ⟨read⟩, of course, we also need the definition of ⟨id-list⟩. This is provided by Rule 6 in Fig. 5.2:

```
⟨id-list⟩  ::= id | ⟨id-list⟩ , id
```

This rule offers two possibilities, separated by the | symbol, for the syntax of an ⟨id-list⟩. The first alternative specifies that an ⟨id-list⟩ may consist simply of a token **id** (the notation **id** denotes an identifier that is recognized by the scanner). The second syntax alternative is an ⟨id-list⟩, followed by the token ",", followed by a token **id**. Note that this rule is recursive, which means the construct ⟨id-list⟩ is defined partially in terms of itself. By trying a few examples you should be able to see that this rule includes in the definition of ⟨id-list⟩ any sequence of one or more **id**'s separated by commas. Thus

```
ALPHA
```

is an ⟨id-list⟩ that consists of a single **id** ALPHA;

```
ALPHA , BETA
```

is an ⟨id-list⟩ that consists of another ⟨id-list⟩ ALPHA, followed by a comma (,), followed by an **id** BETA, and so forth.

It is often convenient to display the analysis of a source statement in terms of a grammar as a tree. This tree is usually called the *parse tree,* or *syntax tree,* for the statement. Figure 5.3(a) shows the parse tree for the statement

```
READ ( VALUE )
```

in terms of the two rules just discussed.

Rule 9 of the grammar in Fig. 5.2 provides a definition of the syntax of an assignment statement:

```
⟨assign⟩  ::= id := ⟨exp⟩
```

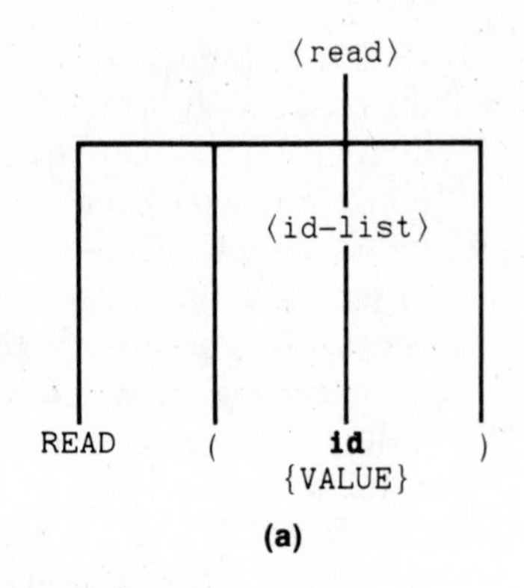

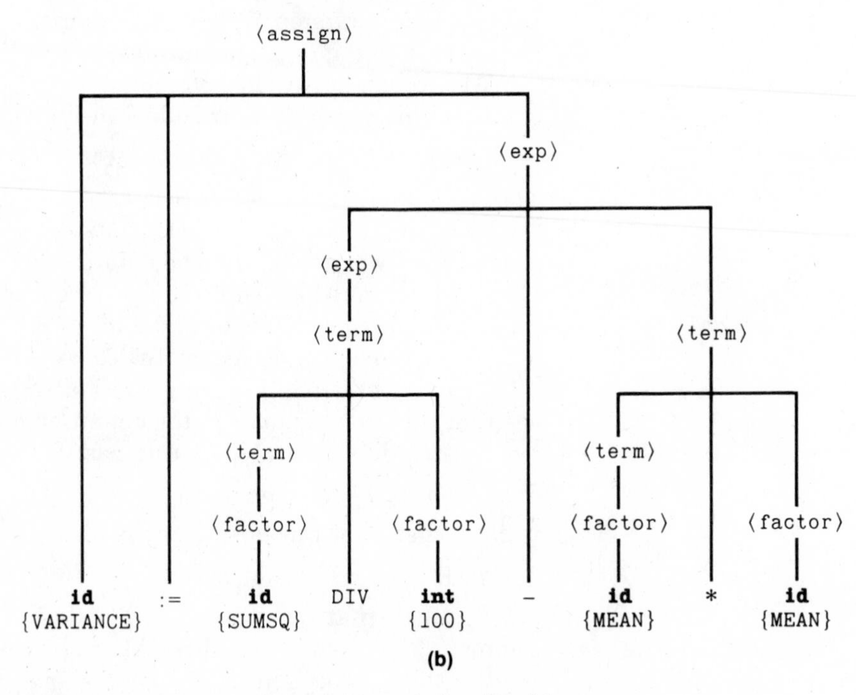

FIGURE 5.3 Parse trees for two statements from Fig. 5.1.

That is, an ⟨assign⟩ consists of an **id**, followed by the token :=, followed by an expression ⟨exp⟩. Rule 10 gives a definition of an ⟨exp⟩:

```
⟨exp⟩  ::= ⟨term⟩ | ⟨exp⟩ + ⟨term⟩ | ⟨exp⟩ - ⟨term⟩
```

By reasoning similar to that applied to ⟨id-list⟩, we can see that this rule defines an expression ⟨exp⟩ to be any sequence of ⟨term⟩s connected by the operators + and −. Similarly, Rule 11 defines a ⟨term⟩ to be any sequence of ⟨factor⟩s connected by * and DIV. Rule 12 specifies that a ⟨factor⟩ may consist of an identifier **id**, or an integer **int**, which is also recognized by the scanner, or an ⟨exp⟩ enclosed in parentheses.

Figure 5.3(b) shows the parse tree for statement 14 from Fig. 5.1, in terms of the rules just described. You should examine this figure carefully to be sure you understand the analysis of the source statement according to the rules of the grammar. In Section 5.1.3, we discuss methods for performing this sort of syntactic analysis in a compiler.

Note that the parse tree in Fig. 5.3(b) implies that multiplication and division are done before addition and subtraction. The terms SUMSQ DIV 100 and MEAN * MEAN must be calculated first since these intermediate results are the operands (left and right subtrees) for the − operation. Another way of saying this is that multiplication and division have higher *precedence* than addition and subtraction. These rules of precedence are implied by the way Rules 10–12 are constructed (see Exercise 5.1.3). In Section 5.1.3 we see a way to make use of such precedence relationships during the parsing process.

The parse trees shown in Fig. 5.3 represent the only possible ways to analyze these two statements in terms of the grammar of Fig. 5.2. For some grammars, this might not be the case. If there is more than one possible parse tree for a given statement, the grammar is said to be *ambiguous*. We prefer to use unambiguous grammars in compiler construction because, in some cases, an ambiguous grammar would leave doubt about what object code should be generated. See, for example, Exercise 5.1.4.

Figure 5.4 shows the parse tree for the entire program in Fig. 5.1. You should examine this figure carefully to see how the form and structure of the program correspond to the rules of the grammar in Fig. 5.2.

5.1.2 Lexical Analysis

Lexical analysis involves scanning the program to be compiled and recognizing the tokens that make up the source statements. Scanners are usually designed to recognize keywords, operators, and identifiers, as well as integers, floating-point numbers, character strings, and other similar items that are written as part of the source program. The exact set of tokens to be recognized, of course, depends upon the programming language being compiled and the grammar being used to describe it.

Items such as identifiers and integers are usually recognized directly as single tokens. As an alternative, these tokens could be defined

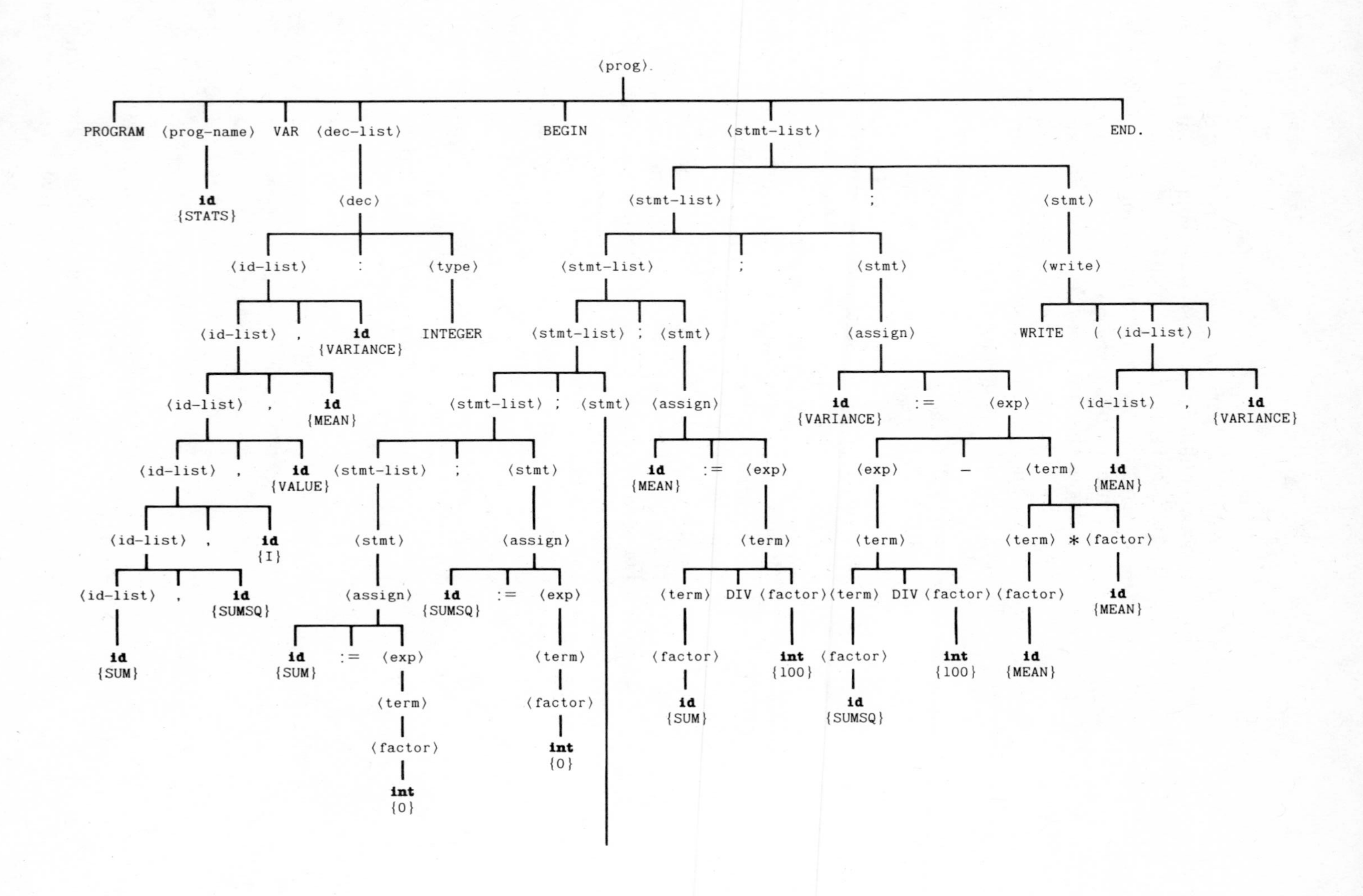

⟨prog⟩
PROGRAM
⟨prog-name⟩
VAR
⟨dec-list⟩
BEGIN
⟨stmt-list⟩
END.
id
{STATS}
⟨dec⟩
⟨stmt-list⟩
;
⟨stmt⟩
⟨id-list⟩
:
⟨type⟩
⟨stmt-list⟩
;
⟨stmt⟩
⟨write⟩
⟨id-list⟩
,
id
{VARIANCE}
INTEGER
⟨stmt-list⟩
;
⟨stmt⟩
⟨assign⟩
WRITE
(
⟨id-list⟩
)
⟨id-list⟩
,
id
{MEAN}
⟨stmt-list⟩
;
⟨stmt⟩
⟨assign⟩
id
{VARIANCE}
:=
⟨exp⟩
⟨id-list⟩
,
id
{VARIANCE}
⟨id-list⟩
,
id
{VALUE}
⟨stmt-list⟩
;
⟨stmt⟩
id
{MEAN}
:=
⟨exp⟩
⟨exp⟩
—
⟨term⟩
id
{MEAN}
⟨id-list⟩
,
id
{I}
⟨stmt⟩
⟨assign⟩
⟨term⟩
⟨term⟩
⟨term⟩
*
⟨factor⟩
⟨id-list⟩
,
id
{SUMSQ}
⟨assign⟩
id
{SUMSQ}
:=
⟨exp⟩
⟨term⟩
DIV
⟨factor⟩
⟨term⟩
DIV
⟨factor⟩
⟨factor⟩
id
{MEAN}
id
{SUM}
id
{SUM}
:=
⟨exp⟩
⟨term⟩
⟨factor⟩
int
{100}
⟨factor⟩
int
{100}
id
{MEAN}
⟨term⟩
⟨factor⟩
int
{0}
id
{SUM}
id
{SUMSQ}
⟨factor⟩
int
{0}

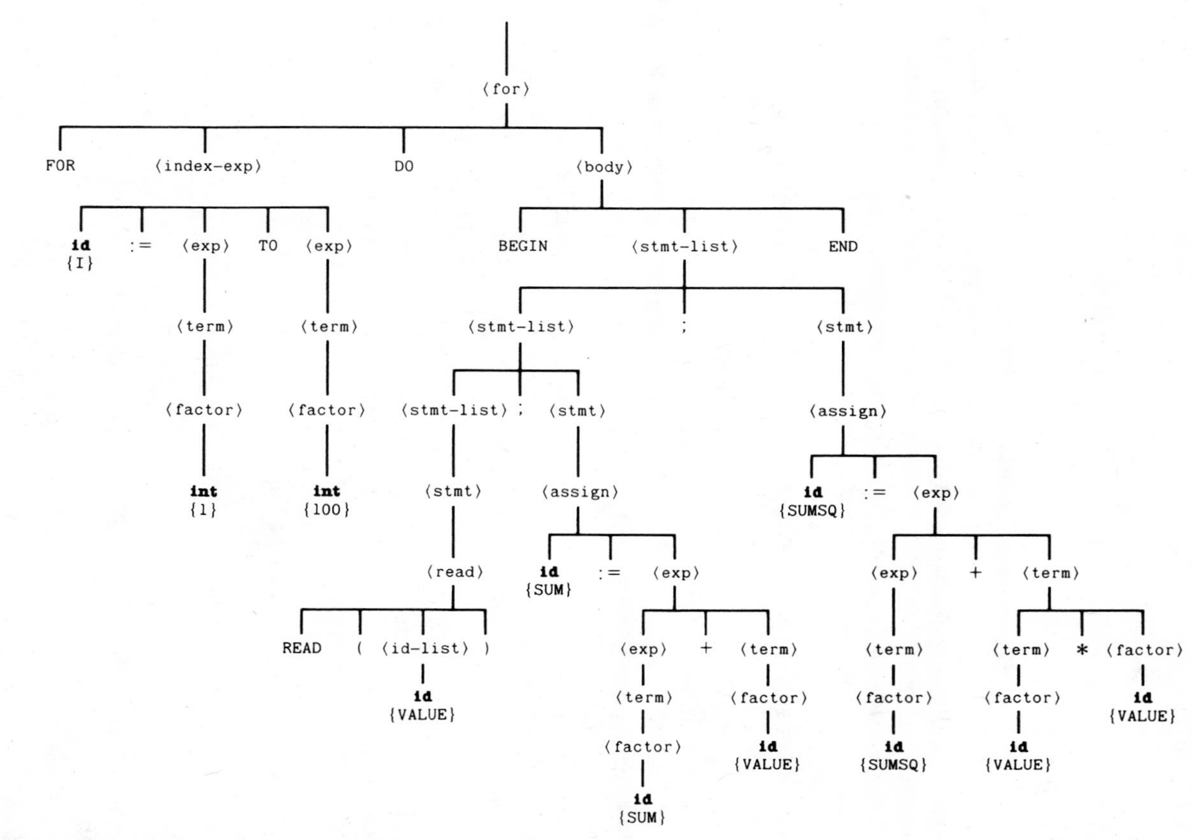

FIGURE 5.4 Parse tree for the program from Fig. 5.1.

as a part of the grammar. For example, an identifier might be defined by the rules

```
⟨ident⟩  ::= ⟨letter⟩ | ⟨ident⟩ ⟨letter⟩ | ⟨ident⟩ ⟨digit⟩
⟨letter⟩ ::= A | B | C | D | ... | Z
⟨digit⟩  ::= 0 | 1 | 2 | 3 | ... | 9
```

In such a case the scanner would recognize as tokens the single characters A, B, 0, 1, and so on. The parser would interpret a sequence of such characters as the language construct ⟨ident⟩. However, this approach would require the parser to recognize simple identifiers using general parsing techniques such as those discussed in the next section. A special-purpose routine such as the scanner can perform this same function much more efficiently. Since a large part of the source program consists of such multiple-character identifiers, this saving in compilation time can be highly significant. In addition, restrictions such as a limitation on the length of identifiers are easier to include in a scanner than in a general-purpose parsing routine.

Similarly, the scanner generally recognizes both single-character and multiple-character tokens directly. For example, the character string READ would be interpreted as a single token rather than as a sequence of four tokens R, E, A, D. The string := would be recognized as a single assignment operator, not as : followed by =. It is, of course, possible to handle multiple-character tokens one character at a time, but such an approach creates considerably more work for the parser.

The output of the scanner consists of a sequence of tokens. For efficiency of later use, each token is usually represented by some fixed-length code, such as an integer, rather than as a variable-length character string. In such a coding scheme for the grammar of Fig. 5.2 (shown in Fig. 5.5) the token PROGRAM would be represented by the integer value 1, an identifier **id** would be represented by the value 22, and so on.

When the token being scanned is a keyword or an operator, such a coding scheme gives sufficient information. In the case of an identifier, however, it is also necessary to specify the particular identifier name that was scanned. The same is true of integers, floating-point values, character-string constants, etc. This can be accomplished by associating a *token specifier* with the type code for such tokens. This specifier gives the identifier name, integer value, etc., that was found by the scanner. Some scanners are designed to enter identifiers directly into a symbol table, similar to the symbol table used by an assembler, when they are first recognized. In that case, the token specifier for an identifier might be a pointer to the symbol-table entry for that identifier.

Token	Code
PROGRAM	1
VAR	2
BEGIN	3
END	4
END.	5
INTEGER	6
FOR	7
READ	8
WRITE	9
TO	10
DO	11
;	12
:	13
,	14
:=	15
+	16
–	17
*	18
DIV	19
(	20
)	21
id	22
int	23

FIGURE 5.5 Token coding scheme for the grammar from Fig. 5.2.

This approach avoids much of the need for table searching during the rest of the compilation process.

Figure 5.6 shows the output from a scanner for the program in Fig. 5.1, using the token coding scheme in Fig. 5.5. For token type 22 (identifier), the token specifier is a pointer to a symbol-table entry (denoted by ^SUM, ^SUMSQ, etc.). For token type 23 (integer), the specifier is the value of the integer (denoted by #0, #100, etc.).

We have shown the output from the scanner as a list of token codings; however, this does not mean the entire program is scanned at one time, before any other processing. More often, the scanner operates as a procedure that is called by the parser when it needs another token. In this case, each call to the scanner would produce the coding (and specifier, if any) for the next token in the source program. The parser

Line	Token type	Token specifier
1	1	
	22	^STATS
2	2	
3	22	^SUM
	14	
	22	^SUMSQ
	14	
	22	^I
	14	
	22	^VALUE
	14	
	22	^MEAN
	14	
	22	^VARIANCE
	13	
	6	
4	3	
5	22	^SUM
	15	
	23	#0
	12	
6	22	^SUMSQ
	15	
	23	#0
	12	
7	7	
	22	^I
	15	
	23	#1
	10	
	23	#100
	11	
8	3	
9	8	
	20	
	22	^VALUE
	21	
	12	
10	22	^SUM
	15	
	22	^SUM
	16	
	22	^VALUE
	12	
11	22	^SUMSQ
	15	
	22	^SUMSQ
	16	
	22	^VALUE
	18	
	22	^VALUE
12	4	
	12	
13	22	^MEAN
	15	
	22	^SUM
	19	
	23	#100
	12	
14	22	^VARIANCE
	15	
	22	^SUMSQ
	19	
	23	#100
	17	
	22	^MEAN
	18	
	22	^MEAN
	12	
15	9	
	20	
	22	^MEAN
	14	
	22	^VARIANCE
	21	
16	5	

FIGURE 5.6 Lexical scan of the program from Fig. 5.1.

would be responsible for saving any tokens that it might require for later analysis. An example of this can be found in the next section.

In addition to its primary function of recognizing tokens, the scanner usually is responsible for reading the lines of the source program as needed, and possibly for printing the source listing. Comments are ignored by the scanner, except for printing on the output listing, so they are effectively removed from the source statements before parsing begins.

The process of lexical scanning, as we have described it, is quite simple. However, many languages have special characteristics that must be considered when programming a scanner. The scanner must take into account any special format required of the source statements. For example, in FORTRAN a number in columns 1–5 of a source statement should be interpreted as a statement number, not as an integer. The scanner must also incorporate knowledge about language-dependent items such as whether blanks function as delimiters for tokens (as in Pascal) or not (as in FORTRAN) and whether statements can be continued freely from one line to the next (as in Pascal) or whether special continuation flags are necessary (as in FORTRAN).

The rules for the formation of tokens may also vary from one part of the program to another. For example, READ should not be recognized as a keyword if it appears within a quoted character string. Blanks are significant within such a quoted string, even if they are not significant in the rest of the program. Similarly, in a FORTRAN program the string F6.3 must be interpreted differently when it appears within a FORMAT statement and when it appears elsewhere in the program.

In some languages, there are unusual cases that must be handled by the scanner. For example, in the FORTRAN statement

```
DO 10 I = 1,100
```

the scanner should recognize DO as a keyword, 10 as a statement number, I as an identifier, etc. However, in the statement

```
DO 10 I = 1
```

the scanner should recognize DO10I as an identifier. Remember that blanks are ignored in FORTRAN statements, so this is an assignment statement that sets the variable DO10I to the value 1. In this case, the scanner must look ahead to see if there is a comma (,) before it can decide on the proper interpretation of the characters DO.

Languages that do not have reserved words create even more difficulties for the scanner. In PL/I, for example, any keyword may also be used as an identifier. Words such as IF, THEN, and ELSE might represent either keywords or variable names defined by the programmer. In fact, the statement

```
IF THEN = ELSE THEN IF = THEN; ELSE THEN = IF;
```

would be legal, although bizarre, in PL/I. In such a case, the scanner might interact with the parser so that it could tell the proper interpretation of each word, or it might simply place identifiers and keywords in the same class, leaving the task of distinguishing between them to the parser.

A number of tools have been developed for automatically constructing lexical scanners from specifications stated in a special-purpose language. A description of one of these tools, which includes provisions for handling some of the special cases previously mentioned, can be found in Aho (1977).

5.1.3 Syntactic Analysis

During syntactic analysis, the source statements written by the programmer are recognized as language constructs described by the grammar being used. We may think of this process as building the parse tree for the statements being translated. Parsing techniques are divided into two general classes—*bottom-up* and *top-down*—according to the way in which the parse tree is constructed. Top-down methods begin with the rule of the grammar that specifies the goal of the analysis (i.e., the root of the tree), and attempt to construct the tree so that the terminal nodes match the statements being analyzed. Bottom-up methods begin with the terminal nodes of the tree (the statements being analyzed), and attempt to combine these into successively higher-level nodes until the root is reached.

A large number of different parsing techniques have been devised, most of which are applicable only to grammars that satisfy certain conditions. In this section we briefly describe one bottom-up method and one top-down method, and show the application of these techniques to our example program. We do not attempt to give all the details of either method. Instead, we illustrate the approach and main concepts involved, and provide references for readers who want to study such techniques further.

The bottom-up parsing technique we consider is called the *operator-precedence* method. This method is based on examining pairs of consecutive operators in the source program, and making decisions about which operation should be performed first. Consider, for example, the arithmetic expression

```
A + B * C - D
```

According to the usual rules of arithmetic, multiplication and division are performed before addition and subtraction—that is, multiplication and division have higher *precedence* than addition and subtraction. If we examine the first two operators (+ and *), we find that + has lower precedence than *. This is often written as

```
+ ⋖ * .
```

Similarly, for the next pair of operators (* and −), we would find that * has higher precedence than −. We may write this as

```
* ⋗ − .
```

The operator-precedence method uses such observations to guide the parsing process. Our previous discussion has led to the following precedence relations for the expression being considered.

```
A + B * C − D
    ⋖   ⋗
```

This implies that the subexpression B * C is to be computed before either of the other operations in the expression is performed. In terms of the parse tree, this means that the * operation appears at a lower level than does either + or −. Thus a bottom-up parser should recognize B * C, by interpreting it in terms of the grammar, before considering the surrounding terms.

The preceding discussion illustrates the fundamental idea behind operator-precedence parsing. During this process, the statement being analyzed is scanned for a subexpression whose operators have higher precedence than the surrounding operators. Then this subexpression is interpreted in terms of the rules of the grammar. This process continues until the root of the tree is reached, at which time the analysis is complete. We now consider the application of this approach to our example program.

The first step in constructing an operator-precedence parser is to determine the precedence relations between the operators of the grammar. In this context, *operator* is taken to mean any terminal symbol (i.e., any token), so we also have precedence relations involving tokens such as BEGIN, READ, **id**, and (. The matrix in Fig. 5.7 shows these precedence relations for the grammar in Fig. 5.2. Each entry in the matrix gives the relation, if any, between the tokens that label the row and column in which it appears. For example, we find from the matrix that

```
PROGRAM ≐ VAR
```

and

```
BEGIN ⋖ FOR .
```

	VAR	BEGIN	END	END	INTEGER	FOR	READ	WRITE	TO	DO	;	:	,	:=	+	−	*	DIV	(	)	**id**	**int**
PROGRAM	≐																				⋖	
VAR		≐									⋖	⋖	⋖								⋖	
BEGIN			≐	≐		⋖	⋖	⋖			⋖										⋖	
END			⋗	⋗							⋗											
INTEGER		⋗									⋗											
FOR										≐											⋖	
READ																			≐			
WRITE																			≐			
TO										⋗					⋖	⋖	⋖	⋖	⋖		⋖	⋖
DO		⋖	⋗	⋗		⋖	⋖	⋖			⋗										⋖	
;		⋗	⋗	⋗		⋖	⋖	⋖			⋗	⋖	⋖								⋖	
:		⋗			⋖						⋗											
,																					≐	
:=			⋗	⋗					≐		⋗				⋖	⋖	⋖	⋖	⋖		⋖	⋖
+			⋗	⋗					⋗	⋗	⋗				⋗	⋗	⋖	⋖	⋖	⋗	⋖	⋖
−			⋗	⋗					⋗	⋗	⋗				⋗	⋗	⋖	⋖	⋖	⋗	⋖	⋖
*			⋗	⋗					⋗	⋗	⋗				⋗	⋗	⋗	⋗	⋖	⋗	⋖	⋖
DIV			⋗	⋗					⋗	⋗	⋗				⋗	⋗	⋗	⋗	⋖	⋗	⋖	⋖
(													⋖		⋖	⋖	⋖	⋖	⋖	≐	⋖	⋖
)			⋗	⋗					⋗	⋗	⋗				⋗	⋗	⋗	⋗		⋗		
id	⋗		⋗	⋗					⋗	⋗	⋗	⋗	⋗	≐	⋗	⋗	⋗	⋗		⋗		
int			⋗	⋗					⋗	⋗	⋗				⋗	⋗	⋗	⋗		⋗		

FIGURE 5.7 Precedence matrix for the grammar from Fig. 5.2.

The relation ≐ indicates that the two tokens involved have equal precedence and should be recognized by the parser as part of the same language construct. Note that the precedence relations do not follow the ordinary rules for comparisons. For example, we have

; ⋗ END

but

END ⋗ ;

Also note that in many cases there is no precedence relation between a pair of tokens. This means that these two tokens cannot appear together in any legal statement. If such a combination occurs during parsing, it should be recognized as a syntax error.

There are algorithmic methods for constructing a precedence matrix like Fig. 5.7 from a grammar (see, for example, Aho, 1977). For the operator-precedence parsing method to be applied, it is necessary for all the precedence relations to be unique. For example, we could not have both ; $\lessdot$ BEGIN and ; $\gtrdot$ BEGIN. This condition holds true for the grammar in Fig. 5.2; however, if seemingly minor changes were made to this grammar, some of the relations would not be unique, and the operator-precedence method could no longer be used.

Figure 5.8 shows the application of the operator-precedence parsing method to two statements from the program in Fig. 5.1. In Fig. 5.8(a), we are examining the READ statement from line 9 of this program. The statement is scanned from left to right, one token at a time. For each pair of operators, the precedence relation between them is determined. In part (i) of Fig. 5.8(a), the parser has identified the portion of the statement delimited by the precedence relations $\lessdot$ and $\gtrdot$ to be interpreted in terms of the grammar. In this case, that portion of the statement consists of the single token **id**. This **id** can be recognized as a ⟨factor⟩ according to Rule 12 of the grammar. Actually, this **id** could also be recognized as a ⟨prog-name⟩ (Rule 2) or an ⟨id-list⟩ (Rule 6). In an operator-precedence parse, it is not really necessary to be concerned with *which* nonterminal symbol is being recognized; we simply interpret **id** as some nonterminal $\langle N_1 \rangle$. Part (ii) of Fig. 5.8 shows

FIGURE 5.8 Operator-precedence parse of two statements from Fig. 5.1.

```
(i)    ... BEGIN   READ   (   id   )
                 ⋖       ≐   ⋖    ⋗

(ii)   ... BEGIN   READ   (   ⟨N1⟩   )   ;                ⟨N1⟩
                 ⋖       ≐        ≐       ⋗                |
                                                           id
                                                         {VALUE}

(iii)  ... BEGIN   ⟨N2⟩   ;                                ⟨N2⟩
                                                            |
                                              +------+------+------+
                                              |      |      |      |
                                             READ    (    ⟨N1⟩     )
                                                            |
                                                            id
                                                          {VALUE}
```

(a)

(i) $\ldots\ \mathbf{id}_1 \quad := \quad \mathbf{id}_2 \quad \mathrm{DIV}$

$\lessdot \qquad \doteq \quad \lessdot \qquad \gtrdot$

(ii) $\ldots\ \mathbf{id}_1 \quad := \quad \langle N_1 \rangle \quad \mathrm{DIV} \quad \mathbf{int} \quad -$

$\lessdot \qquad \doteq \qquad \lessdot \qquad\qquad \lessdot \qquad \gtrdot$

$\langle N_1 \rangle$

$\mathbf{id}_2$
{SUMSQ}

(iii) $\ldots\ \mathbf{id}_1 \quad := \quad \langle N_1 \rangle \quad \mathrm{DIV} \quad \langle N_2 \rangle \quad -$

$\lessdot \qquad \doteq \qquad \lessdot \qquad\qquad\qquad \gtrdot$

$\langle N_1 \rangle$ $\langle N_2 \rangle$

$\mathbf{id}_2$ {SUMSQ} $\mathbf{int}$ {100}

(iv) $\ldots\ \mathbf{id}_1 \quad := \quad \langle N_3 \rangle \quad - \quad \mathbf{id}_3 \quad *$

$\lessdot \qquad \doteq \qquad \lessdot \qquad \lessdot \qquad \gtrdot$

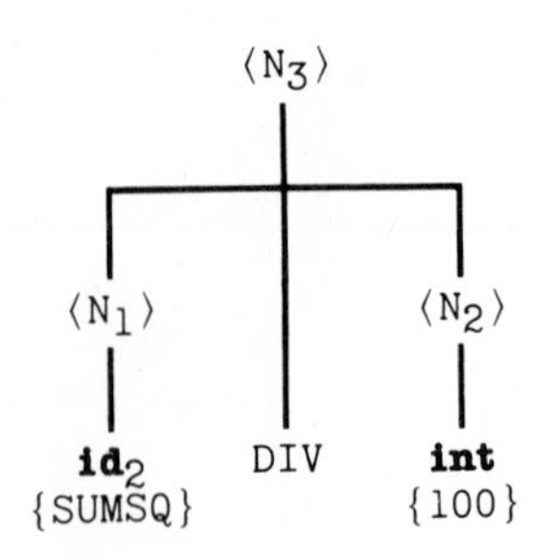

(v) $\ldots\ \mathbf{id}_1 \quad := \quad \langle N_3 \rangle \quad - \quad \langle N_4 \rangle \quad * \quad \mathbf{id}_4 \quad ;$

$\lessdot \qquad \doteq \qquad \lessdot \qquad\qquad \lessdot \qquad \lessdot \qquad \gtrdot$

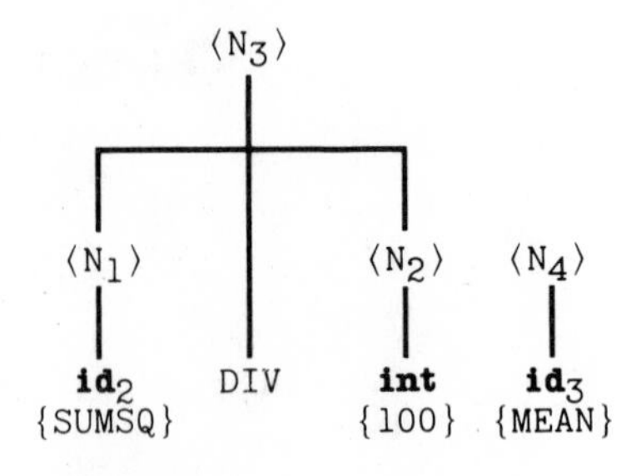

(vi) $\ldots\ \mathbf{id}_1 \quad := \quad \langle N_3 \rangle \quad - \quad \langle N_4 \rangle \quad * \quad \langle N_5 \rangle \quad ;$

$\lessdot \qquad \doteq \qquad \lessdot \qquad\qquad \lessdot \qquad\qquad \gtrdot$

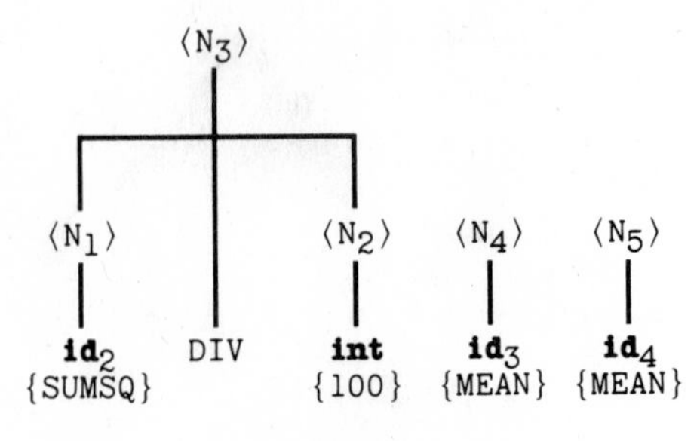

FIGURE 5.8(b)

(vii) ... $\mathbf{id}_1$:= ⟨N_3⟩ – ⟨N_6⟩ ;
⋖ ≐ ⋖ ⋗

⟨N_3⟩ ⟨N_6⟩
⟨N_1⟩ ⟨N_2⟩ ⟨N_4⟩ ⟨N_5⟩
$\mathbf{id}_2$ {SUMSQ} DIV **int** {100} $\mathbf{id}_3$ {MEAN} * $\mathbf{id}_4$ {MEAN}

(viii) ... $\mathbf{id}_1$:= ⟨N_7⟩ ;
⋖ ≐ ⋗

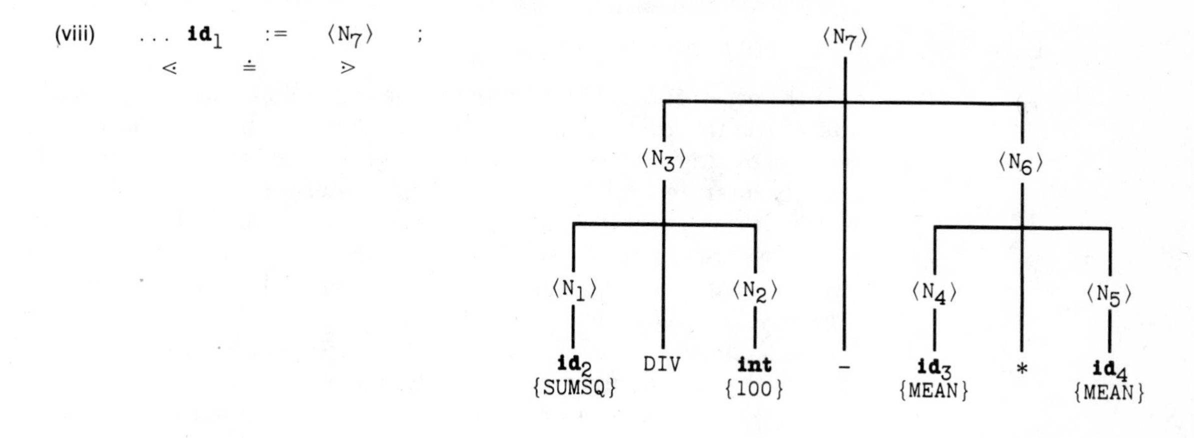

(ix) ... ⟨N_8⟩ ;

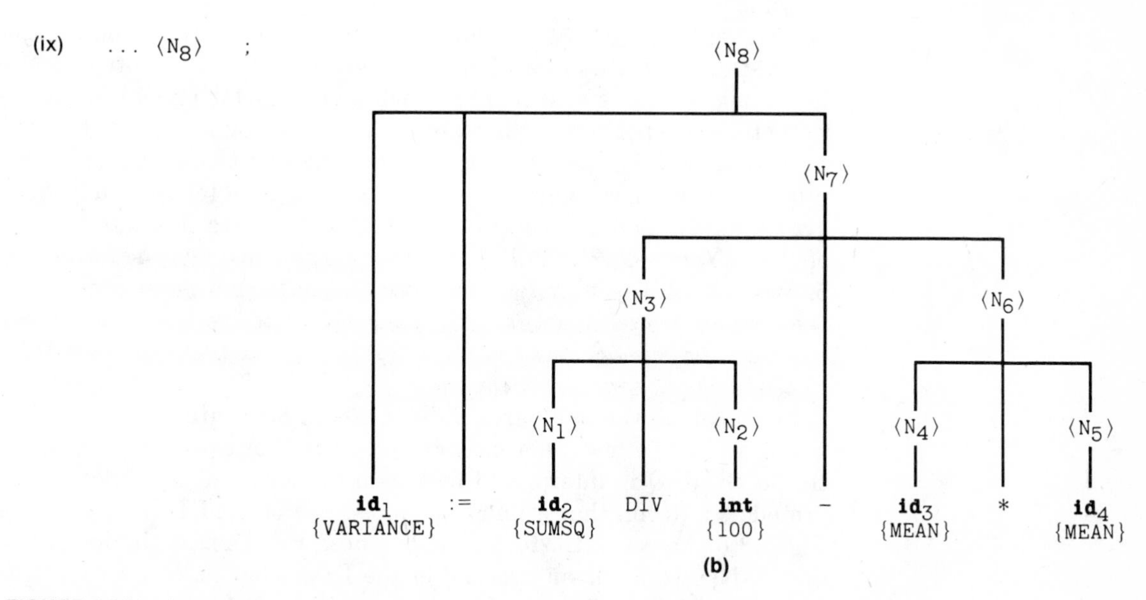

FIGURE 5.8(b) *(cont'd)*

the statement being analyzed with **id** replaced by $\langle N_1 \rangle$. The portion of the parse tree that corresponds to this interpretation appears to the right.

Part (ii) of Fig. 5.8(a) also shows the precedence relations that hold in the new version of the statement. An operator-precedence parser generally uses a stack to save tokens that have been scanned but not yet parsed, so that it can re-examine them in this way. Precedence relations hold only between terminal symbols, so $\langle N_1 \rangle$ is not involved in this process, and a relationship is determined between "(" and ")". This time the portion of the statement to be recognized is

$$\texttt{READ (} \langle N_1 \rangle \texttt{)}$$

which corresponds, except for the name of the nonterminal symbol, to Rule 13 of the grammar. This rule is the only one that could be applied in recognizing this portion of the program. As before, however, we simply interpret the sequence as some nonterminal $\langle N_2 \rangle$.

This completes the parsing of the READ statement. If we compare the parse tree shown in Fig. 5.3(a) with the one just developed, we see that they are the same except for the names of the nonterminal symbols involved. This means that we have correctly identified the *syntax* of the statement, which is the goal of the parsing process. The *names* of the nonterminals are arbitrarily chosen by the person writing the grammar, and have no real bearing on the meaning of the source statement.

Figure 5.8(b) shows a similar step-by-step parsing of the assignment statement from line 14 of the program in Fig. 5.1. Note that the left-to-right scan is continued in each step only far enough to determine the next portion of the statement to be recognized, which is the first portion delimited by $<$ and $>$. Once this portion has been determined, it is interpreted as a nonterminal according to some rule of the grammar. This process continues until the complete statement is recognized. You should follow carefully through the steps shown in Fig. 5.8(b) to be sure you understand how the statement is analyzed with the aid of the precedence matrix in Fig. 5.7. Note that each portion of the parse tree is constructed from the terminal nodes up toward the root, hence the term *bottom-up parsing*.

Comparing the parse tree built in Fig. 5.8(b) with the one in Fig. 5.3(b), we note a few differences in structure. For example, in Fig. 5.3 the **id** SUMSQ is interpreted first as a ⟨factor⟩, which is then interpreted as a ⟨term⟩ that is one of the operands of the DIV operation. In Fig. 5.8(b), however, the **id** SUMSQ is interpreted as the single nonterminal $\langle N_1 \rangle$, which is an operand of the DIV. That is, $\langle N_1 \rangle$ in the tree from Fig. 5.8(b) corresponds to two nonterminals, ⟨factor⟩ and ⟨term⟩, in Fig. 5.3(b). There are other similar differences between the two trees.

These differences are consistent with our use of arbitrary names for the nonterminal symbols recognized during an operator-precedence parse. In Fig. 5.3(b), the interpretation of SUMSQ as a ⟨factor⟩ and then as a ⟨term⟩ is simply a reassignment of names. This renaming is necessary because, according to Rule 11 of the grammar, the first operand in a multiplication operation must be a ⟨term⟩, not a ⟨factor⟩. Since our operator-precedence parse is not concerned with the names of the nonterminals in any case, it is not necessary to perform this additional step in the recognition process. As a matter of fact, the three different names ⟨exp⟩, ⟨term⟩, and ⟨factor⟩ were originally included in the grammar only as a means of specifying the precedence of operators (for example, that multiplication is performed before addition). Since this information is incorporated into our precedence matrix, there is no need to be concerned with the different names during the actual parsing.

Although we have illustrated operator-precedence parsing only on single statements, the same techniques can be applied to an entire program. You may want to work through this process for yourself on the program of Fig. 5.1, using the precedence matrix given in Fig. 5.7. The results should be the same as the parse tree shown in Fig. 5.4, except for differences in the naming of nonterminals such as those previously discussed.

The other parsing technique we discuss in this section is a top-down method known as *recursive descent.* A recursive-descent parser is made up of a procedure for each nonterminal symbol in the grammar. When a procedure is called, it attempts to find a substring of the input, beginning with the current token, that can be interpreted as the nonterminal with which the procedure is associated. In the process of doing this, it may call other procedures, or even call itself recursively, to search for other nonterminals. If a procedure finds the nonterminal that is its goal, it returns an indication of success to its caller, and it also advances the current-token pointer past the substring it has just recognized. If the procedure is unable to find a substring that can be interpreted as the desired nonterminal, it returns an indication of failure, or invokes an error diagnosis and recovery routine.

As an example of this, consider Rule 13 of the grammar in Fig. 5.2. The procedure for ⟨read⟩ in a recursive-descent parser first examines the next two input tokens, looking for READ and (. If these are found, the procedure for ⟨read⟩ then calls the procedure for ⟨id-list⟩. If that procedure succeeds, the ⟨read⟩ procedure examines the next input token, looking for). If all these tests are successful, the ⟨read⟩ procedure returns an indication of success to its caller and advances to the next token following). Otherwise, the procedure returns an indication of failure.

The procedure is only slightly more complicated when there are several alternatives defined by the grammar for a nonterminal. In that case, the procedure must decide which of the alternatives to try. For the recursive-descent technique, it must be possible to decide which alternative to use by examining the next input token. There are other top-down methods that remove this requirement; however, they are not as efficient as recursive descent. Thus the procedure for ⟨stmt⟩ looks at the next token to decide which of its four alternatives to try. If the token is READ, it calls the procedure for ⟨read⟩; if the token is **id**, it calls the procedure for ⟨assign⟩ because this is the only alternative that can begin with the token **id**, and so on.

If we attempted to write a complete set of procedures for the grammar of Fig. 5.2, we would discover a problem. The procedure for ⟨id-list⟩, corresponding to Rule 6, would be unable to decide between its two alternatives since both **id** and ⟨id-list⟩ can begin with **id**. There is, however, a more fundamental difficulty. If the procedure somehow decided to try the second alternative (⟨id-list⟩, **id**), it would immediately call itself recursively to find an ⟨id-list⟩. This could result in another immediate recursive call, which leads to an unending chain. The reason for this is that one of the alternatives for ⟨id-list⟩ begins with ⟨id-list⟩. Top-down parsers cannot be directly used with a grammar that contains this kind of immediate *left recursion.* The same problems also occur with respect to Rules 3, 7, 10, and 11. Methods for eliminating left recursion from a grammar appear in Gries (1971) and Aho (1977).

Figure 5.9 shows the grammar from Fig. 5.2 with left recursion eliminated. Consider, for example, Rule 6a in Fig. 5.9:

```
⟨id-list⟩  ::= id { , id }
```

This notation, which is a common extension to BNF, specifies that the terms between { and } may be omitted, or repeated one or more times. Thus Rule 6a defines ⟨id-list⟩ as being composed of an **id** followed by zero or more occurrences of ", **id**". This is clearly equivalent to Rule 6 of Fig. 5.2. With the revised definition, the procedure for ⟨id-list⟩ simply looks first for an **id**, and then keeps scanning the input as long as the next two tokens are a comma (,) and **id**. This eliminates the problem of left recursion and also the difficulty of deciding which alternative for ⟨id-list⟩ to try.

Similar changes have been made in Rules 3a, 7a, 10a, and 11a in Fig. 5.9. You should compare these rules to the corresponding defini-

```
1     <prog>          ::=  PROGRAM <prog-name> VAR <dec-list> BEGIN <stmt-list> END.
2     <prog-name>     ::=  id
3a    <dec-list>      ::=  <dec> { ; <dec> }
4     <dec>           ::=  <id-list> : <type>
5     <type>          ::=  INTEGER
6a    <id-list>       ::=  id { , id }
7a    <stmt-list>     ::=  <stmt> { ; <stmt> }
8     <stmt>          ::=  <assign> | <read> | <write> | <for>
9     <assign>        ::=  id := <exp>
10a   <exp>           ::=  <term> { + <term> | - <term> }
11a   <term>          ::=  <factor> { * <factor> | DIV <factor> }
12    <factor>        ::=  id | int | ( <exp> )
13    <read>          ::=  READ ( <id-list> )
14    <write>         ::=  WRITE ( <id-list> )
15    <for>           ::=  FOR <index-exp> DO <body>
16    <index-exp>     ::=  id := <exp> TO <exp>
17    <body>          ::=  <stmt> | BEGIN <stmt-list> END
```

FIGURE 5.9 Simplified Pascal grammar modified for recursive-descent parse.

tions in Fig. 5.2 to be sure you understand the changes made. Note that the grammar itself is still recursive: ⟨exp⟩ is defined in terms of ⟨term⟩, which is defined in terms of ⟨factor⟩, and one of the alternatives for ⟨factor⟩ involves ⟨exp⟩. This means that recursive calls among the procedures of the parser are still possible. However, direct left recursion has been eliminated. A chain of calls from ⟨exp⟩ to ⟨term⟩ to ⟨factor⟩ and back to ⟨exp⟩ must always consume at least one token from the input statement.

Figure 5.10 illustrates a recursive-descent parse of the READ statement on line 9 of Fig. 5.1, using the grammar in Fig. 5.9. Figure 5.10(a) shows the procedures for the nonterminals ⟨read⟩ and ⟨id-list⟩, which follow the verbal descriptions just given. It is assumed that TOKEN contains the type of the next input token, using the coding scheme shown in Fig. 5.5. You should examine these procedures carefully to be sure you understand how they were derived from the grammar.

In the procedure IDLIST, note that a comma (,) that is not followed by an **id** is considered to be an error, and the procedure returns an indication of failure to its caller. If a sequence of tokens such as "**id,id,**" could be a legal construct according to the grammar, this recursive-descent technique would not work properly. For such a grammar, it would be necessary to use a more complex parsing method that would allow the top-down parser to backtrack after recognizing that the last comma was not followed by an **id**.

Figure 5.10(b) gives a graphic representation of the recursive-descent parsing process for the statement being analyzed. In part (i),

```
procedure READ
   begin
      FOUND := FALSE
      if TOKEN = 8 {READ} then
         begin
            advance to next token
            if TOKEN = 20 { ( } then
               begin
                  advance to next token
                  if IDLIST returns success then
                     if TOKEN = 21 { ) } then
                        begin
                           FOUND := TRUE
                           advance to next token
                        end {if )}
               end {if (}
         end {if READ}
      if FOUND = TRUE then
         return success
      else
         return failure
   end {READ}


procedure IDLIST
   begin
      FOUND := FALSE
      if TOKEN = 22 {id} then
         begin
            FOUND := TRUE
            advance to next token
            while (TOKEN = 14 {,}) and (FOUND = TRUE) do
               begin
                  advance to next token
                  if TOKEN = 22 {id} then
                     advance to next token
                  else
                     FOUND := FALSE
               end {while}
         end {if id}
      if FOUND = TRUE then
         return success
      else
         return failure
   end {IDLIST}
```

(a)

FIGURE 5.10 Recursive-descent parse of a READ statement.

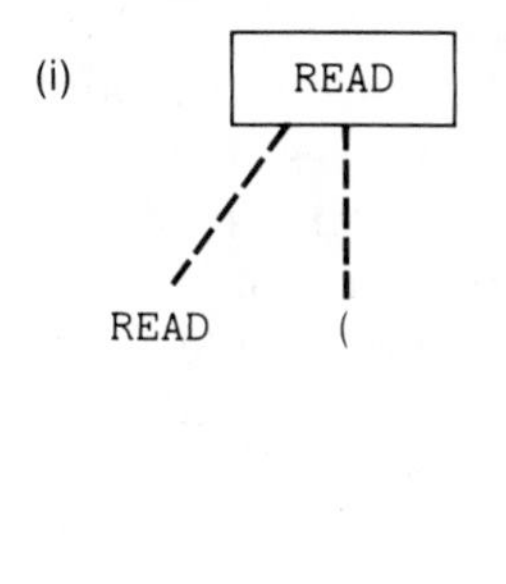

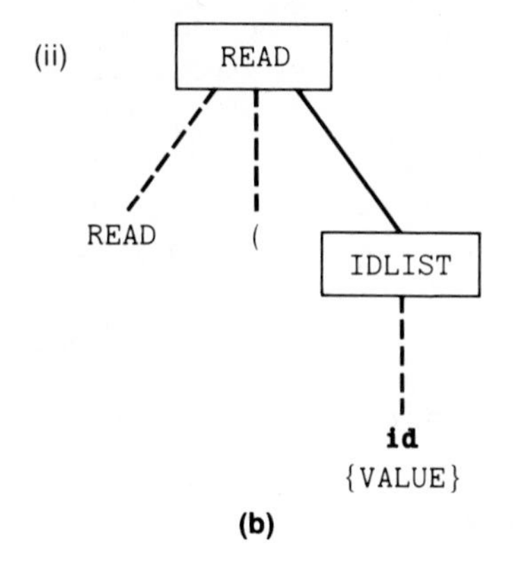

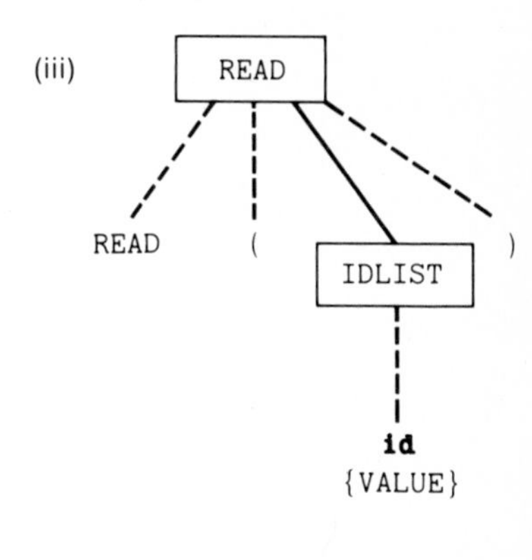

(b)

FIGURE 5.10 *(cont'd)*

the READ procedure has been invoked and has examined the tokens READ and (from the input stream (indicated by the dashed lines). In part (ii), READ has called IDLIST (indicated by the solid line), which has examined the token **id**. In part (iii), IDLIST has returned to READ, indicating success; READ has then examined the input token). This completes the analysis of the source statement. The procedure READ will now return to its caller, indicating that a ⟨read⟩ was successfully found. Note that the sequence of procedure calls and token examinations has completely defined the structure of the READ statement. The representation in part (iii) is the same as the parse tree in Fig. 5.3(a). Note also that the parse tree was constructed beginning at the root, hence the term *top-down parsing*.

Figure 5.11 illustrates a recursive-descent parse of the assignment statement on line 14 of Fig. 5.1. Figure 5.11(a) shows the procedures for the nonterminal symbols that are involved in parsing this statement. You should carefully compare these procedures to the corresponding rules of the grammar. Figure 5.11(b) is a step-by-step representation of the procedure calls and token examinations similar to that shown in Fig. 5.10(b). You are urged to follow through each step of the analysis of this statement, using the procedures in Fig. 5.11(a). Compare the parse tree built in Fig. 5.11(b) to the one in Fig. 5.3(b). Note that the differences between these two trees correspond exactly to the differences between the grammars of Figs. 5.9 and 5.2.

Our examples of recursive-descent parsing have involved only single statements; however, the same technique can be applied to an entire program. In that case, the syntactic analysis would consist simply of a call to the procedure for ⟨prog⟩. The calls from this procedure would create the parse tree for the program. You may want to write the procedures for the other nonterminals, following the grammar of Fig. 5.9, and apply this method to the program in Fig. 5.1. The result should be similar to the parse tree in Fig. 5.4. The only differences should be ones created by the modifications made to the grammar in Fig. 5.9.

Note that there is nothing inherent in a programming language that requires the use of any particular parsing technique. We have used one bottom-up parsing method and one top-down method to parse the same program, using essentially the same grammar. It is even

FIGURE 5.11 Recursive-descent parse of an assignment statement.

```
procedure ASSIGN
   begin
      FOUND := FALSE
      if TOKEN = 22 {id} then
         begin
            advance to next token
            if TOKEN = 15 {:=} then
               begin
                  advance to next token
                  if EXP returns success then
                     FOUND := TRUE
               end {if :=}
         end {if id}
      if FOUND = TRUE then
         return success
      else
         return failure
   end {ASSIGN}


procedure EXP
   begin
      FOUND := FALSE
      if TERM returns success then
         begin
            FOUND := TRUE
            while ((TOKEN = 16 {+}) or (TOKEN = 17 {-}))
             and (FOUND = TRUE) do
               begin
                  advance to next token
                  if TERM returns failure then
                     FOUND := FALSE
               end {while}
         end {if TERM}
      if FOUND = TRUE then
         return success
      else
         return failure
   end {EXP}
```

```
procedure TERM
   begin
      FOUND := FALSE
      if FACTOR returns success then
         begin
            FOUND := TRUE
            while ((TOKEN = 18 {*}) or (TOKEN = 19 {DIV}))
             and (FOUND = TRUE) do
               begin
                  advance to next token
                  if FACTOR returns failure then
                     FOUND := FALSE
               end {while}
         end {if FACTOR}
      if FOUND = TRUE then
         return success
      else
         return failure
   end {TERM}


procedure FACTOR
   begin
      FOUND := FALSE
      if (TOKEN = 22 {id})  or  (TOKEN = 23 {int})  then
         begin
            FOUND := TRUE
            advance to next token
         end {if id or int}
      else
         if TOKEN = 20 { ( } then
            begin
               advance to next token
               if EXP returns success then
                  if TOKEN = 21 { ) } then
                     begin
                        FOUND := TRUE
                        advance to next token
                     end {if )}
            end {if (}
      if FOUND = TRUE then
         return success
      else
         return failure
   end {FACTOR}
```

(a)

FIGURE 5.11 *(cont'd)*

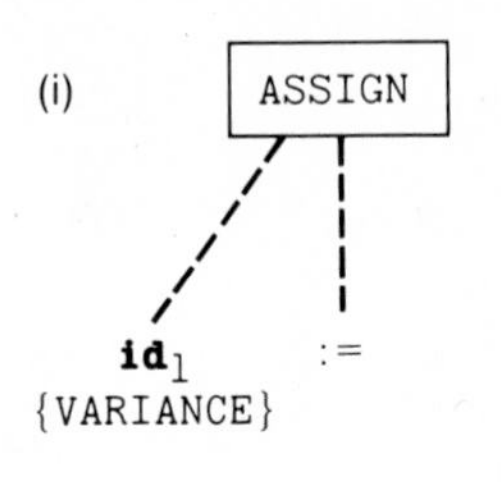

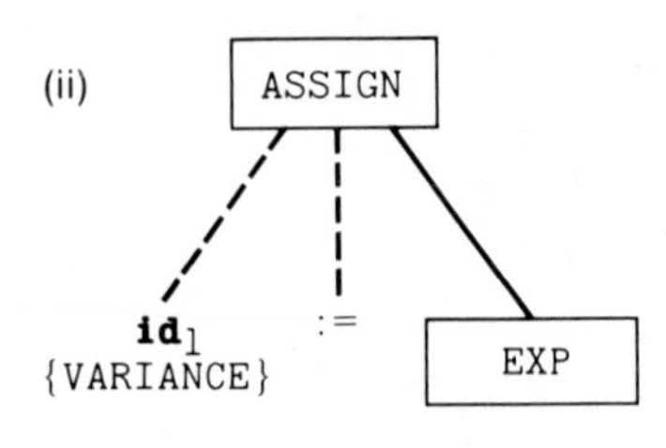

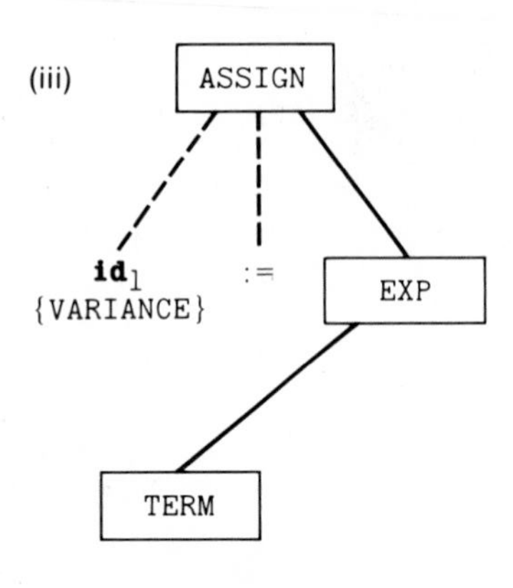

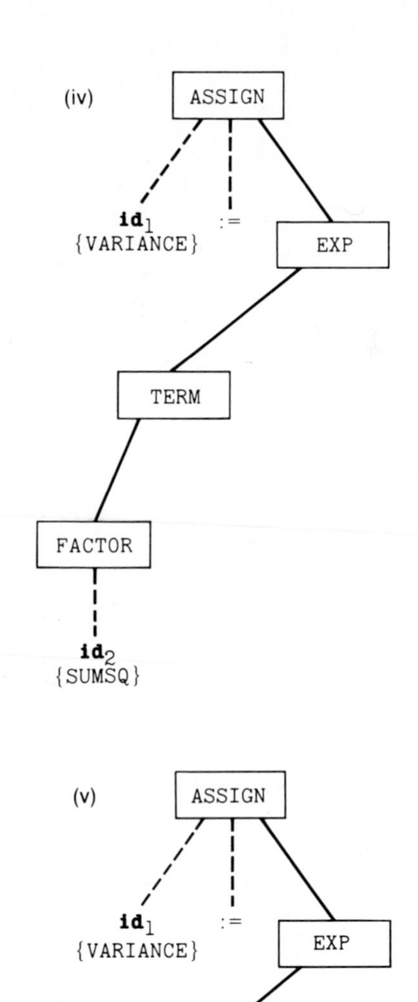

FIGURE 5.11 *(cont'd)*

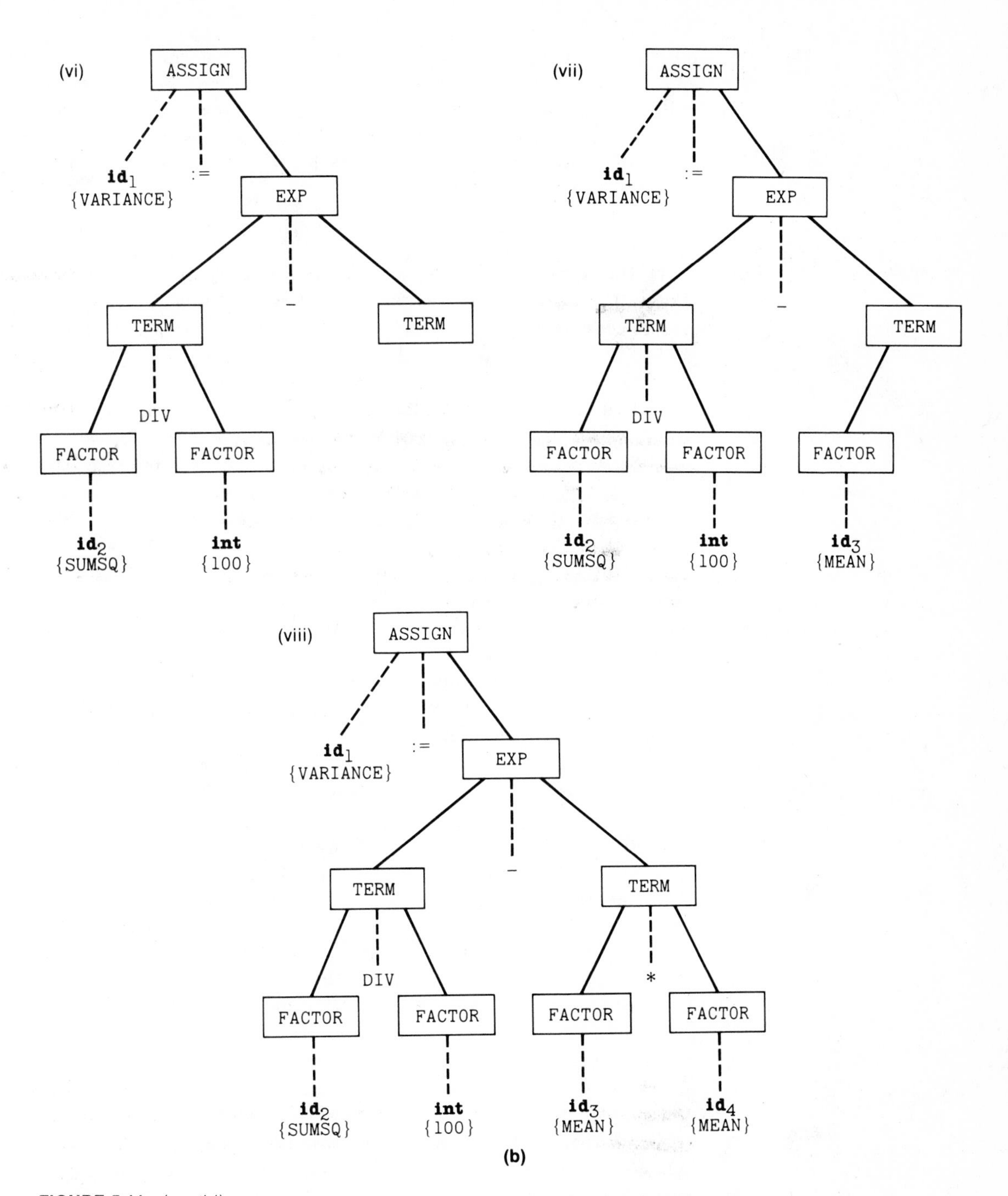

FIGURE 5.11 *(cont'd)*

possible to use a combination of techniques. Some compilers use recursive descent for high-level constructs (for example, down to the statement level), and then switch to a technique such as operator precedence to analyze constructs such as arithmetic expressions. Further discussions of parsing methods can be found in Aho (1977), Lewis (1976), and Gries (1971).

5.1.4 Code Generation

After the syntax of a program has been analyzed, the last task of compilation is the generation of object code. In this section we discuss a simple code-generation technique that creates the object code for each part of the program as soon as its syntax has been recognized.

The code-generation technique we describe involves a set of routines, one for each rule or alternative rule in the grammar. When the parser recognizes a portion of the source program according to some rule of the grammar, the corresponding routine is executed. Such routines are often called *semantic routines* because the processing performed is related to the meaning we associate with the corresponding construct in the language. In our simple scheme, these semantic routines generate object code directly, so we refer to them as *code-generation routines*. In more complex compilers, the semantic routines might generate an intermediate form of the program that would be analyzed further in an attempt to generate more efficient object code. We discuss this possibility in more detail in Sections 5.2 and 5.3.

The code-generation routines we discuss in this section are designed for use with the grammar in Fig. 5.2. As we have seen, neither of the parsing techniques discussed in Section 5.1.3 recognizes exactly the constructs specified by this grammar. The operator-precedence method ignores certain nonterminals, and the recursive-descent method must use a slightly modified grammar. However, there are parsing techniques not much more complicated than those we have discussed that can parse according to the grammar in Fig. 5.2. We choose to use this grammar in our discussion of code generation to emphasize the point that code-generation techniques need not be associated with any particular parsing method.

The specific code to be generated clearly depends upon the computer for which the program is being compiled. In this section we use as an example the generation of object code for a SIC/XE machine.

Our code-generation routines make use of two data structures for working storage: a list and a stack. Items inserted into the list are removed in the order of their insertion, first in–first out. Items pushed onto the stack are removed (popped from the stack) in the opposite order, last in–first out. The variable LISTCOUNT is used to keep a count of the number of items currently in the list. The code-generation routines also make use of the token specifiers described in Section 5.1.2; these specifiers are denoted by S(token). For a token **id**, S(**id**) is the name of the identifier, or a pointer to the symbol-table entry for it. For a token **int**, S(**int**) is the value of the integer, such as #100.

Many of our code-generation routines, of course, create segments of object code for the compiled program. We give a symbolic representation of this code, using SIC assembler language. You should remember, however, that the actual code generated is usually machine language, not assembler language. As each piece of object code is generated, we assume that a location counter LOCCTR is updated to reflect the next available address in the compiled program (exactly as it is in an assembler).

Figure 5.12 illustrates the application of this process to the READ statement on line 9 of the program in Fig. 5.1. The parse tree for this statement is repeated for convenience in Fig. 5.12(a). This tree can be generated with many different parsing methods. Regardless of the technique used, however, the parser always recognizes at each step the leftmost substring of the input that can be interpreted according to a rule of the grammar. In an operator-precedence parse, this recognition occurs when a substring of the input is reduced to some nonterminal $\langle N_i \rangle$. In a recursive-descent parse, the recognition occurs when a procedure returns to its caller, indicating success. Thus the parser first recognizes the **id** VALUE as an ⟨id-list⟩, and then recognizes the complete statement as a ⟨read⟩.

Figure 5.12(c) shows a symbolic representation of the object code to be generated for the READ statement. This code consists of a call to a subroutine XREAD, which would be part of a standard library associated with the compiler. The subroutine XREAD can be called by any program that wants to perform a READ operation. XREAD is linked together with the generated object program by a linking loader or a linkage editor. (The compiler includes in the object program enough information to specify this linking operation, perhaps using Modification records such as those discussed in Chapter 2.) This technique is commonly used for the compilation of statements that perform relatively complex functions. The use of a subroutine avoids the repetitive generation of large amounts of in-line code, which makes the object program smaller.

Since XREAD may be used to perform any READ operation, it must be passed parameters that specify the details of the READ. In this case, the parameter list for XREAD is defined immediately after the JSUB that calls it. The first word in this parameter list contains a value that specifies the number of variables that will be assigned val-

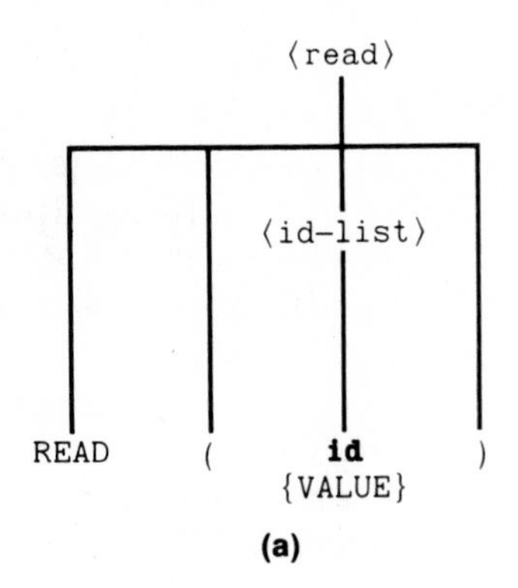

(a)

```
⟨id-list⟩ ::= id

                add S(id) to list
                add 1 to LISTCOUNT


⟨id-list⟩ ::= ⟨id-list⟩ , id

                add S(id) to list
                add 1 to LISTCOUNT


⟨read⟩ ::= READ ( ⟨id-list⟩ )

                generate [        +JSUB   XREAD]
                record external reference to XREAD
                generate [         WORD   LISTCOUNT]
                for each item on list do
                   begin
                      remove S(ITEM) from list
                      generate [         WORD   S(ITEM)]
                   end
                LISTCOUNT := 0
```

(b)

```
+JSUB    XREAD
 WORD    1
 WORD    VALUE
```

(c)

FIGURE 5.12 Code generation for a READ statement.

ues by the READ. The following words give the addresses of these variables. Thus the second line in Fig. 5.12(c) specifies that one variable is to be read, and the third line gives the address of this variable. The address of the first word of the parameter list will automatically be placed in register L by the JSUB instruction. The subroutine XREAD can use this address to locate its parameters, and then add the length of the parameter list to register L to find the true return address.

Figure 5.12(b) shows a set of routines that might be used to accomplish this code generation. The first two routines correspond to alternative structures for ⟨id-list⟩, which are shown in Rule 6 of the grammar in Fig. 5.2. In either case, the token specifier S(**id**) for a new identifier being added to the ⟨id-list⟩ is inserted into the list used by the code-generation routines, and LISTCOUNT is updated to reflect this insertion. After the entire ⟨id-list⟩ has been parsed, the list contains the token specifiers for all the identifiers that are part of the ⟨id-list⟩. When the ⟨read⟩ statement is recognized, these token specifiers are removed from the list and used to generate the object code for the READ.

Remember that the parser, in generating the tree shown in Fig. 5.12(a), recognizes first ⟨id-list⟩ and then ⟨read⟩. At each step, the parser calls the appropriate code-generation routine. You should work through this example carefully to be sure you understand how the code-generation routines in Fig. 5.12(b) create the object code that is symbolically represented in Fig. 5.12(c).

Figure 5.13 shows the code-generation process for the assignment statement on line 14 of Fig. 5.1. Fig. 5.13(a) displays the parse tree for this statement. Most of the work of parsing involves the analysis of the ⟨exp⟩ on the right-hand side of the :=. As we can see, the parser first recognizes the **id** SUMSQ as a ⟨factor⟩ and a ⟨term⟩; then it recognizes the **int** 100 as a ⟨factor⟩; then it recognizes SUMSQ DIV 100 as a ⟨term⟩, and so forth. This is essentially the same sequence of steps shown in Fig. 5.8(b). The order in which the parts of the statement are recognized is the same as the order in which the calculations are to be performed; SUMSQ DIV 100 and MEAN * MEAN are computed, and then the second result is subtracted from the first.

As each portion of the statement is recognized, a code-generation routine is called to create the corresponding object code. For example, suppose we want to generate code that corresponds to the rule

$$\langle\text{term}\rangle_1 \;::=\; \langle\text{term}\rangle_2 \;*\; \langle\text{factor}\rangle$$

The subscripts are used here to distinguish between the two occurrences of ⟨term⟩. Our code-generation routines perform all arithmetic operations using register A, so we clearly need to generate a MUL instruction in the object code. The result of this multiplication, $\langle\text{term}\rangle_1$,

FIGURE 5.13 Code generation for an assignment statement.

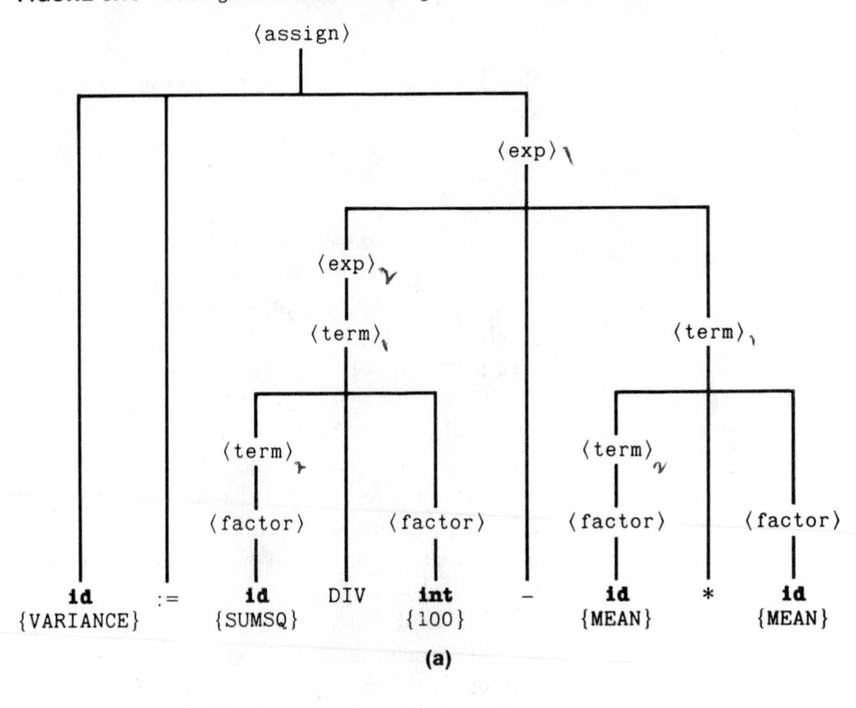

(a)

```
⟨assign⟩ ::= id := ⟨exp⟩

                GETA (⟨exp⟩)
                generate [          STA     S(id)]
                REGA := null


⟨exp⟩ ::= ⟨term⟩

                S(⟨exp⟩) := S(⟨term⟩)
                if S(⟨exp⟩) = rA then
                   REGA := ⟨exp⟩


⟨exp⟩1 ::= ⟨exp⟩2 + ⟨term⟩

                if S(⟨exp⟩2) = rA then
                   generate [          ADD     S(⟨term⟩)]
                else if S(⟨term⟩) = rA then
                   generate [          ADD     S(⟨exp⟩2)]
                else
                   begin
                      GETA (⟨exp⟩2)
                      generate [          ADD     S(⟨term⟩)]
                   end
                S(⟨exp⟩1) := rA
                REGA := ⟨exp⟩1
```

```
⟨exp⟩1 ::= ⟨exp⟩2 - ⟨term⟩

          if S(⟨exp⟩2) = rA then
             generate [          SUB     S(⟨term⟩)]
          else
             begin
                GETA (⟨exp⟩2)
                generate [          SUB     S(⟨term⟩)]
             end
          S(⟨exp⟩1) := rA
          REGA := ⟨exp⟩1


⟨term⟩ ::= ⟨factor⟩

          S(⟨term⟩) := S(⟨factor⟩)
          if S(⟨term⟩) = rA then
             REGA := ⟨term⟩


⟨term⟩1 ::= ⟨term⟩2 * ⟨factor⟩

          if S(⟨term⟩2) = rA then
             generate [          MUL     S(⟨factor⟩)]
          else if S(⟨factor⟩) = rA then
             generate [          MUL     S(⟨term⟩2)]
          else
             begin
                GETA (⟨term⟩2)
                generate [          MUL     S(⟨factor⟩)]
             end
          S(⟨term⟩1) := rA
          REGA := ⟨term⟩1


⟨term⟩1 ::= ⟨term⟩2 DIV ⟨factor⟩

          if S(⟨term⟩2) = rA then
             generate [          DIV     S(⟨factor⟩)]
          else
             begin
                GETA (⟨term⟩2)
                generate [          DIV     S(⟨factor⟩)]
             end
          S(⟨term⟩1) := rA
          REGA := ⟨term⟩1
```

FIGURE 5.13(b) *(cont'd)*

```
⟨factor⟩ ::= id

            S(⟨factor⟩) := S(id)


⟨factor⟩ ::= int

            S(⟨factor⟩) := S(int)


⟨factor⟩ ::= ( ⟨exp⟩ )

            S(⟨factor⟩) := S(⟨exp⟩)
            if S(⟨factor⟩) = rA then
              REGA := ⟨factor⟩
```

(b)

```
procedure GETA (NODE)
   begin
      if REGA = null then
         generate [         LDA     S(NODE)]
      else if S(NODE) ≠ rA then
         begin
            create a new working variable Ti
            generate [         STA     Ti]
            record forward reference to Ti
            S(REGA) := Ti
            generate [         LDA     S(NODE)]
         end {if ≠ rA}
      S(NODE) := rA
      REGA := NODE
   end {GETA}
```

(c)

```
LDA     SUMSQ
DIV     #100
STA     T1
LDA     MEAN
MUL     MEAN
STA     T2
LDA     T1
SUB     T2
STA     VARIANCE
```

(d)

FIGURE 5.13 *(cont'd)*

will be left in register A by the MUL. If either $\langle\text{term}\rangle_2$ or ⟨factor⟩ is already present in register A, perhaps as the result of a previous computation, the MUL instruction is all we need. Otherwise, we must generate a LDA instruction preceding the MUL. In that case we must also save the previous value in register A if it will be required for later use.

Obviously we need to keep track of the result left in register A by each segment of code that is generated. We do this by extending the token-specifier idea to nonterminal nodes of the parse tree. In the example just discussed, the *node specifier* S($\langle\text{term}\rangle_1$) would be set to rA, indicating that the result of this computation is in register A. The variable REGA is used to indicate the highest-level node of the parse tree whose value is left in register A by the code generated so far (i.e., the node whose specifier is rA). Clearly, there can be only one such node at any point in the code-generation process. If the value corresponding to a node is not in register A, the specifier for the node is similar to a token specifier: either a pointer to a symbol table entry for the variable that contains the value, or an integer constant.

As an illustration of these ideas, consider the code-generation routine in Fig. 5.13(b) that corresponds to the rule

```
⟨term⟩1  ::= ⟨term⟩2 * ⟨factor⟩
```

If the node specifier for either operand is rA, the corresponding value is already in register A, so the routine simply generates a MUL instruction. The operand address for this MUL is given by the node specifier for the other operand (the one not in the register). Otherwise, the procedure GETA is called. This procedure, shown in Fig. 5.13(c), generates a LDA instruction to load the value associated with $\langle\text{term}\rangle_2$ into register A. Before the LDA, however, the procedure generates a STA instruction to save the value currently in register A unless REGA is null, indicating that this value is no longer needed. The value is stored in a *temporary* variable. Such variables are created by the code generator (with names T1, T2, . . .) as they are needed for this purpose. The temporary variables used during a compilation will be assigned storage locations at the end of the object program. The node specifier for the node associated with the value previously in register A, indicated by REGA, is reset to indicate the temporary variable used.

After the necessary instructions are generated, the code-generation routine sets S($\langle\text{term}\rangle_1$) and REGA to indicate that the value corresponding to $\langle\text{term}\rangle_1$ is now in register A. This completes the code-generation actions for the $*$ operation.

The code-generation routine that corresponds to the $+$ operation is almost identical to the one just discussed for $*$. The routines for DIV and − are similar, except that for these operations it is necessary for the first operand to be in register A. The code generation for ⟨assign⟩ consists of bringing the value to be assigned into register A (using GETA) and then generating a STA instruction. Note that REGA is then set to null because the code for the statement has been completely generated, and any intermediate results are no longer needed.

The remaining rules shown in Fig. 5.13(b) do not require the generation of any machine instructions since no computation or data movement is involved. The code-generation routines for these rules simply set the node specifier of the higher-level node to reflect the location of the corresponding value.

Figure 5.13(d) shows a symbolic representation of the object code generated for the assignment statement being translated. You should carefully work through the generation of this code to understand the operation of the routines in Figs. 5.13(b) and 5.13(c). You should also confirm that this code will perform the computations specified by the source program statement.

Figure 5.14 shows the other code-generation routines for the grammar in Fig. 5.2. The routine for ⟨prog-name⟩ generates header information in the object program that is similar to that created from the START and EXTREF assembler directives. It also generates instructions to save the return address and jump to the first executable instruction in the compiled program. When the complete ⟨prog⟩ is recognized, storage locations are assigned to any temporary (Ti) variables that have been used. Any references to these variables are then fixed in the object code using the same process performed for forward references by a one-pass assembler. The compiler also generates any Modification records required to describe external references to library subroutines.

Code generation for a ⟨for⟩ statement involves a number of steps. When the ⟨index-exp⟩ is recognized, code is generated to initialize the index variable for the loop and test for loop termination. Information is also saved on the stack for later use. Code is then generated separately for each statement in the body of the loop. When the complete ⟨for⟩ statement has been parsed, code is generated to increment the value of the index variable and to jump back to the beginning of the loop to test for termination. This code-generation routine uses the information saved on the stack by the routine for ⟨index-exp⟩. Using a stack to store this information allows for nested ⟨for⟩ loops.

You are encouraged to trace through the complete code-generation process for this program, using the parse tree in Fig. 5.4 as a guide. The result should be as shown in Fig. 5.15. Once again, it is important to remember that this is merely a symbolic representation of the code

```
⟨prog⟩ ::= PROGRAM ⟨prog-name⟩ VAR ⟨dec-list⟩ BEGIN ⟨stmt-list⟩ END.

                    generate [         LDL     RETADR]
                    generate [         RSUB]
                    for each Ti variable used do
                       generate [Ti       RESW    1]
                    insert [        J      EXADDR]  {jump to first executable
                         instruction}  in bytes 3-5 of object program
                    fix up forward references to Ti variables
                    generate Modification records for external references
                    generate [         END ]


⟨prog-name⟩ ::= id

                    generate [         START    0]
                    generate [         EXTREF   XREAD,XWRITE]
                    generate [         STL      RETADR]
                    add 3 to LOCCTR    {leave room for jump to first
                         executable instruction}
                    generate [RETADR  RESW     1]


⟨dec-list⟩ ::= {either alternative}

                    save LOCCTR as EXADDR  {tentative address of first
                         executable instruction}


⟨dec⟩ ::= ⟨id-list⟩ : ⟨type⟩

                    for each item on list do
                       begin
                          remove S(NAME) from list
                          enter LOCCTR into symbol table as address for NAME
                          generate [S(NAME)  RESW   1]
                       end
                    LISTCOUNT := 0


⟨type⟩ ::= INTEGER

                    {no code generation action}


⟨stmt-list⟩ ::= {either alternative}

                    {no code generation action}
```

FIGURE 5.14 Other code-generation routines for the grammar from Fig. 5.2.

```
⟨stmt⟩ ::= {any alternative}

                    {no code generation action}


⟨write⟩ ::= WRITE (⟨id-list⟩)

                    generate [        +JSUB    XWRITE]
                    record external reference to XWRITE
                    generate [         WORD    LISTCOUNT]
                    for each item on list do
                       begin
                          remove S(ITEM) from list
                          generate [         WORD    S(ITEM)]
                       end
                    LISTCOUNT := 0


⟨for⟩ ::= FOR ⟨index-exp⟩ DO ⟨body⟩

                    pop JUMPADDR from stack {address of jump out of loop}
                    pop S(INDEX) from stack {index variable}
                    pop LOOPADDR from stack {beginning address of loop}
                    generate [         LDA     S(INDEX)]
                    generate [         ADD    #1]
                    generate [         J       LOOPADDR]
                    insert [        JGT    LOCCTR]  at location JUMPADDR


⟨index-exp⟩ ::= id := ⟨exp⟩_1 TO ⟨exp⟩_2

                    GETA (⟨exp⟩_1)
                    push LOCCTR onto stack {beginning address of loop}
                    push S(id) onto stack {index variable}
                    generate [         STA     S(id)]
                    generate [         COMP    S(⟨exp⟩_2)]
                    push LOCCTR onto stack {address of jump out of loop}
                    add 3 to LOCCTR  {leave room for jump instruction}
                    REGA := null


⟨body⟩ ::= {either alternative}

                    {no code generation action}
```

FIGURE 5.14 *(cont'd)*

```
Line    Symbolic Representation of Generated Code

 1      STATS     START   0              {program header}
                  EXTREF  XREAD,XWRITE
                  STL     RETADR         {save return address}
                  J       {EXADDR}
        RETADR    RESW    1
 3      SUM       RESW    1              {variable declarations}
        SUMSQ     RESW    1
        I         RESW    1
        VALUE     RESW    1
        MEAN      RESW    1
        VARIANCE  RESW    1
 5      {EXADDR}  LDA     #0             {SUM := 0}
                  STA     SUM
 6                LDA     #0             {SUMSQ := 0}
                  STA     SUMSQ
 7                LDA     #1             {FOR I := 1 TO 100}
        {L1}      STA     I
                  COMP    #100
                  JGT     {L2}
 9               +JSUB    XREAD          {READ(VALUE)}
                  WORD    1
                  WORD    VALUE
10                LDA     SUM            {SUM := SUM + VALUE}
                  ADD     VALUE
                  STA     SUM
11                LDA     VALUE          {SUMSQ := SUMSQ + VALUE * VALUE}
                  MUL     VALUE
                  ADD     SUMSQ
                  STA     SUMSQ
                  LDA     I              {end of FOR loop}
                  ADD     #1
                  J       {L1}
13      {L2}      LDA     SUM            {MEAN := SUM DIV 100}
                  DIV     #100
                  STA     MEAN
14                LDA     SUMSQ          {VARIANCE := SUMSQ DIV 100 - MEAN * MEAN}
                  DIV     #100
                  STA     T1
                  LDA     MEAN
                  MUL     MEAN
                  STA     T2
                  LDA     T1
                  SUB     T2
                  STA     VARIANCE
15               +JSUB    XWRITE         {WRITE(MEAN,VARIANCE)}
                  WORD    2
                  WORD    MEAN
                  WORD    VARIANCE
                  LDL     RETADR         {return}
                  RSUB
        T1        RESW    1              {working variables used}
        T2        RESW    1
                  END
```

FIGURE 5.15 Symbolic representation of object code generated for the program from Fig. 5.1.

generated. Most compilers would produce machine-language code directly.

5.2 MACHINE-DEPENDENT COMPILER FEATURES

In this section we briefly discuss some machine-dependent extensions to the basic compilation scheme presented in Section 5.1. The purpose of a compiler is to translate programs written in a high-level programming language into the machine language for some computer. Most high-level programming languages are designed to be relatively independent of the machine being used (although the extent to which this goal is realized varies considerably). This means that the process of analyzing the syntax of programs written in these languages should also be relatively machine-independent. It should come as no surprise, therefore, that the real machine dependencies of a compiler are related to the generation and optimization of the object code.

At an elementary level, of course, all code generation is machine-dependent because we must know the instruction set of a computer to generate code for it. However, there are many more complex issues involving such problems as the allocation of registers and the rearrangement of machine instructions to improve efficiency of execution. Such types of code optimization are normally done by considering an *intermediate form* of the program being compiled. In this intermediate form, the syntax and semantics of the source statements have been completely analyzed, but the actual translation into machine code has not yet been performed. It is much easier to analyze and manipulate this intermediate form of the program for the purposes of code optimization than it would be to perform the corresponding operations on either the source program or the machine code.

In Section 5.2.1 we introduce one common way of representing a program in such an intermediate form. The intermediate form used in a compiler, if one is used, is not strictly dependent on the machine for which the compiler is designed. However, such a form is necessary for our discussion of machine-dependent code optimization in Section 5.2.2. There are also many machine-independent techniques for code optimization that use a similar intermediate representation of the program. Some of these techniques are discussed in Section 5.3.

5.2.1 Intermediate Form of the Program

There are many possible ways of representing a program in an intermediate form for code analysis and optimization. See, for example, Aho (1977) and Gries (1971). The method we discuss in this section represents the executable instructions of the program with a sequence of *quadruples*. Each quadruple is of the form

operation, op1, op2, result

where operation is some function to be performed by the object code, op1 and op2 are the operands for this operation, and result designates where the resulting value is to be placed.

For example, the source program statement

```
SUM := SUM + VALUE
```

could be represented with the quadruples

```
+,   SUM, VALUE, i1
:=,  i1 ,      , SUM
```

The entry i_1 designates an intermediate result (SUM + VALUE); the second quadruple assigns the value of this intermediate result to SUM. Assignment is treated as a separate operation (:=) to open up additional possibilities for code optimization.

Similarly, the statement

```
VARIANCE := SUMSQ DIV 100 - MEAN * MEAN
```

could be represented with the quadruples

```
DIV, SUMSQ, #100, i1
*,   MEAN , MEAN, i2
-,   i1   , i2  , i3
:=,  i3   ,     , VARIANCE
```

These quadruples would be created by intermediate code-generation routines similar to those discussed in Section 5.1.4. Many types of analysis and manipulation can be performed on the quadruples for code-optimization purposes. For example, the quadruples can be rearranged to eliminate redundant load and store operations, and the intermediate results i_j can be assigned to registers or to temporary variables to make their use as efficient as possible. We discuss some of these possibilities in later sections. After optimization has been performed, the modified quadruples are translated into machine code.

Note that the quadruples appear in the order in which the corresponding object code instructions are to be executed. This greatly simplifies the task of analyzing the code for purposes of optimization. It also means that the translation into machine instructions will be relatively easy.

Figure 5.16 shows a sequence of quadruples corresponding to the source program in Fig. 5.1. Note that the READ and WRITE statements are represented with a CALL operation, followed by PARAM quadruples that specify the parameters of the READ or WRITE. These PARAM quadruples will, of course, be translated into parameter list

FIGURE 5.16 Intermediate code for the program from Fig. 5.1.

	Operation	Op1	Op2	Result	
(1)	:=	#0		SUM	{SUM := 0}
(2)	:=	#0		SUMSQ	{SUMSQ := 0}
(3)	:=	#1		I	{FOR I := 1 TO 100}
(4)	JGT	I	#100	(15)	
(5)	CALL	XREAD			{READ(VALUE)}
(6)	PARAM	VALUE			
(7)	+	SUM	VALUE	i_1	{SUM := SUM + VALUE}
(8)	:=	i_1		SUM	
(9)	*	VALUE	VALUE	i_2	{SUMSQ := SUMSQ + VALUE * VALUE}
(10)	+	SUMSQ	i_2	i_3	
(11)	:=	i_3		SUMSQ	
(12)	+	I	#1	i_4	{end of FOR loop}
(13)	:=	i_4		I	
(14)	J			(4)	
(15)	DIV	SUM	#100	i_5	{MEAN := SUM DIV 100}
(16)	:=	i_5		MEAN	
(17)	DIV	SUMSQ	#100	i_6	{VARIANCE := SUMSQ DIV 100 -
(18)	*	MEAN	MEAN	i_7	MEAN * MEAN}
(19)	-	i_6	i_7	i_8	
(20)	:=	i_8		VARIANCE	
(21)	CALL	XWRITE			{WRITE(MEAN,VARIANCE)}
(22)	PARAM	MEAN			
(23)	PARAM	VARIANCE			

entries, like those shown in Fig. 5.15, when the machine code is generated. The JGT operation in quadruple 4 compares the values of its two operands and jumps to quadruple 15 if the first operand is greater than the second. The J operation in quadruple 14 jumps unconditionally to quadruple 4.

You should compare Fig. 5.16 with Fig. 5.1 to see the correspondence between source statements and quadruples. You may also want to compare the quadruples in Fig. 5.16 with the object code representation shown in Fig. 5.15.

5.2.2 Machine-Dependent Code Optimization

In this section we briefly describe several different possibilities for performing machine-dependent code optimization. Further details concerning these techniques can be found in many textbooks on compilers, such as Aho (1977) and Gries (1971).

The first problem we discuss is the assignment and use of registers. On many computers there are a number of general-purpose registers that may be used to hold constants, the values of variables, intermediate results, and so on. These same registers can also often be used for addressing (as base or index registers). We concentrate here, however, on the use of registers as instruction operands.

Machine instructions that use registers as operands are usually faster than the corresponding instructions that refer to locations in memory. Therefore, we would prefer to keep in registers all variables and intermediate results that will be used later in the program. Each time a value is fetched from memory, or calculated as an intermediate result, it can be assigned to some register. This value will be available for later use without requiring a memory reference. We used a very simple version of this technique in Section 5.1.4 when we kept track of the value currently in register A.

Consider, for example, the quadruples shown in Fig. 5.16. The variable VALUE is used once in quadruple 7 and twice in quadruple 9. If enough registers are available, it would be possible to fetch this value only once. The value would be retained in a register for use by the code generated from quadruple 9. Likewise, quadruple 16 stores the value of i_5 into the variable MEAN. If i_5 is assigned to a register, this value could still be available when the value of MEAN is required in quadruple 18. Such register assignments can also be used to eliminate much of the need for temporary variables. Consider, for example, the machine code in Fig. 5.15, in which the use of only one register (register A) was sufficient to handle six of the eight intermediate results (i_j) in Fig. 5.16.

Of course, there are rarely as many registers available as we would like to use. The problem then becomes one of selecting which register value to replace when it is necessary to assign a register for some other purpose. One reasonable approach is to scan the program for the next point at which each register value would be used. The value that will not be needed for the longest time is the one that should be replaced. If the register that is being reassigned contains the value of some variable already stored in memory, the value can simply be discarded. Otherwise, this value must be saved using a temporary variable. This is one of the functions performed by the GETA procedure discussed in Section 5.1.4, using the temporary variables Ti.

In making and using register assignments, a compiler must also consider the control flow of the program. For example, quadruple 1 in Fig. 5.16 assigns the value 0 to SUM. This value might be retained in some register for later use. When SUM is next used as an operand in quadruple 7, it might appear that its value can be taken directly from the register; however, this is not necessarily the case. The J operation in quadruple 14 jumps to quadruple 4. If control passes to quadruple 7 in this way, the value of SUM may not be in the designated register. This would happen, for example, if the register were reassigned to hold the value of i_4 in quadruple 12. Thus the existence of Jump instructions creates difficulty in keeping track of register contents.

One way to deal with this problem is to divide the program into *basic blocks*. A basic block is a sequence of quadruples with one entry point, which is at the beginning of the block, one exit point, which is at the end of the block, and no jumps within the block. In other words, each quadruple that is the target of a jump, or that immediately follows a quadruple that specifies a jump, begins a new basic block. Since procedure calls can have unpredictable effects on register contents, a CALL operation is also usually considered to begin a new basic block. The assignment and use of registers within a basic block can follow the method previously described. However, when control passes from one basic block to another, all values currently held in registers are saved in temporary variables.

Figure 5.17 shows the division of the quadruples from Fig. 5.16 into basic blocks. Block A contains quadruples 1–3; block B contains quadruple 4, etc. Figure 5.17 also shows a representation of the control flow of the program: An arrow from block X to block Y indicates that control can pass directly from the last quadruple of X to the first quadruple of Y. (Within a basic block, of course, control must pass sequentially from one quadruple to the next.) This kind of representation is called a *flow graph* for the program. More sophisticated code-optimization techniques can analyze a flow graph and perform register assignments that remain valid from one basic block to another.

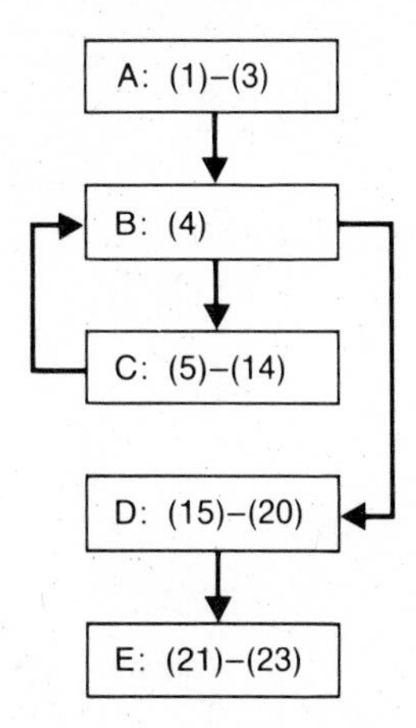

FIGURE 5.17 Basic blocks and flow graph for the quadruples from Fig. 5.16.

Another possibility for code optimization involves rearranging quadruples before machine code is generated. Consider, for example, the quadruples in Fig. 5.18(a). These are essentially the same as quadruples 17–20 in Fig. 5.16; they correspond to source statement 14 of the program in Fig. 5.1. Figure 5.18(a) shows a typical generation of machine code from these quadruples, using only a single register (register A). This is the same as the code shown for source statement 14 in Fig. 5.15.

Note that the value of the intermediate result i_1 is calculated first and stored in temporary variable T1. Then the value of i_2 is calculated. The third quadruple in this series calls for subtracting the value of i_2 from i_1. Since i_2 has just been computed, its value is available in register A; however, this does no good, since the *first* operand for a $-$ operation must be in the register. It is necessary to store the value of i_2 in another temporary variable, T2, and then load the value of i_1 from T1 into register A before performing the subtraction.

With a little analysis, an optimizing compiler could recognize this situation and rearrange the quadruples so the second operand of the subtraction is computed first. This rearrangement is illustrated in Fig. 5.18(b). The first two quadruples in the sequence have been interchanged. The resulting machine code requires two fewer instructions and uses only one temporary variable instead of two. The same technique can be applied to rearrange quadruples that calculate the operands of a DIV operation or any other operation for which the machine code requires a particular placement of operands.

Other possibilities for machine-dependent code optimization in-

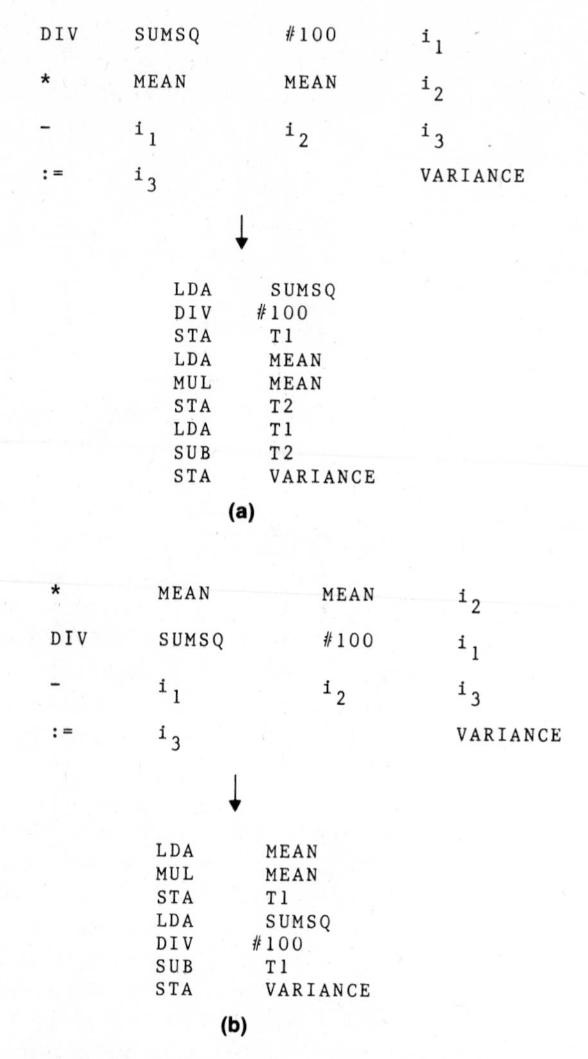

```
DIV     SUMSQ     #100     i1
*       MEAN      MEAN     i2
-       i1        i2       i3
:=      i3                 VARIANCE

            ↓

        LDA     SUMSQ
        DIV     #100
        STA     T1
        LDA     MEAN
        MUL     MEAN
        STA     T2
        LDA     T1
        SUB     T2
        STA     VARIANCE
```

(a)

```
*       MEAN      MEAN     i2
DIV     SUMSQ     #100     i1
-       i1        i2       i3
:=      i3                 VARIANCE

            ↓

        LDA     MEAN
        MUL     MEAN
        STA     T1
        LDA     SUMSQ
        DIV     #100
        SUB     T1
        STA     VARIANCE
```

(b)

FIGURE 5.18 Rearrangement of quadruples for code optimization.

volve taking advantage of specific characteristics and instructions of the target machine. For example, there may be special loop-control instructions or addressing modes that can be used to create more efficient object code. On computers like VAX, there are high-level machine instructions that can perform complicated functions such as calling procedures and manipulating data structures in a single operation. Obviously the use of such features, when possible, can greatly improve the efficiency of the object program.

Some machines, such as certain CYBER models, have a CPU that is made up of several functional units. On such computers, the order in which machine instructions appear can affect the speed of execution. Consecutive instructions that involve different functional units can sometimes be executed at the same time. An optimizing compiler for such a machine could rearrange object code instructions to take advantage of this property. For examples and references, see Gries (1971).

5.3 MACHINE-INDEPENDENT COMPILER FEATURES

In this section we briefly describe some common compiler features that are largely independent of the particular machine being used. As in the preceding section, we do not attempt to give full details of the implementation of these features. Such details may be found in the references cited.

In the compiler design described in Section 5.1, we dealt only with simple variables that were permanently assigned to storage locations within the object program. Section 5.3.1 describes some alternative ways of performing storage allocation for the compiled program. Section 5.3.2 describes methods for handling structured variables such as arrays.

Section 5.3.3 continues the discussion of code optimization begun in Section 5.2.2. This time, we are concerned with machine-independent techniques for optimizing the object code. Section 5.3.4 discusses the problems involved in compiling a block-structured language and indicates some possible solutions for these problems.

5.3.1 Storage Allocation

In the compilation scheme presented in Section 5.1, all programmer-defined variables were assigned storage locations within the object program as their declarations were processed. Temporary variables, including the one used to save the return address, were also assigned fixed addresses within the program. This simple type of storage assignment is usually called *static* allocation. It is often used for languages, such as FORTRAN, that do not allow the recursive use of procedures or subroutines, and do not provide for the dynamic allocation of storage during execution.

If procedures may be called recursively, as in Pascal, static allocation cannot be used. Consider, for example, Fig. 5.19. In Fig. 5.19(a), the program MAIN has been called by the operating system or the loader (invocation 1). The first action taken by MAIN is to store its return address from register L at a fixed location RETADR within MAIN. In Fig. 5.19(b), MAIN has called the procedure SUB (invoca-

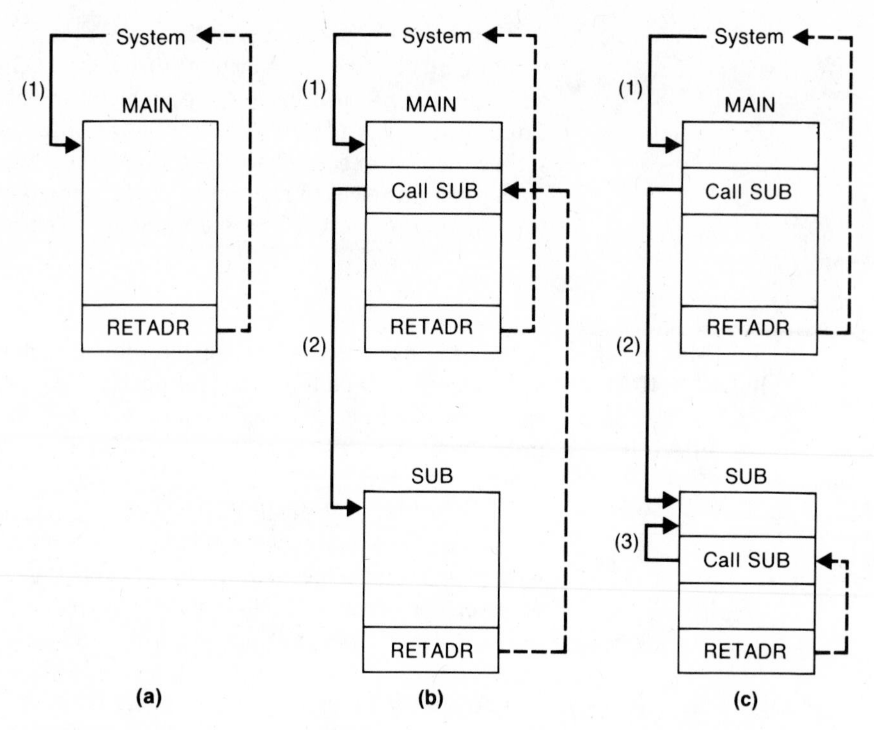

FIGURE 5.19 Recursive invocation of a procedure using static storage allocation.

tion 2). The return address for this call has been stored at a fixed location within SUB. If SUB now calls itself recursively, as in Fig. 5.19(c), a problem occurs. SUB stores the return address for invocation 3 into RETADR from register L. This destroys the return address for invocation 2. As a result, there is no possibility of ever making a correct return to MAIN.

A similar difficulty occurs with respect to any variables used by SUB. When the recursive call is made, variables within SUB may be set to new values; this destroys the previous contents of these variables. However, these previous values may be needed by invocation 2 of SUB after the return from the recursive call. This is the same problem mentioned in Section 4.3.1 in our discussion of recursive macro expansion.

Obviously it is necessary to preserve the previous values of any variables used by SUB, including parameters, temporaries, return addresses, register save areas, etc., when the recursive call is made. This is usually accomplished with a dynamic storage allocation technique. Each procedure call creates an *activation record* that contains storage for all the variables used by the procedure. If the procedure is called recursively, another activation record is created. Each activation record is associated with a particular *invocation* of the procedure, not with the procedure itself. An activation record is not deleted until a return has been made from the corresponding invocation. The starting address for the current activation record is usually contained in a base register, which is used by the procedure to address its variables. In this way, the values of variables used by different invocations of a procedure are kept separate from one another.

Activation records are typically allocated on a stack, with the current record at the top of the stack. This process is illustrated in Fig. 5.20. In Fig. 5.20(a), which corresponds to Fig. 5.19(a), the procedure MAIN has been called; its activation record appears on the stack. The base register B has been set to indicate the starting address of this current activation record. The first word in an activation record would normally contain a pointer PREV to the previous record on the stack.

FIGURE 5.20 Recursive invocation of a procedure using automatic storage allocation.

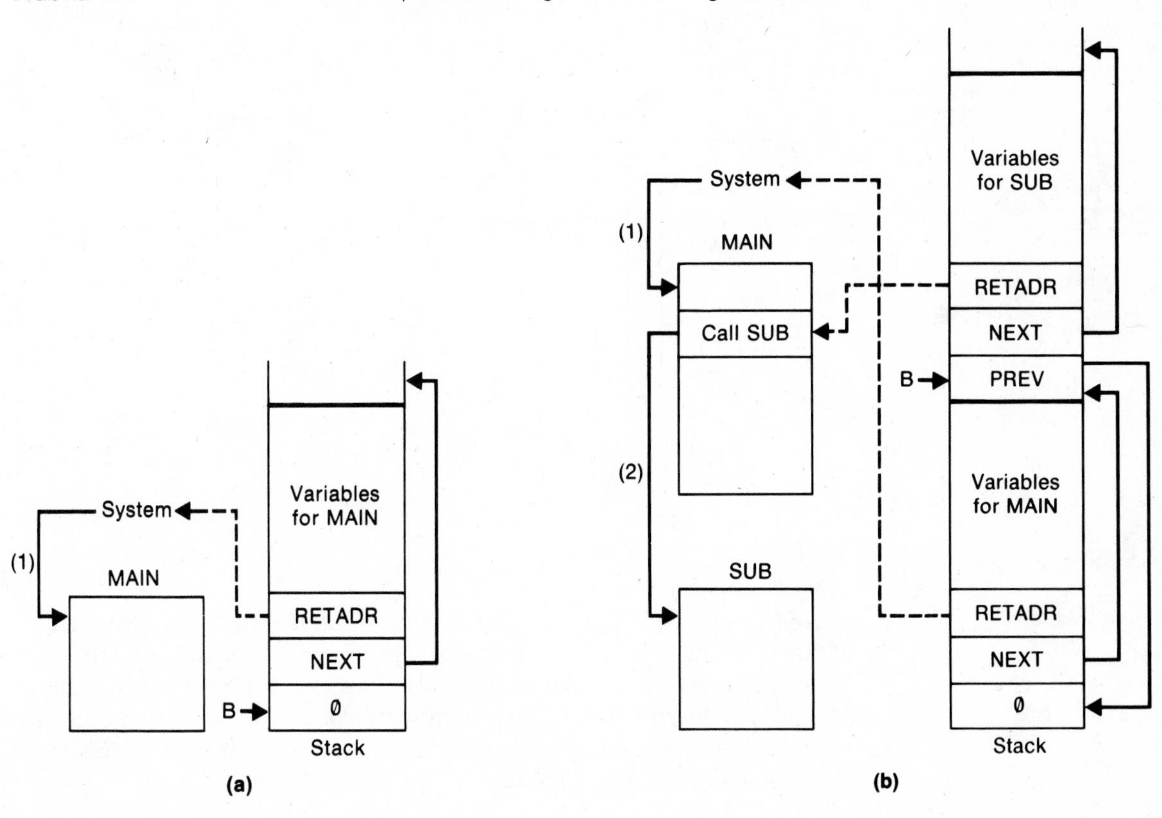

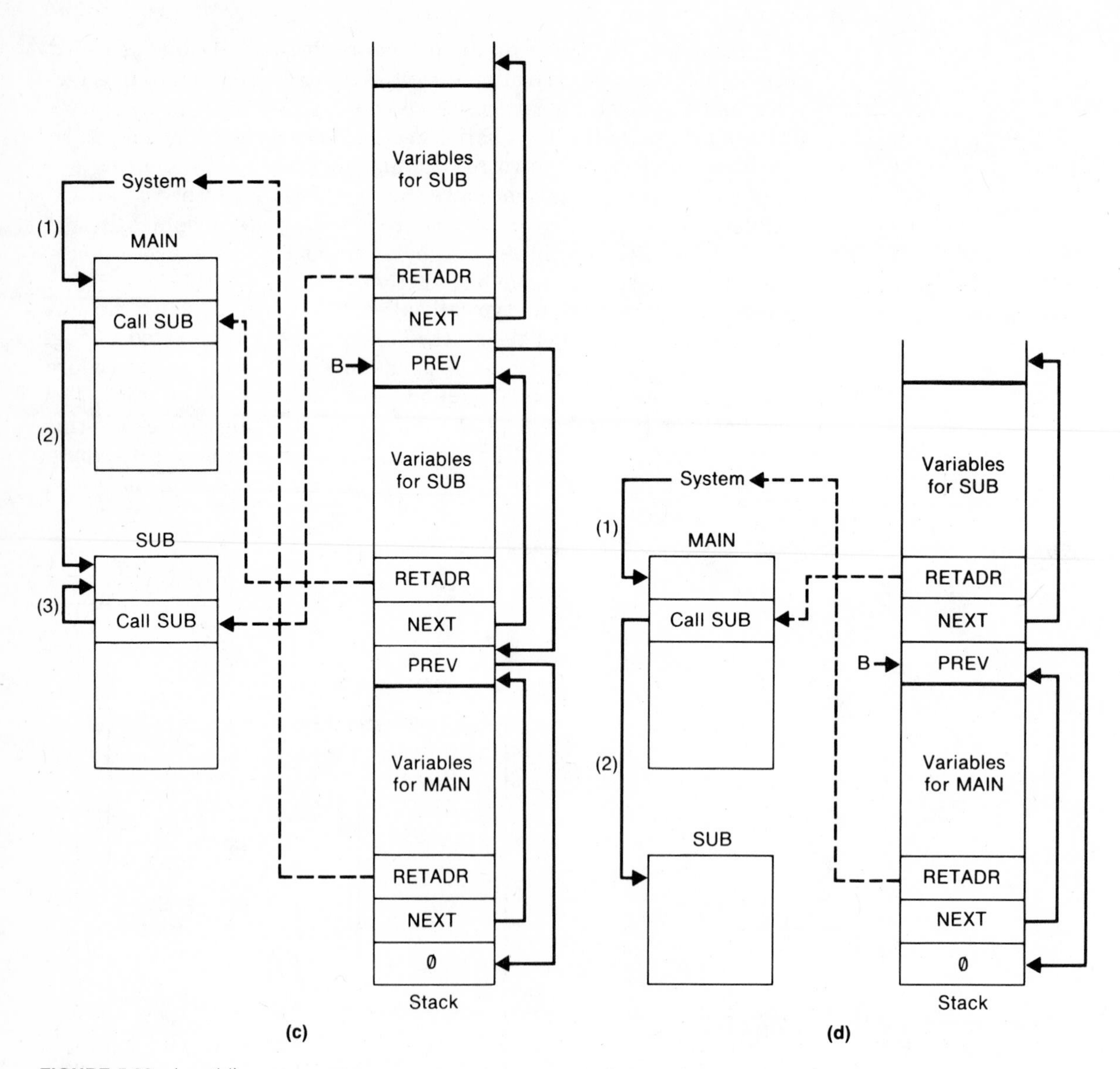

FIGURE 5.20 *(cont'd)*

Since this record is the first, the pointer value is null. The second word of the activation record contains a pointer NEXT to the first unused word of the stack, which will be the starting address for the next activation record created. The third word contains the return address for this invocation of the procedure, and the remaining words contain the values of variables used by the procedure.

In Fig. 5.20(b), MAIN has called the procedure SUB. On the top of the stack, a new activation record has been created with register B set to indicate this new current record. The pointers PREV and NEXT in the two records have been set as shown. In Fig. 5.20(c), SUB has called itself recursively, and another activation record has been created for this current invocation of SUB. Note that the return addresses and variable values for the two invocations of SUB are kept separate by this process.

When a procedure returns to its caller, the current activation record (which corresponds to this invocation) is deleted. The pointer PREV in the deleted record is used to reestablish the previous activation record as the current one and execution continues. Figure 5.20(d) shows the stack as it would appear after SUB returns from the recursive call. Register B has been reset to point to the activation record for the previous invocation of SUB. The return address and all the variable values in this activation record are exactly the same as they were before the recursive call.

This technique is often referred to as *automatic* allocation of storage to distinguish it from other types of dynamic allocation that are under the control of the programmer. When automatic allocation is used, the compiler must generate code for references to variables using some sort of relative addressing. In our example the compiler assigns to each variable an address that is relative to the beginning of the activation record, instead of an actual location within the object program. The address of the current activation record is, by convention, contained in register B, so a reference to a variable is translated as an instruction that uses base relative addressing. The displacement in this instruction is the relative address of the variable within the activation record.

The compiler must also generate additional code to manage the activation records themselves. At the beginning of each procedure there must be code to create a new activation record, linking it to the previous one and setting the appropriate pointers as illustrated in Fig. 5.20. This code is often called a *prologue* for the procedure. At the end of the procedure, there must be code to delete the current activation record, resetting pointers as needed. This code is often called an *epilogue*.

When automatic allocation is used, storage is assigned to all variables used by a procedure when the procedure is called. Other types of dynamic storage allocation allow the programmer to specify when storage is to be assigned. In PL/I, for example, the statement

```
ALLOCATE(A)
```

in the source program is translated into code that allocates storage for the variable A. The statement

```
FREE(A)
```

releases the storage assigned to A by a previous ALLOCATE. This feature is called *controlled* storage allocation in PL/I. When a variable is created by ALLOCATE, it automatically becomes available for use by the program. The previous allocation of the variable, if any, may be saved on a stack and restored after the current allocation is deleted.

Another type of dynamic storage allocation is found in Pascal. The statement

```
NEW(P)
```

allocates storage for a variable and sets the pointer P to indicate the variable just created. The type of the variable created is specified by the way P is declared in the program. The program refers to the created variable by using the pointer P. The statement

```
DISPOSE(P)
```

releases the storage that was previously assigned to the variable pointed to by P. A similar feature is also available in PL/I, where it is called *based* storage allocation.

A variable that is dynamically allocated using NEW does not occupy a fixed location in an activation record, so it cannot be referenced directly using base relative addressing. Such a variable is usually accessed using indirect addressing through the pointer variable P. Since P does occupy a fixed location in the activation record, it can be addressed in the usual way. A similar technique can be used for controlled storage in PL/I. The activation record contains a pointer that is used to indicate the most recent allocation of the controlled variable.

In the preceding discussions we have not described the mechanism by which storage is allocated for a variable. One approach is to let the operating system handle all storage management. A NEW or ALLOCATE statement would be translated into a request to the operating system for an area of storage of the required size. Another method is to handle the required allocation through a library procedure associated with the compiler. With this method, a large block of free storage would be made available at the beginning of the program and would be managed by the library procedure, using some standard allocation strategy. For a discussion and evaluation of such memory management techniques, see Standish (1980).

Dynamic storage allocation, as discussed in this section, provides another example of *delayed binding*. The association of an address with a variable is made when the procedure is executed, not when it is compiled or loaded. This delayed binding allows more flexibility in the use of variables and procedures. However, it also requires more overhead because of the creation of activation records and the use of indirect addressing. (Similar observations were made at the end of Section 3.4.2 with respect to dynamic linking.)

5.3.2 Structured Variables

In this section we briefly consider the compilation of programs that use *structured variables* such as arrays, records, strings, and sets. We are primarily concerned with the allocation of storage for such variables and with the generation of code to reference them. These issues are discussed in a moderate amount of detail for arrays. The same principles can also be applied to the other types of structured variables. Further details concerning these topics can be found in a number of textbooks on compilers, such as Aho (1977) and Gries (1971).

Consider first the Pascal array declaration

```
A : ARRAY[1..10] OF INTEGER
```

If each INTEGER variable occupies one word of memory, then we must clearly allocate 10 words to store this array. More generally, if an array is declared as

```
ARRAY[l..u] OF INTEGER
```

then we must allocate $u-l+1$ words of storage for the array.

Allocation for a multi-dimensional array is not much more difficult. Consider, for example, the two-dimensional array

```
B : ARRAY[0..3,1..6] OF INTEGER
```

Here the first subscript can take on 4 different values (0–3), and the second subscript can take on 6 different values. We need to allocate a total of $4 * 6 = 24$ words to store the array. In general, if the array declaration is

```
ARRAY[l1..u1,l2..u2] OF INTEGER
```

then the number of words to be allocated is given by

$$(u_1 - l_1 + 1) * (u_2 - l_2 + 1)$$

For an array with n dimensions, the number of words required is a product of n such terms.

When we consider the generation of code for array references, it becomes important to know which array element corresponds to each word of allocated storage. For one-dimensional arrays, there is an obvious correspondence. In the array A previously defined, the first word would contain A[1], the second word would contain A[2], and so on. For higher-dimensional arrays, however, the choice of representation is not as clear.

Figure 5.21 shows two possible ways of storing the array B previously defined. In Fig. 5.21(a), all array elements that have the same value of the first subscript are stored in contiguous locations; this is called *row-major* order. In Fig. 5.21(b), all elements that have the same value of the second subscript are stored together; this is called *column-major* order. Another way of looking at this is to scan the words of the array in sequence and observe the subscript values. In row-major order, the rightmost subscript varies most rapidly; in column-major order, the leftmost subscript varies most rapidly. These concepts can be generalized easily to arrays with more than two subscripts.

Compilers for most high-level languages store arrays using row-major order; this is the order we assume in the following discussions. For historical reasons, however, most FORTRAN compilers store arrays in column-major order.

To refer to an array element, we must calculate the address of the referenced element relative to the base address of the array. For a typical computer, the compiler would generate code to place this relative address in an index register. Indexed addressing would then be used to access the desired array element. In the following discussion we assume the base address is the address of the first word allocated to store the array. For a discussion of other possibilities, see Aho (1977).

Consider first the one-dimensional array

```
A : ARRAY[1..10] OF INTEGER
```

and suppose that a statement refers to array element A[6]. There are five array elements preceding A[6]; on a SIC machine, each such ele-

FIGURE 5.21 Storage of B : ARRAY[0..3,1..6] in (a) row-major order and (b) column-major order.

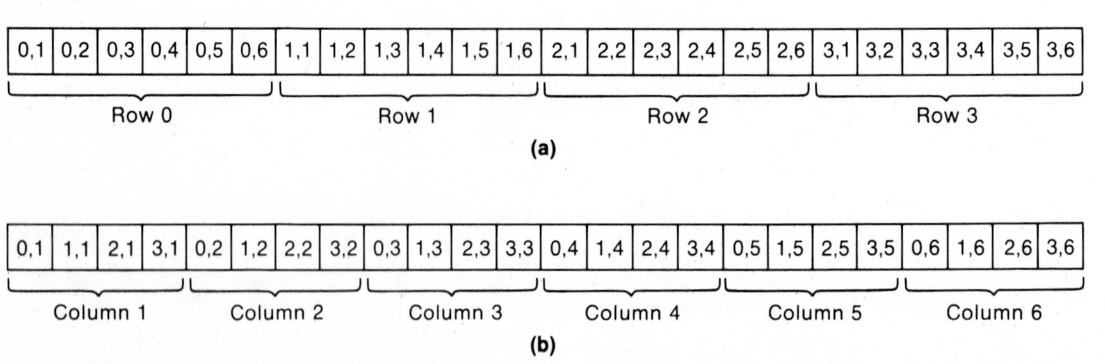

ment would occupy 3 bytes. Thus the address of A[6] relative to the starting address of the array is given by 5 * 3=15.

If an array reference involves only constant subscripts, the relative address calculation can be performed during compilation. If the subscripts involve variables, however, the compiler must generate object code to perform this calculation during execution. Suppose the array declaration is

```
A : ARRAY[l..u] OF INTEGER
```

and each array element occupies w bytes of storage. If the value of the subscript is s, then the relative address of the referenced array element A[s] is given by

$$w * (s - l)$$

The generation of code to perform such a calculation is illustrated in Fig. 5.22(a). The notation A[i_2] in quadruple 3 specifies that the generated machine code should refer to A using indexed addressing, after having placed the value of i_2 in the index register.

For multi-dimensional arrays, the generation of code depends on whether row-major or column-major order is used to store the array. We assume row-major order. Figure 5.21(a) illustrates the storage of the array

```
B : ARRAY[0..3,1..6] OF INTEGER
```

in row-major order. Consider first the array element B[2,5]. If we start at the beginning of the array, we must skip over two complete rows (row 0 and row 1) before arriving at the beginning of row 2 (i.e., element B[2,1]). Each such row contains 6 elements, so this involves 2 * 6=12 array elements. We must also skip over the first 4 elements in row 2 to arrive at B[2,5]. This makes a total of 16 array elements between the beginning of the array and element B[2,5]. If each element occupies 3 bytes, then B[2,5] is located at relative address 48 within the array.

More generally, suppose the array declaration is

```
B : ARRAY[l1..u1,l2..u2] OF INTEGER
```

and we wish to refer to an array element specified by subscripts having values s_1 and s_2. The relative address of B[s_1,s_2] is given by

$$w * [(s_1 - l_1) * (u_2 - l_2 + 1) + (s_2 - l_2)].$$

Figure 5.22(b) illustrates the generation of code to perform such an array reference. You should examine this set of quadruples carefully to be sure you understand the calculations involved.

The methods and formulas discussed above can easily be generalized to higher-dimensional arrays. For details, see Aho (1977).

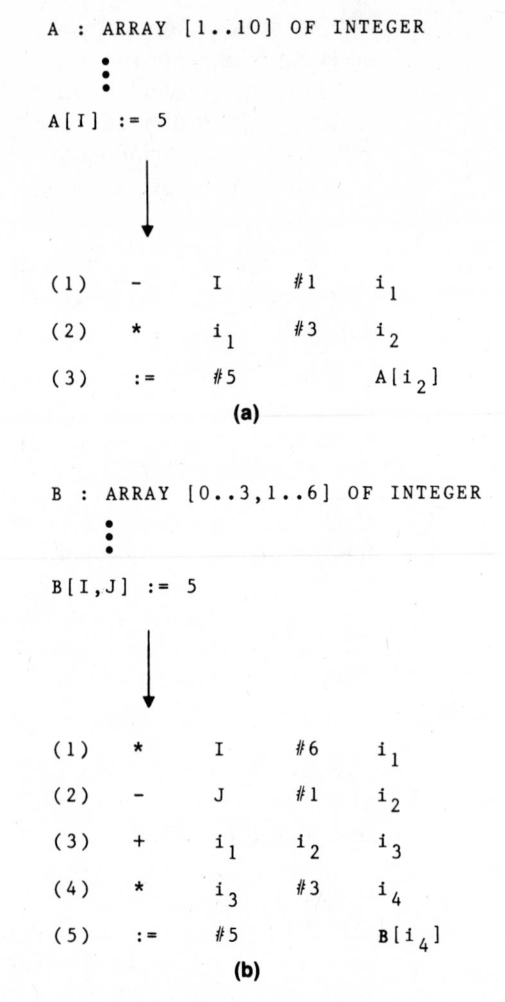

FIGURE 5.22 Code generation for array references.

The symbol-table entry for an array usually specifies the type of the elements in the array, the number of dimensions declared, and the lower and upper limit for each subscript. This information is sufficient for the compiler to generate the code required for array references. In some languages, however, the required information is not known at compilation time. In PL/I, for example, a two-dimensional array could be declared as

```
T (I:J-1,I+1:J)
```

This specifies that the range of the first subscript is from I to J−1, and the range of the second is from I+1 to J. I and J are variable names defined within the program. Obviously, storage must be assigned to such an array using some form of dynamic allocation. In PL/I, this might be accomplished by

```
ALLOCATE T
```

where the variables I and J have previously been assigned values.

Since the values of I and J are not known at compilation time, the compiler cannot directly generate code like that in Fig. 5.22. Instead, the compiler creates a descriptor for the array, called a *dope vector*. This dope vector provides space for storing the lower and upper bounds for each array subscript. When storage is allocated for the array, the values of these bounds are computed and stored in the dope vector. The generated code for an array reference uses the values from the dope vector to calculate relative addresses as required. The dope vector may also include the number of dimensions for the array, the type of the array elements, and a pointer to the beginning of the array. This information can be useful if the allocated array is passed as a parameter to another procedure.

The issues discussed for arrays also arise in the compilation of other structured variables such as records, strings, and sets. The compiler must provide for the allocation of storage for the variable; it must store information concerning the structure of the variable, and use this information to generate code to access components of the structure; and it must construct a dope vector for situations in which the required information is not known at compilation time. For further discussion of these issues as they relate to specific types of structured variables, see Aho (1977) and Gries (1971).

5.3.3 Machine-Independent Code Optimization

In this section we discuss some of the most important types of machine-independent code optimization. As in the previous sections, we do not attempt to give the full details of any of these techniques. Instead, we give an intuitive verbal description and illustrate the main concepts with examples. Algorithms and further details concerning these methods can be found in Aho (1977) and Gries (1971). We assume the source program has already been translated into a sequence of quadruples like those introduced in Section 5.2.1.

One important source of code optimization is the elimination of *common subexpressions,* which are subexpressions that appear at more than one point in the program and that compute the same value. Consider, for example, the statements in Fig. 5.23(a). The term 2 * J is a common subexpression. An optimizing compiler should generate code so that this multiplication is performed only once and the result is used in both places.

```
X,Y : ARRAY[1..10,1..10] OF INTEGER
      .
      .
      .
FOR I := 1 TO 10 DO
   X[I,2*J-1] := Y[I,2*J]
```

(a)

```
 (1)  :=    #1             I        {loop initialization}
 (2)  JGT   I      #10     (20)
 (3)  -     I      #1      i1       {subscript calculation for X}
 (4)  *     i1     #10     i2
 (5)  *     #2     J       i3
 (6)  -     i3     #1      i4
 (7)  -     i4     #1      i5
 (8)  +     i2     i5      i6
 (9)  *     i6     #3      i7
(10)  -     I      #1      i8       {subscript calculation for Y}
(11)  *     i8     #10     i9
(12)  *     #2     J       i10
(13)  -     i10    #1      i11
(14)  +     i9     i11     i12
(15)  *     i12    #3      i13
(16)  :=    Y[i13]         X[i7]    {assignment operation}
(17)  +     #1     I       i14      {end of loop}
(18)  :=    i14            I
(19)  J                    (2)
(20)                                {next statement}
```

(b)

FIGURE 5.23 Code optimization by elimination of common subexpressions and removal of loop invariants.

(1)	:=	#1		I	{loop initialization}
(2)	JGT	I	#10	(16)	
(3)	-	I	#1	i_1	{subscript calculation for X}
(4)	*	i_1	#10	i_2	
(5)	*	#2	J	i_3	
(6)	-	i_3	#1	i_4	
(7)	-	i_4	#1	i_5	
(8)	+	i_2	i_5	i_6	
(9)	*	i_6	#3	i_7	
(10)	+	i_2	i_4	i_{12}	{subscript calculation for Y}
(11)	*	i_{12}	#3	i_{13}	
(12)	:=	Y[i_{13}]		X[i_7]	{assignment operation}
(13)	+	#1	I	i_{14}	{end of loop}
(14)	:=	i_{14}		I	
(15)	J			(2)	
(16)					{next statement}

(c)

FIGURE 5.23 *(cont'd)*

Common subexpressions are usually detected through the analysis of an intermediate form of the program. Such an intermediate form is shown in Fig. 5.23(b). If we examine this sequence of quadruples, we see that quadruples 5 and 12 are the same except for the name of the intermediate result produced. Note that the operand J is not changed in value between quadruples 5 and 12. It is not possible to reach quadruple 12 without passing through quadruple 5 first because the quadruples are part of the same basic block. Therefore, quadruples 5 and 12 compute the same value. This means we can delete quadruple 12 and replace any reference to its result (i_{10}) with a reference to i_3, the result of quadruple 5. This modification eliminates the duplicate calculation of 2 $*$ J, which we identified previously as a common subexpression in the source statement.

After the substitution of i_3 for i_{10} is performed, quadruples 6 and 13 are the same, except for the name of the result. Thus we can remove quadruple 13 and substitute i_4 for i_{11} wherever i_{11} is used. Similarly,

(1)	*	#2	J	i_3	{computation of invariants}
(2)	-	i_3	#1	i_4	
(3)	-	i_4	#1	i_5	
(4)	:=	#1		I	{loop initialization}
(5)	JGT	I	#10	(16)	
(6)	-	I	#1	i_1	{subscript calculation for X}
(7)	*	i_1	#10	i_2	
(8)	+	i_2	i_5	i_6	
(9)	*	i_6	#3	i_7	
(10)	+	i_2	i_4	i_{12}	{subscript calculation for Y}
(11)	*	i_{12}	#3	i_{13}	
(12)	:=	Y[i_{13}]		X[i_7]	{assignment operation}
(13)	+	#1	I	i_{14}	{end of loop}
(14)	:=	i_{14}		I	
(15)	J			(5)	
(16)					{next statement}

(d)

FIGURE 5.23 *(cont'd)*

quadruples 10 and 11 can be removed because they are equivalent to quadruples 3 and 4.

The result of applying this technique is shown in Fig. 5.23(c). The quadruples have been renumbered in this figure. However, the intermediate result names i_j have been left unchanged, except for the substitutions just described, to make the comparison with Fig. 5.23(b) easier. Note that the total number of quadruples has been reduced from 19 to 15. Each of the quadruple operations used here will probably take approximately the same length of time to execute on a typical machine, so there should be a corresponding reduction in the overall execution time of the program.

Another common source of code optimization is the removal of *loop invariants*. These are subexpressions within a loop whose values do not change from one iteration of the loop to the next. Thus their values can be computed once, before the loop is entered, rather than be recalculated for each iteration. Because most programs spend a large majority of their running time in the execution of loops, the time savings from this sort of optimization can be highly significant. We assume the existence of algorithms that can detect loops by analyzing the control flow of the program. One example of such an algorithm is the method for constructing a program flow graph that is described in Section 5.2.2.

An example of a loop-invariant computation is the term 2 * J in Fig. 5.23(a) (see quadruple 5 of Fig. 5.23c). The result of this computation depends only on the operand J, which does not change in value during the execution of the loop. Thus we can move quadruple 5 in Fig. 5.23(c) to a point immediately before the loop is entered. A similar argument can be applied to quadruples 6 and 7.

Figure 5.23(d) shows the sequence of quadruples that results from these modifications. The total number of quadruples remains the same as in Fig. 5.23(c); however, the number of quadruples within the body of the loop has been reduced from 14 to 11. Each execution of the FOR statement in Fig. 5.23(a) causes 10 iterations of the loop, which means the total number of quadruple operations required for one execution of the FOR is reduced from 141 to 114.

Our modifications have reduced the total number of quadruple operations for one execution of the FOR from 181 (Fig. 5.23b) to 114 (Fig. 5.23d), which saves a substantial amount of time. There are methods for handling common subexpressions and loop invariants that are considerably more sophisticated than the techniques we used here. As might be expected, such methods can produce more highly optimized code. For examples and discussions of these, see Aho (1977).

Some optimization, of course, can be obtained by rewriting the source program. For example, the statements in Fig. 5.23(a) could have been written as

```
T1 := 2 * J;
T2 := T1 - 1;
FOR I := 1 TO 10 DO
   X[I,T2] := Y[I,T1];
```

However, this would achieve only a part of the benefits realized by the optimization process just described. The rest of the optimizations are related to the process of calculating a relative address from subscript values; these details are inaccessible to the source programmer. For example, the optimizations involving quadruples 3, 4, 10, and 11 in Fig. 5.23(b) could not be achieved with any rewriting of the source statement. It could also be argued that the original statements in Fig. 5.23(a) are preferable because they are clearer than the modified version involving T1 and T2. An optimizing compiler should allow the programmer to write source code that is clear and easy to read, and it should compile such a program into machine code that is efficient to execute.

Another source of code optimization is the substitution of a more efficient operation for a less efficient one. Consider, for example, the FORTRAN program segment in Fig. 5.24(a). This DO loop creates a table that contains the first 20 powers of 2. In each iteration of the loop, the constant 2 is raised to the power I. Figure 5.24(b) shows a represen-

FIGURE 5.24 Code optimization by reduction in strength of operations.

```
   DO 10 I = 1,20
10 TABLE(I) = 2**I
```

(a)

```
(1)  :=    #1            I           {loop initialization}
(2)  EXP   #2     I      i1          {calculation of 2**I}
(3)  -     I      #1     i2          {subscript calculation}
(4)  *     i2     #3     i3
(5)  :=    i1            TABLE[i3]   {assignment operation}
(6)  +     I      #1     i4          {end of loop}
(7)  :=    i4            I
(8)  JLE   I      #20    (2)
```

(b)

```
(1)  :=    #1            i1          {initialization of temporaries}
(2)  :=    #(-3)         i3
(3)  :=    #1            I           {loop initialization}
(4)  *     i1     #2     i1          {calculation of 2**I}
(5)  +     i3     #3     i3          {subscript calculation}
(6)  :=    i1            TABLE[i3]   {assignment operation}
(7)  +     I      #1     i4          {end of loop}
(8)  :=    i4            I
(9)  JLE   I      #20    (4)
```

(c)

tation of these statements as a series of quadruples. Exponentiation is represented with the operation EXP. At the machine-code level, EXP would involve either a loop that performs a series of multiplications, or a call to a subroutine that uses logarithms to arrive at the result.

On closer examination, we can see that there is a more efficient way to perform the computation. For each iteration of the loop, the value of I increases by 1. Therefore, the value of 2 ** I for the current iteration can be found by multiplying the value for the previous iteration by 2. Clearly this method of computing 2 ** I is much more efficient than performing a series of multiplications or using a logarithmic technique. Such a transformation is called *reduction in strength* of an operation.

A similar transformation can be applied in the calculation of the relative address for the array element TABLE(I). Assuming that each array element occupies one word, a SIC object program would calculate this displacement as 3 * (I − 1). This calculation appears in quadruples 3 and 4 of Fig. 5.24(b). Thus the object program would need to perform one multiplication for each array reference.

The same sort of reduction in strength can be applied to this situation. Each iteration of the loop refers to the next array element in sequence, so the calculation of the required displacement can be accomplished by adding 3 to the previous displacement. If addition is faster than multiplication on the target machine, this reduction in strength will result in more efficient object code.

Figure 5.24(c) shows a modification of the quadruples from Fig. 5.24(b) that implements these two reductions in strength. An algorithm for performing this sort of transformation can be found in Gries (1971). As in our previous examples, a part of this optimization could have been performed at the source-code level. However, the strength reduction in the subscript calculation process could not be accomplished by the programmer, who has no access to the details of code generation for array references.

There are a number of other possibilities for machine-independent code optimization. For example, computations whose operand values are known at compilation time can be performed by the compiler. This optimization is known as *folding*. Other optimizations include converting a loop into straight-line code (*loop unrolling*) and merging of the bodies of loops (*loop jamming*). For details on these and other optimization techniques, see Gries (1971) and Aho (1977).

5.3.4 Block-Structured Languages

In some languages, such as ALGOL, a program can be divided into units called *blocks*. A block is a portion of a program that has the ability to declare its own identifiers. The definition of a block is also met by units such as procedures and functions in Pascal. In this section we consider some of the issues involved in compiling and executing programs written in such block-structured languages.

Figure 5.25(a) shows the outline of a block-structured program in a Pascal-like language. Each procedure corresponds to a block. In the following discussion, therefore, we use the terms *procedure* and *block* interchangeably. Note that blocks may be nested within other blocks. For example, procedures B and D are nested within procedure A, and procedure C is nested within procedure B. Each block may contain a declaration of variables, as shown. A block may also refer to variables that are defined in any block that contains it, provided the same names are not redefined in the inner block.

Consider, for example, the INTEGER variables X, Y, and Z that are declared in procedure A on line 2. Procedure B contains declarations of X and Y as REAL variables on line 4. Within procedure B, a use of the name X refers to the REAL variable declared within B. However, a use of the name Z refers to the INTEGER variable declared by A because the name Z is not redefined within B. Similarly, within procedure C the name W refers to the variable declared by C; the names X and Y refer to the variables declared by B; and the name Z refers to the variable declared by A. Variables cannot be used outside the block in which they are declared. For example, the name W cannot be referred to outside of procedure B, and V cannot be referred to outside of procedure C.

In compiling a program written in a block-structured language, it is convenient to number the blocks as shown in Fig. 5.25(a). As the beginning of each new block is recognized, it is assigned the next block number in sequence. The compiler can then construct a table that describes the block structure, as illustrated in Fig. 5.25(b). The block-level entry gives the nesting depth for each block. The outermost block has a level number of 1, and each other block has a level number that is one greater than that of the surrounding block.

Since a name can be declared more than once in a program (by different blocks), each symbol-table entry for an identifier must contain the number of the declaring block. A declaration of an identifier is legal if there has been no previous declaration of that identifier by the current block, so there can be several symbol-table entries for the same name. The entries that represent declarations of the same name by different blocks can be linked together in the symbol table with a chain of pointers.

When a reference to an identifier appears in the source program, the compiler must first check the symbol table for a definition of that

```
1     PROCEDURE A;
2        VAR  X,Y,Z : INTEGER;
              .
              .
3           PROCEDURE B;
4              VAR  W,X,Y : REAL;
                 .
                 .
5                 PROCEDURE C;
6                    VAR  V,W : INTEGER;
                        .
                        .
                        .
7                 END {C};
                 .
                 .
8           END {B};
              .
              .
9           PROCEDURE D;
10             VAR  X,Z : CHAR;
                 .
                 .
                 .
11          END {D};
              .
              .
12    END  { A };
```

(Brackets at right mark blocks: 1 = A, lines 1–12; 2 = B, lines 3–8; 3 = C, lines 5–7; 4 = D, lines 9–11.)

(a)

Block name	Block number	Block level	Surrounding block
A	1	1	—
B	2	2	1
C	3	3	2
D	4	2	1

(b)

FIGURE 5.25 Nesting of blocks in a source program.

identifier by the current block. If no such definition is found, the compiler looks for a definition by the block that surrounds the current one, then by the block that surrounds that, and so on. If the outermost block is reached without finding a definition of the identifier, then the reference is an error.

The search process just described can easily be implemented within a symbol table that uses hashed addressing. The hashing function is used to locate one definition of the identifier. The chain of definitions for that identifier is then searched for the appropriate entry. There are other symbol-table organizations that store the definitions of identifiers according to the nesting of the blocks that define them. This kind of structure can make the search for the proper definition more efficient. See, for example, Aho (1977).

Most block-structured languages make use of automatic storage allocation, as described in Section 5.3.1. That is, the variables that are defined by a block are stored in an activation record that is created each time the block is entered. If a statement refers to a variable that is declared within the current block, this variable is present in the current activation record, so it can be accessed in the usual way. However, it is possible for a statement to refer to a variable that is declared in some surrounding block. In that case, the most recent activation record for that block must be located to access the variable.

One common method for providing access to variables in surrounding blocks uses a data structure called a *display*. The display contains pointers to the most recent activation records for the current block and for all blocks that surround the current one in the source program. When a block refers to a variable that is declared in some surrounding block, the generated object code uses the display to find the activation record that contains this variable.

The use of a display is illustrated in Fig. 5.26. We assume that procedure A has been invoked by the system, A has then called procedure B, and B has called procedure C. The resulting situation is shown in Fig. 5.26(a). The stack contains activation records for the invocations of A, B, and C. The display contains pointers to the activation records for C and for the surrounding blocks (A and B).

Now let us assume procedure C calls itself recursively. Another activation record for C is created on the stack as a result of this call. Any reference to a variable declared by C should use this most recent activation record; the display pointer for C is changed accordingly. Variables that correspond to the previous invocation of C are not accessible for the moment, so there is no display pointer to this activation record. This situation is illustrated in Fig. 5.26(b).

Suppose now that procedure C calls D. (This is allowed because the

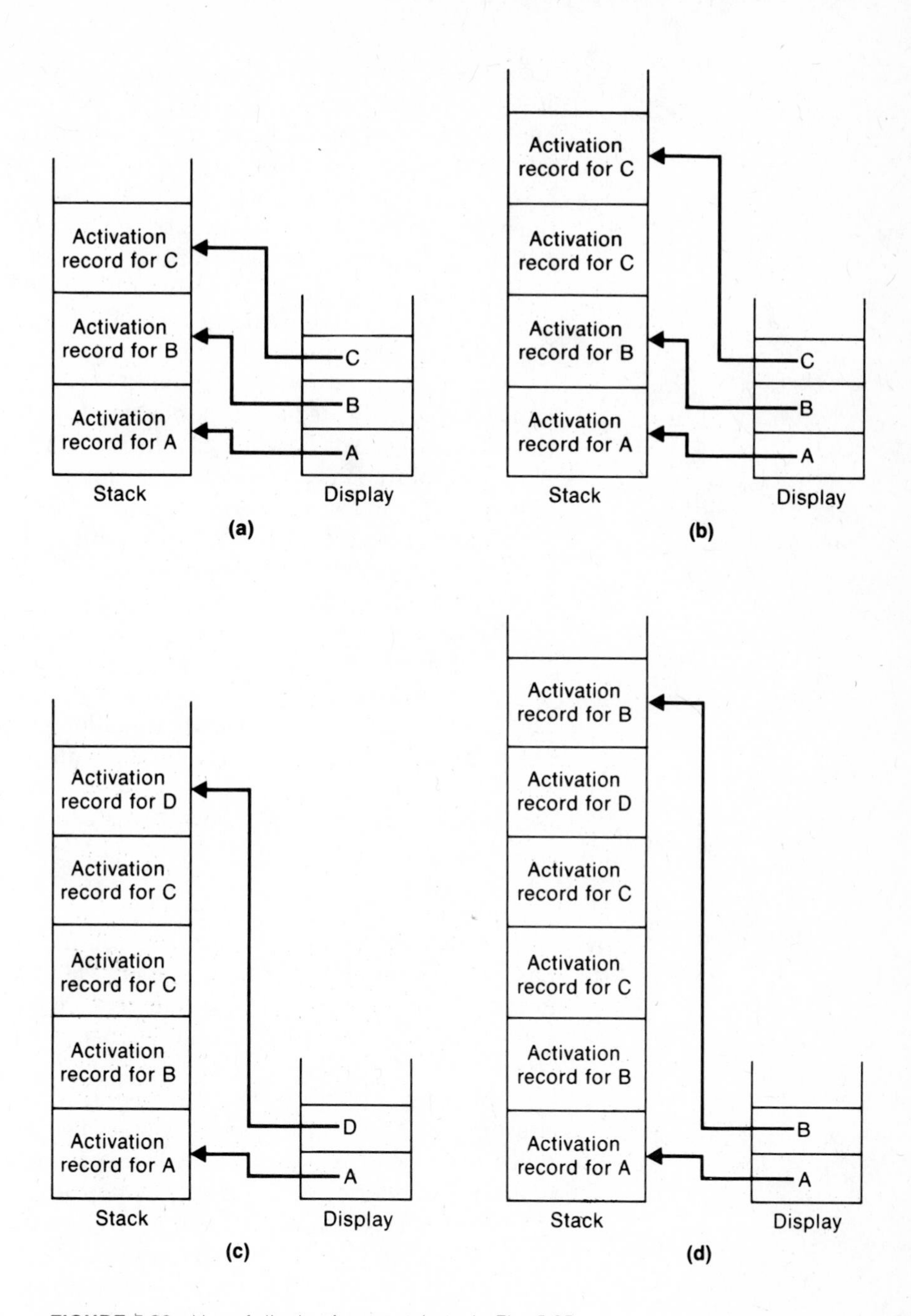

FIGURE 5.26 Use of display for procedures in Fig. 5.25.

identifier D is defined in procedure A, which contains C.) The resulting stack and display are shown in Fig. 5.26(c). An activation record for D has been created in the usual way and added to the stack. Note, however, that the display now contains only two pointers: one each to the activation records for D and A. This is because procedure D cannot refer to variables in B or C, except through parameters that are passed to it, even though it was called from C. According to the rules for the scope of names in a block-structured language, procedure D can refer only to variables that are declared by D or by some block that contains D in the source program (in this case, procedure A).

A similar situation, illustrated in Fig. 5.26(d), occurs if procedure D now calls B. Procedure B is allowed to refer only to variables declared by either B or A, which is reflected in the contents of the display. After procedure B returns to D, the contents of the stack and display will again appear as they were in Fig. 5.26(c).

It is important to be aware of the difference between the runtime allocation of variables, as represented by the stack of activation records, and the rules for referring to variables in the block-structured program, as represented by the display. You should carefully examine Figs. 5.25 and 5.26 to be sure you understand why the stack and display appear as they do in each situation.

The compiler for a block-structured language must include code at the beginning of a block to initialize the display for that block. At the end of the block, it must include code to restore the previous display contents. For the details of how this is accomplished, see Aho (1977).

5.4 COMPILER DESIGN OPTIONS

In this section we consider some of the possible alternatives for the design and construction of a compiler. The discussions in this section are necessarily very brief. Our purpose is to introduce terms and concepts rather than to give a comprehensive discussion of any of these topics.

The compilation scheme presented in Section 5.1 was a simple one-pass design. Sections 5.2 and 5.3 described features that usually require more than one pass to implement. In Section 5.4.1 we briefly discuss the general question of dividing a compiler into passes, and consider the advantages of one-pass and multi-pass designs.

Section 5.4.2 discusses interpreters, which execute an intermediate form of the program instead of translating it into machine code. Section 5.4.3 introduces the related topic of P-code systems, which compile high-level language programs into object code for a hypothetical machine.

Finally, Section 5.4.4 describes compiler-writing systems, which use software tools to automate much of the process of compiler construction.

5.4.1 Division into Passes

In Section 5.1 we presented a simple one-pass compilation scheme for a subset of the Pascal language. In this design, the compiler was driven by the parsing process. The lexical scanner was called when the parser needed another input token, and a code-generation routine was invoked as each language construct was recognized by the parser. The object code produced was not highly efficient. Most of the code-optimization techniques discussed in Sections 5.2 and 5.3 could not be applied in such a one-pass compiler. However, the compilation process itself, which required only one pass over the program and no intermediate code-generation step, was quite efficient.

Not all languages can be translated by such a one-pass compiler. In Pascal, declarations of variables must appear in the program before the statements that use these variables. In FORTRAN, declarations may appear at the beginning of the program; any variable that is not declared is assigned characteristics by default. However, in some languages, such as PL/I, the declaration of an identifier may appear after it has been used in the program. One-pass compilers must have the ability to fix up forward references in jump instructions, using techniques like those discussed for one-pass assemblers. Forward references to data items, however, present a much more serious problem.

Consider, for example, the assignment statement

```
X := Y * Z
```

If all of the variables X, Y, and Z are of type INTEGER, the object code for this statement might consist of a simple integer multiplication followed by storage of the result. If the variables are a mixture of REAL and INTEGER types, one or more conversion operations will need to be included in the object code, and floating-point arithmetic instructions may be used. Obviously the compiler cannot decide what machine instructions to generate for this statement unless information about the operands is available. The statement may even be illegal for certain combinations of operand types. Thus a language that allows forward references to data items cannot be compiled in one pass.

Some programming languages, because of other characteristics, require more than two passes to compile. For example Hunter (1981) shows that ALGOL 68 requires at least three passes.

There are a number of factors that should be considered in deciding between one-pass and multi-pass compiler designs (assuming that the language in question can be compiled in one pass). If speed of compilation is important, a one-pass design might be preferred. For example, computers running student jobs tend to spend a large amount of time performing compilations. The resulting object code is usually executed only once or twice for each compilation; these test runs are normally very short. In such an environment, improvements in the speed of compilation can lead to significant benefits in system performance and job turnaround time.

If programs are executed many times for each compilation, or if they process large amounts of data, then speed of execution becomes more important than speed of compilation. In such a case, we might prefer a multi-pass compiler design that could incorporate sophisticated code-optimization techniques. Multi-pass compilers are also used when the amount of memory, or other system resources, is severely limited. The requirements of each pass can be kept smaller if the work of compilation is divided into several passes.

Other factors may also influence the design of the compiler. If a compiler is divided into several passes, each pass becomes simpler and, therefore, easier to understand, write, and test. Different passes can be assigned to different programmers and can be written and tested in parallel, which shortens the overall time required for compiler construction.

For further discussion of the problem of dividing a compiler into passes, see Hunter (1981) and Aho (1977). Examples of actual compilers, detailing the division of the work into passes, can be found in Gries (1971) and Aho (1977).

5.4.2 Interpreters

An *interpreter* processes a source program written in a high-level language, just as a compiler does. The main difference is that interpreters execute a version of the source program directly, instead of translating it into machine code.

An interpreter usually performs lexical and syntactic analysis functions like those we have described for a compiler, and then translates the source program into an internal form. Many different internal forms can be used. One possibility is a sequence of quadruples like those discussed in Section 5.2. More often, an extension to Polish postfix notation is used (see Gries, 1971). It is even possible to use the original source program itself as the internal form; however, it is generally much more efficient to perform some preprocessing of the program before execution.

After translating the source program into an internal form, the interpreter executes the operations specified by the program. During this phase, an interpreter can be viewed as a set of subroutines. The execution of these subroutines is driven by the internal form of the program.

The process of translating a source program into some internal form is simpler and faster than compiling it into machine code. However, execution of the translated program by an interpreter is much slower than execution of the machine code produced by a compiler. Thus an interpreter would not normally be used if speed of execution is important. If speed of translation is of primary concern, and execution of the translated program will be short, then an interpreter may be a good choice.

The real advantage of an interpreter over a compiler, however, is in the debugging facilities that can easily be provided. The symbol table, source line numbers, and other information from the source program are usually retained by the interpreter. During execution, these can be used to produce symbolic dumps of data values, traces of program execution related to the source statements, etc. Thus interpreters are especially attractive in an educational environment where the emphasis is on learning and program testing. Details concerning the implementation of debugging tools in an interpreter can be found in Gries (1971).

Most programming languages can be either compiled or interpreted successfully. However, some languages are particularly well suited to the use of an interpreter. As we have seen, compilers usually generate calls to library routines to perform functions such as I/O and complex conversion operations. For some languages, such as SNOBOL and APL, a large part of the compiled program would consist of calls to such routines. In such cases, an interpreter might be preferred because of its speed of translation. Most of the execution time for the translated program would be consumed by the standard library routines. These routines would be the same, regardless of whether a compiler or an interpreter were used.

Certain languages also have features that lend themselves naturally to interpretation. In APL and SNOBOL, for example, the type of a variable can change during the execution of a program. In APL, the variables that can be referred to by a function or a subroutine are determined by the sequence of calls made during execution, not by the nesting of blocks in the source program. (See Figs. 5.25 and 5.26 for an illustration of this difference.) It would be very difficult to compile such languages efficiently and allow for dynamic changes in the types of variables and the scope of names. These features can be more easily handled by an interpreter, which provides delayed binding of symbolic variable names to data types and locations.

For further discussions of the construction and use of interpreters, see Gries (1971).

5.4.3 P-Code Compilers

P-code compilers are very similar in concept to interpreters. In both cases, the source program is analyzed and converted into an intermediate form, which is then executed interpretively. With a P-code compiler, however, this intermediate form is the machine language for a hypothetical computer, often called a *pseudo-machine* or *P-machine*. The process of using such a P-code compiler is illustrated in Fig. 5.27. The source program is compiled, with the resulting object program being in P-code. This P-code program is then read and executed under the control of a P-code interpreter.

The main advantage of this approach is portability of software. It is not necessary for the compiler to generate different code for different computers, because the P-code object programs can be executed on any machine that has a P-code interpreter. Even the compiler itself can be transported if it is written in the language that it compiles. To accomplish this, the source version of the compiler is compiled into P-code; this P-code can then be interpreted on another computer. In this way, a P-code compiler can be used without modification on a wide variety of systems if a P-code interpreter is written for each different machine. Although writing such an interpreter is not a trivial task, it is certainly easier than writing a new compiler for each different machine. The same approach can also be used to transport other types of system

FIGURE 5.27 Translation and execution using a P-code compiler.

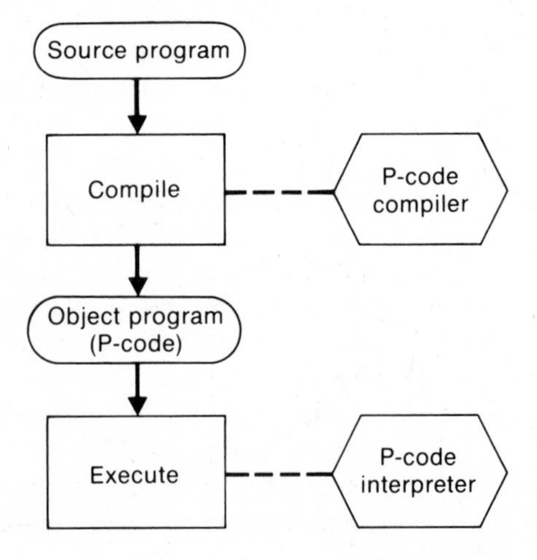

software without rewriting. See, for example, the description of the UCSD Pascal® system in Section 6.5.

The design of a P-machine and the associated P-code is often related to the requirements of the language being compiled. For example, the P-code for a Pascal compiler might include single P-instructions that perform array-subscript calculations, handle the details of procedure entry and exit, and perform elementary operations on sets. This simplifies the code-generation process, leading to a smaller and more efficient compiler. In addition, the P-code object program is often much smaller than a corresponding machine-code program would be. This is particularly useful on machines with severely limited memory size.

Obviously the interpretive execution of a P-code program may be much slower than the execution of the equivalent machine code. Depending upon the environment, however, this may not be a problem. Many P-code compilers are designed for a single user running on a dedicated microcomputer system. In that case, speed of execution may be relatively insignificant because the limiting factor in system performance may be the response time and "think time" of the user. If execution speed is important, some P-code compilers support the use of machine-language subroutines. By rewriting a small number of commonly used routines in machine language, rather than P-code, it is often possible to achieve substantial improvements in performance. Of course, this approach sacrifices some of the portability associated with the use of P-code compilers.

The most widely known P-code compiler is the UCSD Pascal compiler described in Section 5.5.2 of this text. The terms P-code, P-machine, etc., are often used as though they referred specifically to the UCSD compiler. However, they are actually more general terms and have been used in connection with a variety of other software such as the P-compiler described in Nori (1981).

5.4.4 Compiler-Compilers

The process of writing a compiler usually involves a great deal of time and effort. In some areas, particularly the construction of scanners and parsers, it is possible to perform much of this work automatically. A *compiler-compiler* is a software tool that can be used to help in the task of compiler construction. Such tools are also often called *compiler generators* or *translator-writing systems.*

The process of using a typical compiler-compiler is illustrated in Fig. 5.28. The user (i.e., the compiler writer) provides a description of the language to be translated. This description may consist of a set of lexical rules for defining tokens and a grammar for the source lan-

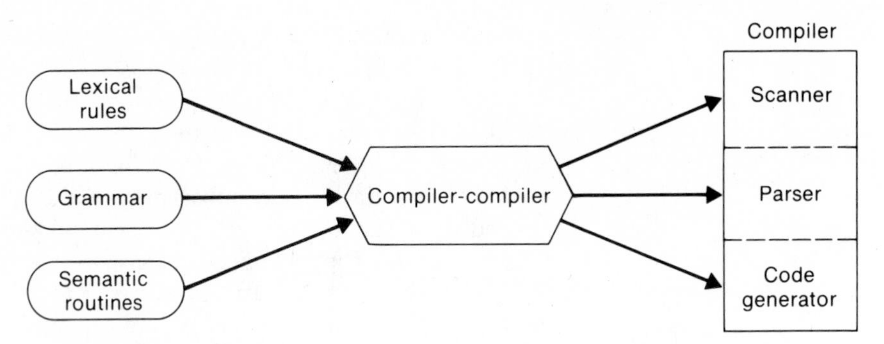

FIGURE 5.28 Automated compiler construction using a compiler-compiler.

guage. Some compiler-compilers use this information to generate a scanner and a parser directly. Others create tables for use by standard table-driven scanning and parsing routines that are supplied by the compiler-compiler.

In addition to the description of the source language, the user provides a set of semantic or code-generation routines. Often there is one such routine for each rule of the grammar, as we discussed in Section 5.1. This routine is called by the parser each time it recognizes the language construct described by the associated rule. However, some compiler-compilers can parse a larger section of the program before calling a semantic routine. In that case, an internal form of the statements that have been analyzed, such as a portion of the parse tree, may be passed to the semantic routine. This latter approach is often used when code optimization is to be performed. Compiler-compilers frequently provide special languages, notations, data structures, and other similar facilities that can be used in the writing of semantic routines.

The main advantage in using a compiler-compiler is, of course, ease of compiler construction and testing. However, the amount of work required from the user varies considerably from one compiler-compiler to another depending upon the degree of flexibility provided. Compilers that are generated in this way tend to require more memory and compile programs more slowly than handwritten compilers. However, the object code generated by the compiler may actually be better when a compiler-compiler is used. Because of the automatic construction of scanners and parsers, and the special tools provided for writing semantic routines, the compiler writer is freed from many of the mechanical details of compiler construction. The writer can therefore focus more attention on good code generation and optimization.

A brief description of one compiler-compiler (YACC) is given in

Section 5.5.4. Further discussion and examples of compiler-compilers can be found in Gries (1971) and Hopgood (1969).

5.5 IMPLEMENTATION EXAMPLES

In this section we briefly describe the design of several real compilers. Section 5.5.1 describes the ETH Zurich Pascal compiler, which is in widespread use on CDC CYBER systems. Section 5.5.2 discusses the UCSD Pascal compiler, one of the best known P-code compilers. Both of these are one-pass compilers that attempt little in the way of code optimization.

Section 5.5.3 describes the IBM FORTRAN H compiler, which is in use on IBM 370 systems. This is a multi-pass compiler that is designed to produce highly optimized object code.

Section 5.5.4 presents a description of the YACC compiler-compiler, developed at Bell Laboratories for use with the UNIX™ operating system. We also briefly describe LEX, a scanner generator that is commonly used with YACC.

As in our previous discussions of real systems, we do not attempt to give a complete description of any of these compilers. References are provided for those readers who want more information.

5.5.1 ETH Zurich Pascal Compiler*

The programming language Pascal was developed by Niklaus Wirth at ETH Zurich in 1968–70, and the first Pascal compiler was written in 1970. The experience gained led to a revised Pascal specification in 1972 and the development of a new compiler for the revised language in 1972–74. This compiler is now in use at over 150 installations throughout the world, and its community of users is steadily growing. It is this second ETH compiler we describe in this section. Further details can be found in Ammann (1981).

The ETH compiler is a one-pass compiler that produces machine code for the CDC 6000 and CYBER series computers. It is a *self-compiler,* which means it is written in the language that it compiles. One benefit of this approach is the ability to bootstrap the compiler onto a new machine. The original version of the compiler, which runs on machine A, is modified to produce object code for machine B. When this version of the compiler is used to compile itself, the result is a compiler that runs on machine B. Another benefit of a self-compiler is the feedback of improvements in code generation. If the compiler is modified to produce more efficient object code, and then used to compile itself, the

* Adapted from "The Zurich Implementation," by Urs Ammann. *Pascal: The Language and Its Implementation,* copyright 1981 by John Wiley & Sons, Ltd. Reprinted by permission.

result is a more efficient compiler. Such self-compilers are common. Coincidentally, the other two compilers described in Section 5.5 are also self-compilers.

The ETH compiler contains a lexical scanner of the type we discussed previously. This scanner uses a table of reserved words, stored in a linear list in order of length, to aid in the identification of tokens. Syntactic analysis is performed by a recursive-descent parser.

The symbol table for the compiler is organized as a binary search tree. Storage for this symbol table, and for the other data structures used, is allocated dynamically to make effective use of the available memory. When the compilation of a procedure is completed, the associated symbol-table entries are no longer needed; therefore, the storage allocated for them is released for later use.

Run-time storage is assigned to data items using the automatic allocation technique described in Section 5.3.1. Activation records are created on a stack as each procedure is entered. These activation records are chained together, in the order of their appearance on the stack, as illustrated in Fig. 5.20. There is also a separate chain that includes only those activation records that are currently accessible according to the Pascal rules for the scope of names. This chain is used for the same purpose as the display we discussed in Section 5.3.4 (see Fig. 5.26).

To improve code generation, the compiler keeps track of the contents of all 24 CYBER registers as they will appear during execution. When a data value or an address pointer is needed during code generation, these register content descriptions are examined. If the required value or pointer is already in a register, no load operation is necessary. When it is necessary to assign an X register to contain some data value, the compiler first searches for a free register. If no free register is available, the compiler selects a register whose value is to be replaced. This selection is based on the length of time since the last previous reference to the register contents and on the kind of value contained. The replacement of a constant is preferred to the replacement of a variable value since the constant is easier to restore when it is needed. This selection technique seems to be quite effective. When the ETH compiler was used to compile itself, the resulting object code was nearly 30 percent smaller than that produced by a previous version that did not keep track of register contents.

The ETH compiler produces relocatable object code in a format compatible with the system's linking loader. All forward references within the procedures being compiled are fixed up by the compiler before the object program is written. To achieve this, all object code for each procedure is retained in memory until compilation of that procedure is completed. Storage is dynamically allocated for this purpose.

When the end of the procedure is reached, the object code is written out and the allocated storage is released.

5.5.2 UCSD Pascal Compiler

The UCSD Pascal system is a complete program development and execution environment designed for small computers. Programs developed under this system are portable across a large number of different machines. In this section we describe the UCSD Pascal compiler. Section 6.5 presents a brief discussion of the entire UCSD Pascal system.

The UCSD Pascal compiler was modeled on the P-code compiler developed at ETH Zurich, which is a modified version of the ETH compiler described in Section 5.5.1. See Nori (1981) for a description. The UCSD Pascal compiler is a one-pass compiler, written in Pascal, that produces P-code for a pseudo-machine (P-machine). The language compiled is standard Pascal with a number of extensions.

The UCSD Pascal P-machine has a stack-oriented architecture that incorporates a number of high-level instructions and data structures. There are no general-purpose registers on the P-machine. Most arithmetic, logical, and other operations take their operands from the stack. The stack is also used to store parameters and bookkeeping information about procedure and function calls. There are several special-purpose registers that are used by the operating system and the P-code interpreter. For example, two of these registers are a program counter and a pointer to the current activation record.

The design of the P-machine grew out of an analysis of Pascal object-program instructions. An effort was made to identify the instruction sequences that occupied the most space in the object program and to provide direct support for these operations in P-code. Single P-code instructions can access data items from the current activation record and from any other activation record that may be referred to by the current procedure. There are single instructions that can calculate array subscripts, perform array and string manipulations, and evaluate the results of operations such as the union and intersection of sets. There are also instructions that support the calling of procedures. In a single P-machine operation, these instructions take care of building activation records, modifying other data structures, and properly linking together all these structures. Because of these special instructions, a P-code object program is usually much shorter than an equivalent program for a machine with a more conventional architecture.

The UCSD Pascal compiler allows programmers to write P-code instructions directly as a part of the Pascal program. This feature is useful in certain special cases, such as low-level systems programming. It is also possible to generate machine code for the host computer, using an optional step in the compilation process. A *native-code generator* accepts as input a complete P-code program created by the compiler, and translates selected procedures from P-code into actual machine code. The translated machine code usually occupies more space than the corresponding P-code; however, the speed of execution of the translated procedures may be improved by a factor of 10 or more.

More information about the UCSD Pascal compiler and the P-machine can be found in SofTech Microsystems (1983) and Overgaard (1980).

5.5.3 IBM FORTRAN H Compiler

The FORTRAN H compiler is designed for use on IBM 360 and 370 series computers. The language translated is FORTRAN IV; the output is an object module that can be processed by the system's linkage editor. One of the primary goals of this compiler is the production of efficient object code. The user can select three different levels of optimization for each compilation.

The FORTRAN H compiler consists of a *system director* that controls the compilation process, four logical processing phases (designated as phases 10, 15, 20, and 25), and an error-handling phase (phase 30). The exact processing performed by some of the phases, and, therefore, the number of passes required for compilation, varies according to the level of optimization selected. The compiler is structured as an overlay program with 13 segments. The root segment of the overlay structure is the system director. Each of the other segments is a phase or a logical portion of a phase.

The system director performs initialization, calls the various phases for execution, and allocates storage for use during the compilation. It also receives input/output requests from the other phases and passes them to the operating system for execution.

Phase 10 reads the source program and performs the lexical analysis function. It makes entries in the appropriate tables for all variables, constants, statement numbers, etc., that appear in the source statements. The output from phase 10 is a series of operator–operand pairs in the order in which they appear in the source statements. In this context, the term *operator* includes such elements as parentheses and commas in addition to the usual arithmetic, logical, and relational operators. The term *operand* includes such elements as variables, constants, and literals. In addition to performing lexical analysis, phase 10 also prints the listing of the source program and the cross-reference listing, if these are requested by the programmer.

Phase 15 is divided into two parts. The first part translates the output from phase 10 into quadruples. An operator-precedence technique is used in the analysis of the source statements. If optimization is requested, phase 15 also divides the program into basic blocks and gathers information on branching and the usage of constants and variables for each block. The second part of phase 15 assigns relative addresses to constants and variables. It also assigns address constants to be used in addressing these data items.

Phase 20 performs a variety of code-optimization functions. The processing depends upon the level of optimization requested by the programmer. If no optimization is requested, phase 20 assigns registers to operands from the quadruple form of the program. However, it does not take full advantage of the available registers, and it does not attempt to retain operand values in registers for later use. This process is called *basic register assignment* and is accomplished in a single pass.

If the first level of optimization is selected, phase 20 performs a process called *full register assignment.* This is similar in purpose to the basic register assignment; however, it takes greater advantage of the available registers and attempts to keep operand values in registers for subsequent use. An attempt is also made to keep the most frequently used operands in registers throughout the execution of the object program. Full register assignment requires a number of passes over the intermediate form of the program. Phase 20 also performs optimization of branch instructions to avoid unnecessary loading of registers with branch addresses.

If the highest level of optimization is selected, phase 20 performs optimization on a loop-by-loop basis. Phase 20 first determines the structure of the program in terms of the loops within it and the order in which the loops are executed. It then performs optimizations such as the elimination of common subexpressions and the removal of loop invariants. Phase 20 also performs full register assignment and branch optimization as just described.

Phase 25 generates an object program from the information produced by the previous phases. Phase 30 is called after phase 25 processing is completed only if errors were detected by the previous phases. The purpose of this phase is printing the appropriate error messages.

Further information about the FORTRAN H compiler can be found in IBM (1972b).

5.5.4 The YACC Compiler-Compiler*

YACC (Yet Another Compiler-Compiler) is a parser generator that is available on UNIX systems. YACC has been used in the production of compilers for Pascal, RATFOR, APL, C, and a number of other pro-

* Adapted from "Language Development Tools on the Unix System" by S. C. Johnson, from the IEEE publication *Computer,* Vol. 13, No. 8, pp. 16–21, August 1980. © 1980 IEEE.

gramming languages. It has also been used for several less conventional applications, including a typesetting language and a document-retrieval system. In this section we give brief descriptions of YACC and LEX, the scanner generator that is related to YACC. Further information about these software tools can be found in Johnson (1980), Johnson (1975), and Lesk (1975).

A lexical scanner must be supplied for use with YACC. This scanner is called by the parser whenever a new input token is needed. It returns an integer that identifies the type of token found, as described in Section 5.1. The scanner may also make entries in a symbol table for the identifiers that are processed.

LEX is a scanner generator that can be used to create scanners of the type required by YACC. A portion of an input specification for LEX is shown in Fig. 5.29(a). Each entry in the left-hand column is a pattern to be matched against the input stream. When a pattern is matched, the corresponding action routine in the right-hand column is invoked. These action routines are written in the programming language C. The routines usually return an indication of the token that was recognized. They may also make entries in tables and perform other similar tasks.

In the example shown in Fig. 5.29(a), the first pattern has no associated action; the effect of this is to delete blanks as the input is scanned. The actions for the next three patterns simply return a token type: the token LET for the keyword *let,* MUL for the operator *, and ASSIGN for the operator =. As discussed above, the internal representations of LET, MUL, and the other tokens are integers. The fifth pattern specifies the form of identifiers to be recognized. The first character must be in the range a–z or A–Z. This may be followed by any number of characters in the ranges a–z, A–Z, or 0–9. The * in this pattern indicates that an arbitrary number of repetitions of the preceding item are allowed. The action routine for this pattern makes entries in the appropriate tables to describe the identifier found and then returns the token type ID.

According to the specifications given in Fig. 5.29(a), the input

```
let x = y * z
```

would be scanned as the sequence of tokens

```
LET  ID  ASSIGN  ID  MUL  ID
```

Note that the first pattern that matches the input stream is selected, so the keyword *let* is recognized as the token LET, not as ID.

LEX can be used to produce quite complicated scanners. Some languages, such as FORTRAN, however, have lexical analyzers that must still be generated by hand.

```
•
•
•
" "                         ; /* ignore blanks */
let                         return(LET);
"*"                         return(MUL);
"="                         return(ASSIGN);
[a-zA-Z] [a-zA-Z0-9]*       {make entries in tables; return(ID)};
•
•
•
```

(a)

```
%token   ASSIGN ID LET MUL ...
•
•
•
statement :   LET  ID  ASSIGN expr
                   { ... }
expr      :   expr  MUL  expr
                   { $$ = build(MUL,$1,$3);}
expr      :   ID
                   { ... }
•
•
•
```

(b)

FIGURE 5.29 Example of input specifications for LEX and YACC.

The YACC parser generator accepts as input a grammar for the language being compiled and a set of actions corresponding to rules of the grammar. A portion of such an input specification appears in Fig. 5.29(b). The first line shown is a declaration of the token types used. The other entries are rules of the grammar. The YACC parser calls the semantic routine associated with each rule as the corresponding language construct is recognized. Each routine may return a value by assigning it to the variable $$. Values returned by previous routines, or by the scanner, may be referred to as *$1, $2,* etc. These variables designate the values returned for the components on the right-hand side of the corresponding rule, reading from left to right.

An example of the use of such values is shown in Fig. 5.29(b). The semantic routine associated with the rule

expr : expr MUL expr

constructs a portion of the parse tree for the statement, using a tree-building function *build.* This subtree is returned from the semantic routine by assigning the subtree to $$. The arguments passed to the build function are the operator MUL and the values (i.e., the subtrees) returned when the operands were recognized. These values are denoted by *$1* and *$3*.

It is sometimes useful to perform semantic processing as each part of a rule is recognized. YACC permits this by allowing semantic routines to be written in the middle of a rule as well as at the end. The value returned by such a routine is available to any of the routines that appear later in the rule. It is also possible for the user to define global variables that can be used by all of the semantic routines and by the lexical scanner.

The parsers generated by YACC use a bottom-up syntactic analysis technique called LALR(1). This parsing method can handle a large and interesting class of grammars. It is not necessary to avoid left recursion. It is even possible to use ambiguous grammars by specifying auxiliary rules to resolve the ambiguities. The parsers produced by YACC have very good error detection properties. Error handling permits the reentry of the items in error or a continuation of the input process after the erroneous entries are skipped.

EXERCISES

Section 5.1

1. Draw parse trees, according to the grammar in Fig. 5.2, for the following ⟨id-list⟩s:

   a) `ALPHA`

   b) `ALPHA, BETA, GAMMA`

2. Draw parse trees, according to the grammar in Fig. 5.2, for the following ⟨exp⟩s:

   a) `ALPHA + BETA`

   b) `ALPHA - BETA * GAMMA`

   c) `ALPHA DIV (BETA + GAMMA) - DELTA`

3. Suppose Rules 10 and 11 of the grammar in Fig. 5.2 were changed to

```
⟨exp⟩  ::= ⟨term⟩ | ⟨exp⟩ * ⟨term⟩ | ⟨exp⟩ DIV ⟨term⟩

⟨term⟩ ::= ⟨factor⟩ | ⟨term⟩ + ⟨factor⟩ | ⟨term⟩ - ⟨factor⟩
```

   Draw the parse trees for the ⟨exp⟩s in Exercise 2 according to this modified grammar. How has the change in the grammar affected the precedence of the arithmetic operators?

4. Assume that Rules 10 and 11 of the grammar in Fig. 5.2 are deleted and replaced with the single rule

```
⟨exp⟩ ::= ⟨factor⟩ | ⟨exp⟩ + ⟨factor⟩ | ⟨exp⟩ - ⟨factor⟩
            | ⟨exp⟩ * ⟨factor⟩ |⟨exp⟩ DIV ⟨factor⟩
```

Draw the parse trees for the ⟨exp⟩s in Exercise 2 according to this modified grammar. What problems arise?

5. Modify the grammar in Fig. 5.2 to include exponentiation operations of the form X↑Y. Be sure that exponentiation has higher priority than any other arithmetic operation.

6. Modify the grammar in Fig. 5.2 to include statements of the form

```
IF condition THEN statement-1 ELSE statement-2
```

where the ELSE clause may be omitted. Assume that the condition must be of the form $a < b$, $a = b$, or $a > b$, where a and b are single identifiers or integers. You do not need to allow for nested IFs—that is, statement-1 and statement-2 may not be IF statements.

7. Modify the grammar in Fig. 5.2 so that the I/O list for a WRITE statement may include character strings enclosed in quotation marks, as well as identifiers.

8. Write an algorithm that scans an input stream, recognizing operators and identifiers. An identifier may be up to 10 characters long. It must start with a letter, and the remaining characters, if any, must be letters and digits. The operators to be recognized are +, −, *, DIV, and :=. Your algorithm should return an integer that represents the type of token found, using the coding scheme of Fig. 5.5. If an illegal combination of characters is found, the algorithm should return the value −1.

9. Modify the scanner you wrote in Exercise 8 so that it recognizes integers as well as identifiers. Integers may begin with a sign (+ or −); however, they may not begin with the digit 0.

10. Select a high-level programming language with which you are familiar and write a lexical scanner for it.

11. Parse the following statements from the example program in Fig. 5.1, using the operator-precedence technique and the precedence matrix in Fig. 5.7:

    a) the assignment statement on line 11

    b) the declaration on line 3

    c) the FOR statement beginning on line 7

12. Parse the entire program for Fig. 5.1, using the operator-precedence techniques and the precedence matrix in Fig. 5.7.

13. Parse the assignment statement on line 11 of Fig. 5.1, using the method of recursive descent and the procedures given in Fig. 5.11.

14. Write recursive-descent parsing procedures that correspond to the rules for ⟨dec-list⟩, ⟨dec⟩, and ⟨type⟩ in Fig. 5.9. Use these procedures to parse the declaration on line 3 of Fig. 5.1.

15. Write recursive-descent parsing procedures for the remaining nonterminals in the grammar of Fig. 5.9. Parse the entire program in Fig. 5.1, using the method of recursive descent.

16. Use the routines in Figs. 5.12–5.14 to generate code for the following statements from the example program in Fig. 5.1:

    a) the assignment statement on line 11

    b) the WRITE statement on line 15

    c) the FOR statement beginning on line 7.

    Refer to the parse tree in Fig. 5.4 to see the order in which the parser recognizes the various constructs involved in these statements.

17. Use the routines in Figs. 5.12–5.14 to generate code for the entire program in Fig. 5.1.

18. Write code-generation routines for the new rules that you added to the grammar in Exercise 6 to define the IF statement.

19. Suppose that the grammar in Fig. 5.2 is modified to allow floating-point variables (i.e., the ⟨type⟩ REAL) as well as integers. How would the code-generation routines given in the text need to be changed? Assume that mixed-mode arithmetic expressions are allowed according to the usual rules of Pascal.

20. The code-generation routines in the text use immediate addressing for integers written by the programmer in arithmetic expressions (for example, the 100 in the expression SUM DIV 100). How could such constants be handled by a compiler for a machine that does not have immediate addressing?

21. What kinds of source program errors would be detected during lexical analysis?

22. What kinds of source program errors would be detected during syntactic analysis?

23. What kinds of source program errors would be detected during code generation?

24. In what ways might the symbol table used by a compiler be different from the symbol table used by an assembler?

Section 5.2

1. Rewrite the code-generation routines given in Figs. 5.12 and 5.13 to produce quadruples instead of object code.

2. Write a set of routines to generate object code from the quadruples produced by your routines in Exercise 1. (Hint: You will need a routine that is similar in function to the GETA procedure in Fig. 5.13.)

3. Use the routines you wrote in Exercise 1 to produce quadruples for the following program fragment:

```
READ(X,Y);

Z := 3 * X - 5 * Y + X * Y;
```

4. Use the routines you wrote in Exercise 2 to produce object code from the quadruples generated in Exercise 3.

5. Rewrite the code-generation routines given in Fig. 5.14 to produce quadruples instead of object code.

6. Use the routines you wrote in Exercises 1 and 5 to produce quadruples for the program in Fig. 5.1.

7. Divide the quadruples you produced in Exercise 6 into basic blocks and draw a flow graph for the program.

8. Assume that you are generating SIC/XE object code from the quadruples produced in Exercise 6. Show one way of performing register assignments to optimize the object code, using registers S and T to hold variable values and intermediate results.

Section 5.3

1. Write an algorithm for the prologue of a procedure, assuming the activation record format shown in Fig. 5.20.

2. Write an algorithm for the epilogue of a procedure, assuming the activation record format shown in Fig. 5.20.

3. Suggest a way of using the activation record stack to perform dynamic storage allocation for controlled variables. What would be the advantages and disadvantages of such a technique as compared to using a separate area of free storage to perform these allocations?

4. Assuming the array declaration

```
C : ARRAY[5..20] OF INTEGER
```

generate quadruples for the statement

```
C[I] := 0
```

5. Assuming the array declaration

```
D : ARRAY[-10..10,2..12] OF INTEGER
```

generate quadruples for the statement

```
D[I,J] := 0
```

6. Generalize the methods given in Section 5.3.2 to the storage of three-dimensional arrays in row-major order. Assuming the array declaration

```
E : ARRAY[1..5,1..10,0..8] OF INTEGER
```

generate quadruples for the statement

```
E[I,J,K] := 0
```

7. How could the base address for the array A defined in Fig. 5.22(a) be modified to avoid the need for subtracting 1 from the subscript value (quadruple 1)?
8. How could the technique derived in Exercise 7 be extended to two-dimensional arrays?
9. Assume the array declaration

```
T : ARRAY[1..5,1..100] of INTEGER
```

Translate the following statements into quadruples and perform elimination of common subexpressions on the result.

```
K := J - 1;
FOR I := 1 TO 5 DO
   BEGIN
      T[I,J] := K * K;
      J := J + K;
      T[I,J] := K * K - 1;
   END
```

10. Modify the quadruples produced in Exercise 9 to remove loop invariants.
11. Write an algorithm to construct the proper display when a procedure is invoked. Your algorithm may use the old display (i.e., the current display before the call), the address of the activation record created for the procedure being called, and the block-nesting level of the procedure being called.

CHAPTER 6

OPERATING SYSTEMS

In this chapter we discuss the functions and design of operating systems. This very large topic has been the subject of many entire textbooks. We do not attempt to cover such a topic thoroughly in a single chapter. Instead we discuss the most important concepts and issues related to operating systems, giving examples and providing references for further reading.

Operating systems vary widely in purpose and design. Some are very simple systems designed to support a single user on a personal computer. Others are extremely complex systems that support many concurrent users and manage highly sophisticated hardware and software resources. Section 6.1 discusses the basic features of an operating system that should be found in almost any such piece of software. Because of the large variety of operating systems, this list of basic features is surprisingly brief. It consists of only a few generic functions that could almost be taken as a definition of the term *operating system*.

Section 6.2 describes some important machine-dependent operating system features. Section 6.3 describes a number of machine-independent features. Many of the functions discussed in these two sections are actually required in operating systems that support more than one user at a time. Such functions include, for example, the allocation of system resources and the management of communication among different users.

Section 6.4 briefly presents some design alternatives for operating systems. Section 6.5 describes a number of actual operating systems, illustrating some of the variety of form and function in such software.

6.1 BASIC OPERATING SYSTEM FUNCTIONS

In this section we discuss the fundamental functions common to all operating systems. The main purpose of an operating system is to make the computer easier to use. That is, the software provides an interface that is more user-friendly than the underlying hardware. As a part of this process, the operating system manages the resources of the computer in an attempt to meet overall system goals such as efficiency. The details of this resource management can be quite complicated; however, the operating system usually hides such complexities from the user.

Our discussion of basic features is much shorter and more general than those in previous chapters. In Chapter 2, for example, we were able to give a common framework that could be applied to all assemblers—those for microprocessors as well as those for multi-user supercomputers. However, the operating systems for two such dissimilar computers would be quite different. Except for the general approaches described in this section, these operating systems would have little in common.

We may visualize the basic functions of an operating system as illustrated in Fig. 6.1. The operating system supports a *user interface* that governs the interactions with programmers, operators, etc. The user interface is what is usually described in response to the question, "What kind of operating system is this?". This interface may, for example, provide a control language. The users of such a language could enter a command such as RUN P to invoke the system loader to load and execute a program. Section 6.1.1 introduces some operating system terminology and describes a classification of operating systems based on the kind of user interface they provide. Section 6.1.2 briefly discusses some of the possible functions of the user interface.

The operating system also provides programs with a set of services that can aid in the performance of many common tasks. For example,

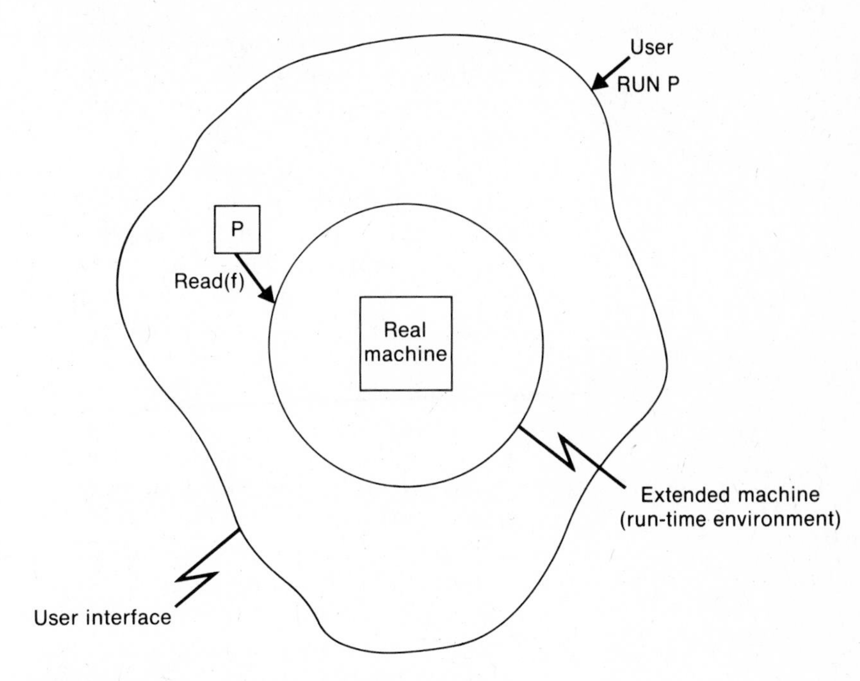

FIGURE 6.1 Basic concept of an operating system.

suppose program P wants to read data sequentially from a file. An operating system might provide a service routine that could be invoked with a command such as *read(f)*. With such a command, the program would specify a file name; the operating system would take care of the details of performing the actual machine-level I/O.

Such service routines can be thought of as providing a *run-time environment* for the programs being executed. Section 6.1.3 gives a general discussion of such a run-time environment. More detailed descriptions of some common service functions and routines are contained in Sections 6.2 and 6.3.

Throughout this chapter we describe operating system functions and services as though they were provided by software. Many of the same functions may also be performed by *firmware,* which consists of microprogrammed routines. For further information and references, see Deitel (1984).

6.1.1 Types of Operating Systems

The most common ways of classifying operating systems are based on the kind of user interface provided. Much operating system terminology arises from the way the system appears to a user. In this section we introduce terms commonly used to describe operating systems. The types of systems mentioned are not always distinct. Some of the classifications overlap and many real operating systems fall into more than one category.

One way of classifying operating systems is concerned with the number of users the system can support at one time. A *single-job* system is one that runs one user job at a time. Single-job systems, which are commonly found today on microcomputers and personal computers, were the earliest type of operating system. A single-job operating system would probably be used on a standard SIC computer. Because of the limited memory size and lack of data channels and other resources, it would be difficult to support more than one user on such a machine.

A *multiprogramming* system permits several user jobs to be executed concurrently. The operating system takes care of switching the CPU among the various user jobs. It also provides a suitable run-time environment and other support functions so the jobs do not interfere with each other. A *multiprocessing* system is similar to a multiprogramming system except there is more than one CPU available.

The basic purpose of multiprogramming is to improve performance by allowing the resources of the computing system to be shared among several jobs. For example, one job might be executing instructions on the CPU while another job is waiting for the completion of an I/O operation. We discuss this further in Section 6.2.

Operating systems are also classified by the type of access provided to a user. In a *batch processing* system, a job is described by a sequence of control statements stored in a machine-readable form (for example, on cards or disk). The operating system can read and execute a series of such jobs without human intervention except for such functions as tape and disk mounting. The order in which the jobs are executed can be selected in several different ways. This job scheduling problem is discussed in Section 6.3. A *timesharing* system provides *interactive, or conversational,* access to a number of users. The operating system executes commands as they are entered, attempting to provide each user with a reasonably short response time to each command. A *real-time* system is designed to respond quickly to external signals such as those generated by data sensors. Real-time systems are used, for example, on computers that monitor and control time-critical processes such as nuclear reactor operation or spacecraft flight.

In general, the goal of a multiprogramming batch processing system is to make the most efficient use of the computer. On the other hand, the goal of a timesharing system is to provide good response time to the interactive users. To provide good response time, it may be necessary to accept less efficient machine utilization. The goal of a real-time system is to provide a *guaranteed* response time to time-critical external events. It is quite common for these goals to be mixed in a single operating system. For example, many batch processing systems also support timesharing users, and some may also provide support for real-time applications.

Further discussions concerning these types of operating systems and descriptions of their historical background can be found in Deitel (1984).

6.1.2 User Interface

The user interface provided by an operating system is designed to serve the needs of the various groups of people who must deal with the computer. In a very simple operating system, such as one designed for a personal computer, the user interface is also relatively simple. Such a system typically provides commands that allow the user to access system programs such as editors, compilers, and loaders. The system also usually provides a capability for managing external files. The command language is generally designed to be easy to use: it may include interactive prompting for commands, or there may be a *menu* from which the user selects various options.

For more complex systems, there may be a number of different user-interface languages. An interactive *command language* is sometimes provided for occasional users of the system. There may also be a more complex and more powerful language, often called a *job control language,* that is intended for use by professional programmers. In addition, there is usually a special language that is used to communicate with the operators of the computer. This operator interface provides facilities for starting and stopping the execution of jobs, inquiring about the status of jobs or system resources, and handling external operations such as the mounting of tapes and disks.

The user interface of an operating system does not usually present major technical problems. However, the design of this interface is extremely important because it is the part of the system that is experienced by most users. Ideally the user interface should be designed to match the needs of all the various types of users, and to be compatible with the goals and purposes of the computing system. Further discussions of user interfaces can be found in Deitel (1984) and Peterson (1983).

An operating system must also contain service routines to support the user interface. For a personal computer, these might be simple routines to handle input from the keyboard and output to a display. In more complicated systems, there may be interfaces to remote timesharing terminals, remote card readers and printers, or remote sensing devices. There may also be interfaces to other computers that are linked into a network with the local system. As a part of the operator interface, many operating systems maintain logs of system activity that can be used for performance analysis and for error recovery.

6.1.3 Run-Time Environment

One of the most important functions of an operating system is supporting a run-time environment for user programs. This run-time environment contains a set of service routines that are available for use during program execution. It also provides facilities for managing the resources of the computing system, assigning these resources to user programs as needed.

As an illustration of services in the run-time environment, consider the input/output function. Nearly all operating systems contain routines that help in performing I/O operations. Suppose that a program is running under a single-job operating system on a SIC computer. To perform a read operation without operating system assistance, the program would need to contain a loop that tests the device status and issues an RD instruction for each byte to be read. (See Fig. 2.1 for an example of this.) The program would also need to test for I/O errors and provide error-recovery routines.

With the support of an operating system, the task of the user program would be much easier. The program could simply invoke a service routine and specify the device to be used. The operating system would take care of all details such as status testing and counting of bytes transferred, and it would also handle any necessary error checking.

A service routine such as the one just described can be thought of as providing an extension to the underlying machine. In a typical operating system, there are many such routines. Taken together, these service routines can be thought of as defining an *extended machine* for use by programs during execution. Programs deal with the functions and capabilities provided by this extended machine; they have no need to be concerned with the underlying real machine. The extended machine is easier to use than the real machine would be. For example, the details of performing an I/O operation are much simpler. The extended machine may also be more attractive in other ways. For example, I/O operations on the extended machine may appear to be less error prone than on the real machine, because the operating system takes care of error detection and recovery.

The extended machine is sometimes referred to as a *virtual machine*. However, the term virtual machine is also used in a different, although related context. This alternative usage of the term is described in Section 6.4.

In a multiprogramming operating system, the run-time environment also contains routines that manage the resources of the computer, allocating them to user jobs as needed. For example, the central memory of the machine is divided between the jobs being multiprogrammed, and the CPU is switched among the user jobs according to some predefined policy. Except for making certain requests of the operating system, the user jobs do not need to be concerned with resource management. The run-time environment provides each user job with the illusion of having its own separate extended machine, even though the underlying real machine is being shared among several users.

On some systems, the user programs can call operating system functions by referring directly to fixed locations in memory. The user documentation for the operating system gives a description of the entry points and data areas provided, along with their actual addresses. For example, the entry point to an I/O service routine might be located at memory address 238. After setting up appropriate parameters in registers, the user program could invoke this service function with an instruction such as JSUB 238. Alternatively, there could be a single service-request entry point in memory, and the exact service needed could be specified with a request code.

This method of communicating with the operating system through fixed memory locations is used on some minicomputers and personal computers. However, this method tends to be inconvenient and error prone, and it also may allow the user to bypass certain safeguards built into the operating system. On more advanced systems, the users generally request operating system functions by means of some special hardware instruction such as a *supervisor call* (SVC). Execution of an SVC instruction generates an *interrupt* that transfers control to an operating system service routine. A code supplied by the SVC instruction specifies the type of request. The handling of interrupts by an operating system is discussed in Section 6.2.1.

On a typical machine, the generation of an interrupt also causes the CPU to switch from *user mode* to *supervisor mode*. In supervisor mode, all machine instructions and features can be used. Most parts of the operating system are designed to run in supervisor mode. In user mode, however, some instructions are not available. These might include, for example, instructions that perform I/O functions, set memory protection flags, or switch the CPU from one mode to another. We discuss examples of such instructions later in this chapter. Restricting the use of such *privileged instructions* forces programs to make use of the services provided by the run-time environment. That is, user programs must deal with the extended machine interface, rather than utilizing the underlying hardware functions directly. This restriction also prevents user programs from interfering, either deliberately or accidentally, with the resource management functions of the operating system. Privileged instructions and user/supervisor modes (or their equivalents) are a practical necessity for a system that supports more than one user at a time.

In Sections 6.2 and 6.3 we discuss many different functions and services that are commonly provided by the run-time environment. At this level, there is much similarity between operating systems that might appear very different at the user interface. Most of the techniques discussed can be applied, with a few modifications, to all types of operating systems: batch processing, timesharing, real-time, etc.

6.2 MACHINE-DEPENDENT OPERATING SYSTEM FEATURES

One of the most important functions of an operating system is managing the resources of the computer on which it runs. Many of these resources are directly related to hardware units such as central memory, I/O channels, and the CPU. Thus many operating system functions are closely related to the machine architecture.

Consider, for example, a standard SIC computer. This machine has a small central memory, no I/O channels, no supervisor-call instruction, and no interrupts. Such a machine might be suitable as a personal computer for a single user; however, it could not reasonably be shared among several concurrent users. Thus an operating system for a standard SIC machine would probably be a single-job system, providing a simple user interface and a minimal set of functions in the run-time environment. It would probably provide few, if any, capabilities beyond the simple ones discussed in Section 6.1.

On the other hand, a SIC/XE computer has a much larger central memory, I/O channels, and many other hardware features not present on the standard SIC machine. A SIC/XE computer might well have a multiprogramming operating system. Such a system would allow several concurrent users to share the expanded machine resources that are available, and would take better advantage of the more advanced hardware. Of course, the sharing of a computing system between several users creates many problems, such as resource allocation, that must be solved by the operating system. In addition, the operating system must provide support for the more advanced hardware features such as I/O channels and interrupts.

In this section we discuss some important machine-dependent operating system functions. This discussion is presented in terms of a SIC/XE computer; however, the same principles can easily be applied to other machines that have architectural features similar to those of SIC/XE. We describe a number of significant SIC/XE hardware features as a part of our discussion. For ease of reference, these features are also summarized in Appendix C.

Section 6.2.1 introduces fundamental concepts of interrupts and interrupt processing that are used throughout the remainder of this chapter. Section 6.2.2 discusses the problem of switching the CPU among the several user jobs being multiprogrammed. Section 6.2.3 describes a method for managing input and output using I/O channels in a multiprogramming operating system. Sections 6.2.4 and 6.2.5 discuss the problem of dividing the central memory between user jobs. Section 6.2.4 presents techniques for managing real memory, and Section 6.2.5 introduces the important topic of virtual memory.

6.2.1 Interrupt Processing

An *interrupt* is a signal that causes a computer to alter its normal flow of instruction execution. Such signals can be generated by many different conditions, such as the completion of an I/O operation, the expiration of a preset time interval, or an attempt to divide by zero.

The sequence of events that occurs in response to an interrupt is illustrated in Fig. 6.2. Suppose program A is being executed when an interrupt signal is generated by some source. The interrupt automatically transfers control to an *interrupt-processing routine* (also called an *interrupt handler*) that is usually a part of the operating system. This interrupt-processing routine is designed to take some action in response to the condition that caused the interrupt. After completion of the interrupt processing, control can be returned to program A at the point at which its execution was interrupted.

In the sequence of events just described, the generation and processing of the interrupt may be completely unrelated to program A. For example, the interrupt might be generated by the completion of an I/O operation requested by another program. In general, it is impossible to predict when, and for what reason, program A will be interrupted in this way. Another way of expressing this is to say that the interrupts may be *asynchronous* with respect to program A. The hardware and software take care of saving the status of the computer when A is interrupted, and restoring it when A is resumed. Because of this,

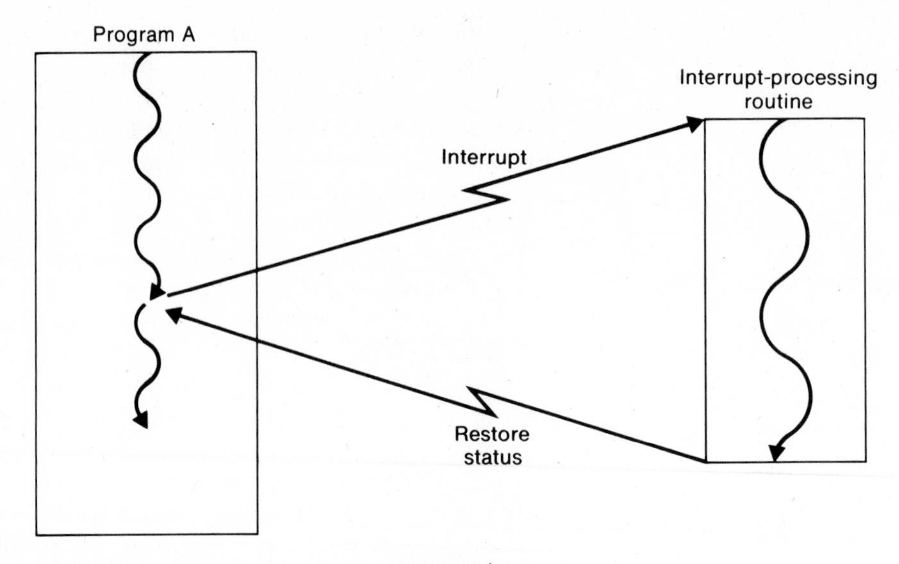

FIGURE 6.2 Basic concept of interrupt processing.

the execution of A is unaffected, except for timing, by the occurrence of the interrupt. Indeed, there may be no direct way for A even to detect that an interrupt has occurred.

Figure 6.3 describes the four classes of interrupts on a SIC/XE computer. An *SVC interrupt* (Class I) is generated when a supervisor-call instruction is executed by the CPU. This instruction is used by programs to request operating system functions. A *program interrupt* (Class II) is generated by some condition, such as an attempt to divide by zero or an attempt to execute an illegal machine instruction, that occurs during program execution. Appendix C contains a complete list of the conditions that can cause a program interrupt.

A *timer interrupt* (Class III) is generated by an *interval timer* within the CPU. This timer contains a register that can be set to an initial positive value by the privileged instruction STI. The value in this register is automatically decremented by 1 for each millisecond of

FIGURE 6.3 SIC/XE interrupt types.

Class	Interrupt type
I	SVC
II	Program
III	Timer
IV	I/O

CPU time that is used. When the value reaches zero, a timer interrupt occurs. The interval timer is used by the operating system to govern how long a user program can remain in control of the machine.

An *I/O interrupt* (Class IV) is generated by an I/O channel or device. Most such interrupts are caused by the normal completion of some I/O operation; however, an I/O interrupt may also signal a variety of error conditions.

When an interrupt occurs, the status of the CPU is saved, and control is transferred to an interrupt-processing routine. We describe the method used by SIC/XE to accomplish this. The mechanism described is typical of the one used on most computers.

On a SIC/XE machine, there is a fixed *interrupt work area* corresponding to each class of interrupt, as illustrated in Fig. 6.4. For example, the area assigned to the timer interrupt begins at memory address 160. When a timer interrupt occurs, the contents of all registers are stored in this work area, as shown in Fig. 6.4(a). Then the status word SW and the program counter PC are loaded with values that are prestored in the first two words of the area. This storing and loading of registers is done automatically by the hardware of the machine.

The loading of the program counter PC with a new value automatically causes a transfer of control. The next instruction to be executed is taken from the address given by the new value of PC. This address, which is prestored in the interrupt work area, is the starting address of the interrupt-handling routine for a timer interrupt. The loading of the status word SW also causes certain changes, described later in this section, in the state of the CPU.

After taking whatever action is required in response to the interrupt, the interrupt-handling routine returns control to the interrupted program by executing a Load Processor Status (LPS) instruction. This action is illustrated in Fig. 6.4(b). LPS causes the stored contents of SW, PC, and the other registers to be loaded from consecutive words beginning at the address specified in the instruction. This restores the CPU status and register contents that existed at the time of the interrupt, and transfers control to the instruction following the one that was being executed when the interrupt occurred. The saving and restoring of the CPU status and register contents are often referred to as *context switching* operations.

The status word SW contains several pieces of information that are important in the handling of interrupts. We discuss the contents of SW for a SIC/XE machine. Most computers have a similar register, which is often called a *program status word* or a *processor status word.*

Figure 6.5 shows the contents of the status word SW. The first bit, MODE, specifies whether the CPU is in user mode or supervisor mode.

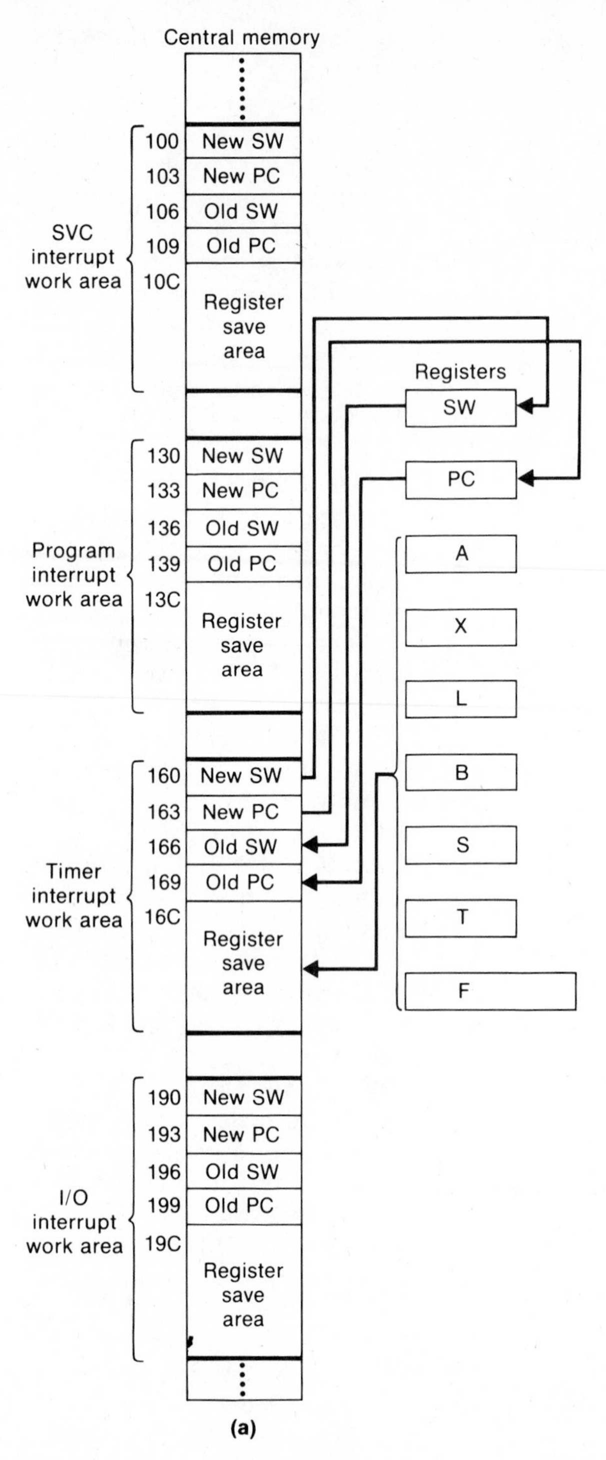

FIGURE 6.4 Context switching operations caused by (a) timer interrupt and (b) LPS 166.

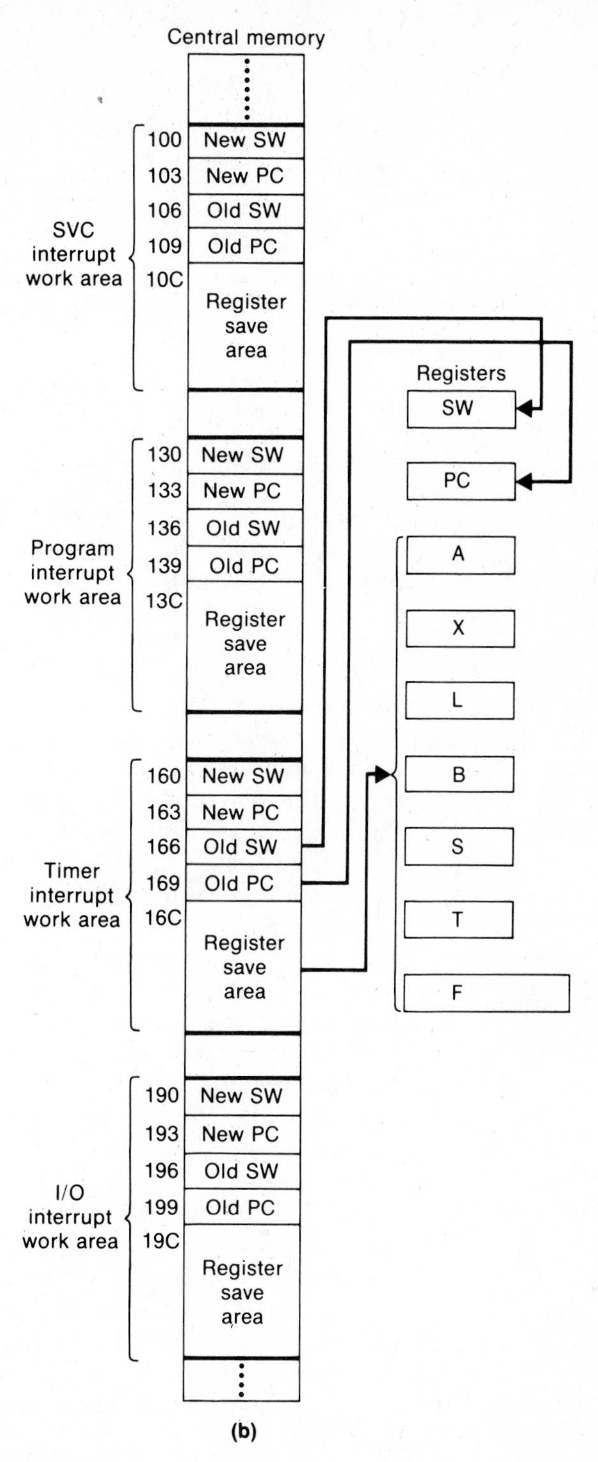

(b)

FIGURE 6.4 *(cont'd)*

Bit position	Field name	Use
0	MODE	0 = user mode, 1 = supervisor mode
1	IDLE	0 = running, 1 = idle
2–5	ID	Process identifier
6–7	CC	Condition code
8–11	MASK	Interrupt mask
12–15		Unused
16–23	ICODE	Interruption code

FIGURE 6.5 SIC/XE status word contents.

Ordinary programs are executed in user mode (MODE=0). When an interrupt occurs, the new SW contents that are loaded have MODE=1, which automatically switches the CPU to supervisor mode so that privileged instructions may be used. Before the old value of SW is saved, the ICODE field is automatically set to a value that indicates the cause of the interrupt. For an SVC interrupt, ICODE is set to the value supplied by the user in the SVC instruction. This value specifies the type of service request being made. For a program interrupt, ICODE indicates the type of condition, such as division by zero, that caused the interrupt. For an I/O interrupt, ICODE gives the number of the I/O channel that generated the interrupt. Further information about the possible values of ICODE can be found in Appendix C.

The status word also contains the condition code CC. Saving SW automatically preserves the condition code value that was being used by the interrupted process. The use of the fields IDLE and ID will be described later in this chapter. IDLE specifies whether the CPU is executing instructions or is idle. ID contains a four-bit value that identifies the user program currently being executed.

The remaining status word field, MASK, is used to control whether interrupts are allowed. This control is necessary to prevent loss of the stored processor status information. Suppose, for example, that an I/O interrupt occurs. The values of SW, PC, and the other registers would be stored in the I/O interrupt work area as just described, and the CPU would begin to execute the I/O-interrupt handler. If another I/O interrupt occurred before the processing of the first had been completed, another context switch would take place. This time, however, the register contents stored in the interrupt work area would be the values currently being used by the interrupt handler. The values that were saved by the original interrupt would be destroyed, so it would be impossible to return control to the user program that was executing at the time of the first interrupt.

To avoid such a problem, it is necessary to prevent certain interrupts from occurring while the first one is being processed. This is accomplished by using the MASK field in the status word. MASK contains one bit that corresponds to each class of interrupt. If a bit in MASK is set to 1, interrupts of the corresponding class are allowed to occur. If the bit is set to 0, interrupts of the corresponding class are not allowed. When interrupts are prohibited, they are said to be *masked* (also often called *inhibited* or *disabled*). Interrupts that are masked are not lost, however, because the hardware saves the signal that would have caused the interrupt. An interrupt that is being temporarily delayed in this way is said to be *pending*. When interrupts of the appropriate class are again permitted, because MASK has been reset, the signal is recognized and an interrupt occurs.

The masking of interrupts on a SIC/XE machine is under the control of the operating system. It depends upon the value of MASK in the SW that is prestored in each interrupt work area. One approach is to set all the bits in MASK to 0, which prevents the occurrence of any other interrupt. However, it is not really necessary to inhibit all interrupts in this way.

Each class of interrupt on a SIC/XE machine is assigned an *interrupt priority*. SVC interrupts (Class I) have the highest priority, program interrupts (Class II) have the next highest priority, and so on. The MASK field in the status word corresponding to each interrupt class is set so that all interrupts of equal or lower priority are inhibited; however, interrupts of higher priority are allowed to occur. For example, the status word that is loaded in response to a program interrupt would have the MASK bits for program, timer, and I/O interrupts set to 0; these classes of interrupt would be inhibited. The MASK bit for SVC interrupts would be set to 1, so these interrupts would be allowed. When interrupts are enabled at the end of an interrupt-handling routine, there may be more than one type of interrupt pending (for example, one timer interrupt and one I/O interrupt). In such a case, the pending interrupt with the highest priority is recognized first.

With this type of priority scheme, it is possible for an interrupt-processing routine itself to be interrupted. Such a *nested* interrupt situation is illustrated in Fig. 6.6, which shows program A in control of the CPU when an I/O interrupt occurs. The status of A is then saved, and control passes to the I/O-interrupt handler. During the execution of this routine, a timer interrupt occurs, and control is transferred to the timer-interrupt handler. After the processing of the timer interrupt is completed, an LPS instruction is used to reload the processor status from the timer-interrupt work area. This returns control to the I/O-interrupt handler. Because the previous value of MASK is re-

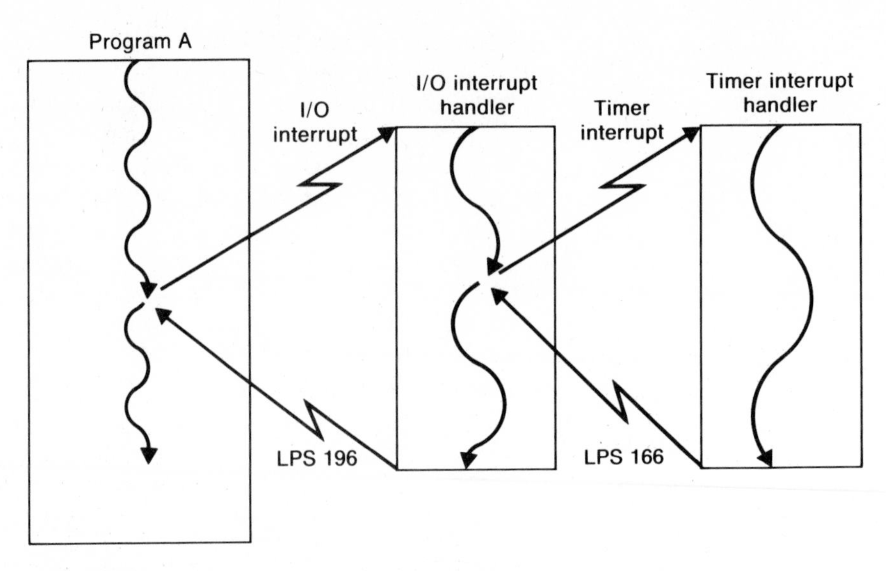

FIGURE 6.6 Example of nested interrupt processing.

loaded, timer interrupts, which had been inhibited, are once again allowed. I/O interrupts, however, are still inhibited. After the I/O-interrupt processing is completed, another LPS returns control to program A, restoring the CPU status as it was at the time of the original interrupt. At this time, all interrupts are allowed because the status word being used by program A has all MASK bits set to 1.

In later sections of this chapter, we see how interrupts can be used in such operating system functions as process scheduling, I/O management, and memory allocation.

6.2.2 Process Scheduling

A *process,* sometimes called a *task,* is most often defined as a program in execution. Many other possible definitions of the term *process* can be found in Deitel (1984). The CPU is assigned to processes by the operating system in order to perform computing work. In a single-job operating system, there is only one user process at a time. In a multiprogramming system, however, there may be many independent processes competing for control of the CPU. *Process scheduling* is the management of the CPU by switching control among the various competing processes according to some scheduling policy.

In most cases, a process corresponds to a user job. However, some operating systems allow one user job to create several different processes that are executed concurrently. In addition, some systems allow one program to be shared by several independent processes. Further information about such topics can be found in Deitel (1984). In our discussions we assume that each process corresponds to exactly one program and one user job.

A process is created when a user job begins execution, and this process is destroyed when the job terminates. During the period of its existence, the process can be considered to be in one of three states. A process is *running* when it is actually executing instructions using the CPU. A process is *blocked* if it must wait for some event to occur before it can continue execution. For example, a process might be blocked because it must wait for the completion of an I/O operation before proceeding. Processes that are neither blocked nor running are said to be *ready*. These processes are candidates to be assigned the CPU when the currently running process gives up control.

Figure 6.7 shows the possible transitions between these three process states. At any particular time, there can be no more than one process in the running state (i.e., in control of the CPU). When the operating system transfers control to a user process, it sets the interval timer to specify a *time-slice,* which is a maximum amount of CPU time the process is allowed to use before giving up control. If this time expires, the process is removed from the running state and placed in the ready state. The operating system then selects some process from the ready state, according to its scheduling policy. This process is placed in the running state and given control of the CPU. The selection of a process, and the transfer of control to it, is usually called *dispatching*. The part of the operating system that performs this function is known as the *dispatcher*.

Before it has used all its assigned time-slice, a running process may find that it must wait for the occurrence of some event such as the completion of an I/O operation. In such a case, the running process enters the blocked state, and a new process is dispatched. When an

FIGURE 6.7 Process state transitions.

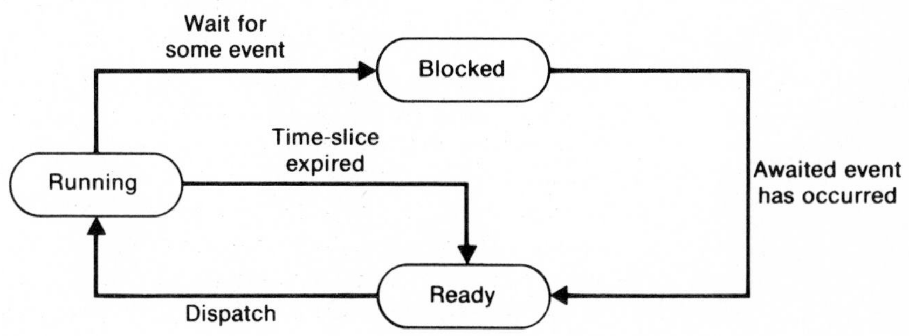

awaited event occurs, the blocked process associated with that event is moved to the ready state, where it is again a candidate for dispatching. The operations of waiting for an event, and of signalling that an event has occurred, are implemented as operating system service requests (using SVC). A mechanism often used to associate processes with awaited events is described later in this section.

A process is usually switched between the running, ready, and blocked states many times before completing its execution. Each time a process leaves the running state, its current status must be saved. This status must be restored the next time the process is dispatched so that the switching will have no effect on the results of the computation being performed. The status information for each process is saved by the operating system in a *process status block* (PSB) for that process. A process status block is created when a process first begins execution and is deleted when that process terminates. The PSB contains an indication of the process state (running, ready, or blocked), an area that is used to save all machine registers (including SW and PC), and a variety of other information (for example, an indication of the system resources used by the process).

An outline of the algorithm used for dispatching is shown in Fig. 6.8. Since the CPU is about to be switched from one process to another, it is first necessary to save status information for the previously running process. If that process lost control of the CPU because its time-slice expired, the status information can be found in the timer-interrupt work area. If the process gave up control (via an SVC request) because it needed to wait for the occurrence of some event, the status information can be found in the SVC-interrupt work area. Of course, it is possible that there was no previously running process. This would happen, for example, if all processes in the system were in the blocked state. In that case, there is no need to save status information.

FIGURE 6.8 Algorithm for dispatcher.

```
procedure DISPATCH

   update PSB of previously Running process (if any)
   select next Ready process to receive control
   if a Ready process was found then
      begin
         mark selected process as Running
         set interval timer for desired time slice using STI
         switch control to selected process using LPS
      end
   else
      place CPU in idle status using LPS
```

After saving the status of the previously running process, the dispatcher selects a new process to receive control. The dispatcher sets the interval timer to specify the time-slice to be given to the selected process. It then switches control by using the LPS instruction to load the status information saved in the PSB for that process. If there is no process in the ready state, the dispatcher uses LPS to place the CPU in an *idle* status by loading a status word that has IDLE=1 (see Fig. 6.5).

There are several different methods for selecting the next process to be dispatched. One common technique, known as *round robin,* treats all processes equally. The dispatcher cycles through the PSBs, selecting the next process that is in the ready state. Each process dispatched is given the same length time-slice as all other processes.

More complicated dispatching methods may select processes based on a priority scheme. In some systems, the priorities are predefined, based on the nature of the user job. The goal of such a system is to provide the desired level of service for each class of job. On other systems, the priorities may be assigned by the operating system itself. In this case, the assignment of priorities is made in an effort to improve the overall system performance. Priorities may be allowed to vary dynamically, depending on the system load and performance. It is also possible to assign different time-slices to different processes in conjunction with the priority system. Further discussion of these more sophisticated dispatching techniques can be found in Deitel (1984) and Lorin (1981).

When a running process reaches a point at which it must wait for some event to occur, the process informs the operating system by making a WAIT (SVC 0) service request. The occurrence of an event on which other processes may be waiting is communicated to the operating system by a SIGNAL (SVC 1) request. In Section 6.2.3 we present examples of the use of WAIT and SIGNAL; at this time, we are concerned with how these requests are related to the process-scheduling function.

Figure 6.9 gives the sequence of logical steps that is performed by the operating system in response to such service requests. The event to be awaited or signalled is specified by giving the address of an *event status block* (ESB) that is associated with the event. The event status block contains a flag bit ESBFLAG that records whether or not the associated event has occurred. The ESB also contains a pointer to ESBQUEUE, a list of all processes currently waiting for the event. Further information about ESBs, and examples of their creation and use, are presented in Section 6.2.3.

The WAIT request is issued by a running process and indicates that the process cannot proceed until the event associated with ESB

```
procedure WAIT(ESB)

   if ESBFLAG = 1 then    {event has already occurred}
      return control to requesting process using LPS
   else
      begin
         mark requesting process as Blocked
         enter requesting process on ESBQUEUE
         DISPATCH
      end
```

(a)

```
procedure SIGNAL(ESB)

   ESBFLAG := 1    {indicate that event has occurred}
   for each process on ESBQUEUE do
      begin
         mark process as Ready
         remove process from ESBQUEUE
      end
   return control to requesting process using LPS
```

(b)

FIGURE 6.9 Algorithms for WAIT (SVC 0) and SIGNAL (SVC 1).

has occurred. Thus the algorithm for WAIT first examines ESBFLAG, and if the event has already occurred, control is returned to the requesting process. If the event has not yet occurred, the running process is placed in the blocked state and is entered on ESBQUEUE. The dispatcher is then called to select the next process to be run.

The SIGNAL request is made by a process that detects that some event corresponding to ESB has occurred. The algorithm for SIGNAL therefore records the event occurrence by setting ESBFLAG. It then scans ESBQUEUE, the list of processes waiting for this event. Each process on the list is moved from the blocked state to the ready state. Control is then returned to the process that made the SIGNAL request.

If the dispatching method being used is based on priorities, a slightly different SIGNAL algorithm is often used. On such systems, it may happen that one or more of the processes that were made ready has a higher priority than the currently running process. To take this into account, the SIGNAL algorithm would invoke the dispatcher instead of returning control directly to the requesting process. The dispatcher would then transfer control to the highest-priority process that is currently ready. This scheme is known as *preemptive* process scheduling. It permits a process that becomes ready to seize control from a lower-priority process that is currently running, without waiting for the time-slice of the lower-priority process to expire.

6.2.3 I/O Supervision

On a typical small computer, such as a standard SIC machine, input and output are usually performed one byte at a time. For example, a program that needs to read data might enter a loop that tests the status of the I/O device and executes a series of *read-data* instructions. On such systems, the CPU is involved with each byte of data being transferred to or from the I/O device. An example of this type of I/O programming can be found in Fig. 2.1.

More advanced computers, such as SIC/XE, have *I/O channels* to take care of the details of transferring data and controlling I/O devices. Figure 6.10 shows a typical I/O configuration for SIC/XE. There may be as many as 16 channels, and up to 16 devices may be connected to each channel. The identifying number assigned to an I/O device also reflects the channel to which it is connected. For example, the devices numbered 20–2F are connected to channel 2.

FIGURE 6.10 Typical I/O configuration for SIC/XE.

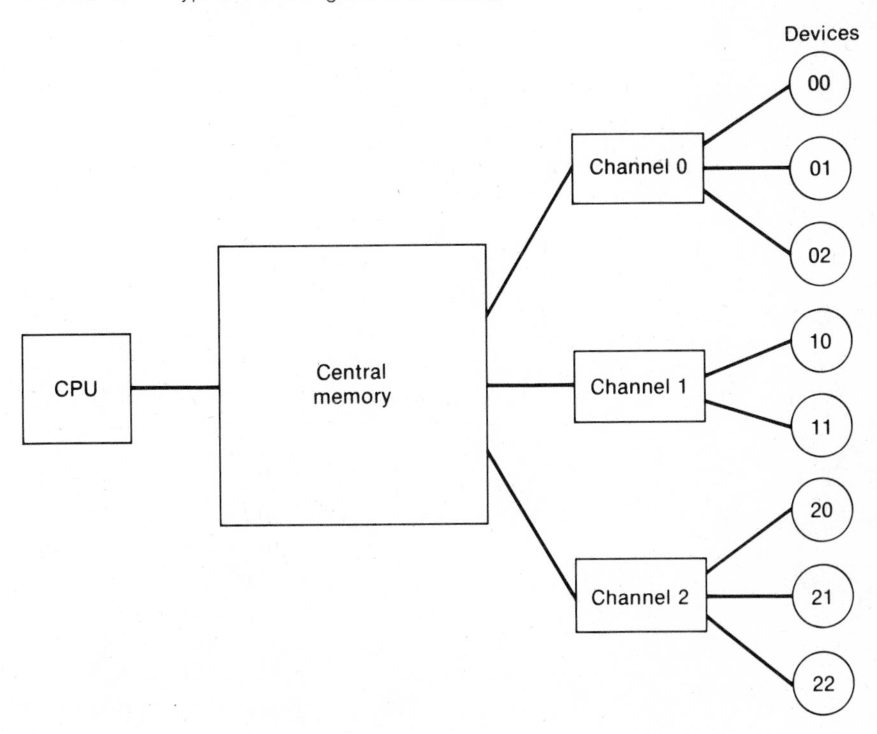

The sequence of operations to be performed by a channel is specified by a *channel program,* which consists of a series of *channel commands.* To perform an I/O operation, the CPU executes a Start I/O (SIO) instruction, specifying a channel number and the beginning address of a channel program. The channel then performs the indicated I/O operation without further assistance from the CPU. After completing its program, the channel generates an I/O interrupt. Several channels can operate simultaneously, each executing its own channel program, so several different I/O operations can be in progress at the same time. Each channel operates independently of the CPU, so the CPU is free to continue computing while the I/O operations are carried out.

The operating system for a computer like SIC/XE is involved with the I/O process in several different ways. The system must accept I/O requests from user programs and inform these programs when the requested operations have been completed. It must also control the operation of the I/O channels and handle the I/O interrupts generated by the channels. In the remainder of this section we discuss how these functions are performed and illustrate the process with several examples.

A SIC/XE program requests an I/O operation by executing an SVC 2 instruction. Parameters specify the channel number, the beginning address of a channel program, and the address of an event status block that is used to signal completion of the I/O operation. When the program must wait for the results of an I/O operation, it executes an SVC 0 (WAIT) instruction. This instruction specifies the address of the event status block that corresponds to the I/O operation being awaited. Thus the general pattern for performing an I/O operation is

```
SVC 2  {request I/O operation}
.
.
.
SVC 0  {wait for result}
```

In some cases, the WAIT may come immediately after the I/O request. However, because computing and I/O can be performed at the same time, it may be possible for the program to continue processing while awaiting the results of the I/O operation.

This procedure is illustrated in more detail by the program in Fig. 6.11. This program first loads the beginning address of a channel program, a channel number, and the address of an event status block into registers. The program then executes an SVC instruction to request

```
P1      START   0
        .
        .                           {initialization}
        .
        LDA     #READ               ADDRESS OF CHANNEL PROGRAM
        LDS     #1                  CHANNEL NUMBER
        LDT     #ESB                ADDRESS OF EVENT STATUS BLOCK
        SVC      2                  ISSUE READ REQUEST
LOOP    LDA     #ESB                ADDRESS OF ESB
        SVC      0                  WAIT FOR COMPLETION OF READ
        .
        .                           {move data to program's work area}
        .
        LDA     #0                  INITIALIZE ESB
        STA      ESB
        LDA     #READ
        LDS     #1
        LDT     #ESB
        SVC      2                  ISSUE NEXT READ REQUEST
        .
        .                           {process data}
        .
        J       LOOP
.
.                               CHANNEL PROGRAM FOR READ
.                                   FIRST COMMAND --
READ    BYTE    X'11'                       COMMAND CODE = READ, DEVICE = 1
        BYTE    X'0100'                     BYTE COUNT = 256
        WORD    BUFIN                       ADDRESS OF INPUT BUFFER
.                                   SECOND COMMAND --
        BYTE    X'000000000000'             HALT CHANNEL
.
ESB     BYTE    X'000000'           EVENT STATUS BLOCK FOR READ
BUFIN   RESB    256                 BUFFER AREA FOR READ
        .
        .
        .
        END
```

FIGURE 6.11 Example of performing I/O using SVC requests.

the I/O operation. The channel program, defined as a sequence of data items, contains two channel commands. The first command specifies that a read operation is to be executed on device number 1 connected to the channel; 256 bytes of data are to be transferred into memory beginning at address BUFIN. The second command causes the channel to halt, which generates an I/O interrupt. The event status block consists of a 3-byte data area. The first bit of this ESB is a flag that is used to indicate whether or not the associated event has already occurred (0 = no, 1 = yes). The rest of the ESB is used to store a pointer to the queue of processes that are waiting for this event. If no processes are currently waiting, the pointer value is zero. Thus an initial ESB value of X'000000' indicates that the associated event has not yet occurred and that no processes are currently waiting for it. Further details concerning the format of SIC/XE channel commands can be found in Appendix C.

After issuing the I/O request, the program in Fig. 6.11 executes an SVC 0 instruction. Register A contains the address of the ESB that corresponds to the event being awaited, which in this case is the I/O operation just requested. After the read operation has been completed, the program moves the input data to a work area. It then re-initializes the ESB and requests an I/O operation to read the next 256 bytes of data. While this I/O operation is being performed, the program can process the data that has previously been read, thus overlapping the computation and input functions. After completing the processing of the previous data, the program returns to the top of its main loop to await the completion of the next read operation.

A slightly more complicated example is shown in Fig. 6.12. This program copies 4096-byte data records from device 22 to device 14. There are two channel programs, one for the read operation and one for the write, and two event status blocks. The main loop of this program first issues a read request, and then waits for the completion of this read and for the completion of the previous write. After both operations are completed, the program builds the output record and issues the write request. It then returns to the top of the loop to read the next input record. On the first iteration of the loop, there has been no previous write request. At this time, however, the ESB for the write operation has its initially defined value of X'800000'. The first bit of this ESB has the value 1, indicating that the corresponding event has already occurred, so control is returned directly to the user program when the operating system WAIT routine is called (see Fig. 6.9).

Because the input and output operations for the program in Fig. 6.12 use different channels, these two operations are performed independently of each other. Either operation might be completed before the other. It is also possible that the two operations might actually be performed at the same time. The program is able to coordinate the related I/O operations because there is a different ESB corresponding to each operation. This program illustrates how I/O channels can be used to perform several overlapped I/O operations. Later in this section we consider a detailed example of this kind of overlap.

The programs in Figs. 6.11 and 6.12 illustrate I/O requests from the user's point of view. Now we are ready to discuss how such requests are actually handled by the operating system and the machine. The SIC/XE hardware provides a *channel work area* in memory corresponding to each I/O channel. This work area contains the starting address of the channel program currently being executed, if any, and the ad-

```
P2      START   0
        .                          {initialization}
        .
        .
LOOP    LDA     #0                 INITIALIZE ESB FOR READ
        STA      RDESB
        LDA     #READ              ISSUE READ REQUEST FOR DEVICE 22
        LDS     #2
        LDT     #RDESB
        SVC      2
        LDA     #RDESB             WAIT FOR COMPLETION OF READ
        SVC      0
        LDA     #WRESB             WAIT FOR COMPLETION OF PREVIOUS WRITE
        SVC      0
        .
        .                          {build output record}
        .
        LDA     #0                 INITIALIZE ESB FOR WRITE
        STA      WRESB
        LDA     #WRITE             ISSUE WRITE REQUEST FOR DEVICE 14
        LDS     #1
        LDT     #WRESB
        SVC      2
        J        LOOP
.
.                                CHANNEL PROGRAM FOR READ
.                                   FIRST COMMAND --
READ    BYTE    X'12'                        COMMAND = READ, DEVICE = 2
        BYTE    X'1000'                      BYTE COUNT = 4096
        WORD    BUFIN                        ADDRESS OF INPUT BUFFER
.                                   SECOND COMMAND --
        BYTE    X'000000000000'              HALT CHANNEL
.
.                                CHANNEL PROGRAM FOR WRITE
.                                   FIRST COMMAND --
WRITE   BYTE    X'24'                        COMMAND = WRITE, DEVICE = 4
        BYTE    X'1000'                      BYTE COUNT = 4096
        WORD    BUFOUT                       ADDRESS OF OUTPUT BUFFER
.                                   SECOND COMMAND --
        BYTE    X'000000000000'              HALT CHANNEL
.
RDESB   BYTE    X'000000'          EVENT STATUS BLOCK FOR READ
WRESB   BYTE    X'800000'          EVENT STATUS BLOCK FOR WRITE
BUFIN   RESB    4096               INPUT BUFFER
BUFOUT  RESB    4096               OUTPUT BUFFER
        .
        .
        .
        END
```

FIGURE 6.12 Program illustrating multiple I/O requests.

dress of the ESB corresponding to the current operation. When an I/O operation is completed, the outcome is indicated by status flags, such as normal completion, I/O error, or device unavailable, that are stored in the channel work area. The channel work area also contains a pointer to a queue of I/O requests for the channel. This queue is maintained by the operating system routines. Appendix C contains additional details on the location and contents of the channel work areas for SIC/XE.

Figure 6.13 outlines the actions taken by the operating system in response to an I/O request from a user program. If the channel on which I/O is being requested is busy performing another operation, the operating system inserts the request into the queue for that channel. If the channel is not busy, the operating system stores the current request in the channel work area and starts the channel. In either case, control is then returned to the process that made the I/O request so that it can continue to execute while the I/O is being performed.

Figure 6.14 describes the actions taken by the operating system in response to an I/O interrupt. The number of the I/O channel that generated the interrupt can be found in the status word that is stored in the I/O-interrupt work area. The interrupt-handling routine then examines the status flags in the work area for this channel to determine the cause of the interrupt.

If the channel status flags indicate normal completion of the I/O operation, the interrupt handler signals this completion via the event status block that was specified in the I/O request. This may be done either by making an SVC request, which results in a nested interrupt situation, or by directly invoking the part of the operating system that processes SIGNAL requests. In either case, the ESB is marked to indicate completion of the I/O operation. Any process that had previously been awaiting this completion is returned to the ready state (see Section 6.2.2). The I/O-interrupt handler then examines the queue of pending requests for this channel and starts the channel performing the next request, if any.

If the channel status flags indicate some abnormal condition, the operating system initiates the appropriate error-recovery action. This

FIGURE 6.13 Algorithm for processing an I/O request (SVC 2).

```
procedure IOREQ(CHAN,CP,ESB)

   test channel using TIO
   if channel is busy then
      insert (CP,ESB) on queue for channel
   else
      begin
         store (CP,ESB) in channel work area
         start channel using SIO
      end
   return control to requesting process using LPS
```

```
procedure IOINTERRUPT(CHAN)

   examine status flags in channel work area
   if normal completion of operation then
      begin
         get ESB address from channel work area
         use SVC to signal occurrence of event for ESB
         if request queue for channel is not empty then
            begin
               get (CP,ESB) for next request from queue
               store (CP,ESB) in channel work area
               start channel using SIO
            end {if not empty}
      end {if normal completion}
   else
      take appropriate error recovery action
   if CPU was in idle state when the interrupt occurred then
      DISPATCH
   else
      return control to interrupted process using LPS
```

FIGURE 6.14 Algorithm for processing an I/O interrupt.

action, of course, depends upon the nature of the I/O device and the error detected. For example, a parity error on a magnetic tape device is normally handled by backspacing the tape and restarting the I/O operation (up to some maximum number of times). On the other hand, an indication that a line printer is out of paper is handled by issuing a message to the computer operator before attempting any further recovery. If the operating system determines that an I/O error is uncorrectable, it may terminate the process that made the I/O request and send an appropriate message to the user. Alternatively, it might signal completion of the operation and store an error code in the ESB. The requesting process could then make its own decision about whether or not to continue.

After its processing is complete, the interrupt handler ordinarily returns control by restoring the status of the interrupted process. However, if the CPU was idle at the time of the interrupt, the dispatcher must be invoked. This is because the interrupt processing may have caused a process to become ready to execute. If this is the case, the CPU should not be restored to an idle status. Likewise, the dispatcher would be invoked if preemptive process scheduling were being used (see Section 6.2.2).

Figure 6.15 provides an illustration of the process-scheduling and I/O-supervision functions we have described. Two user processes, designated P1 and P2, are being executed concurrently. These are the

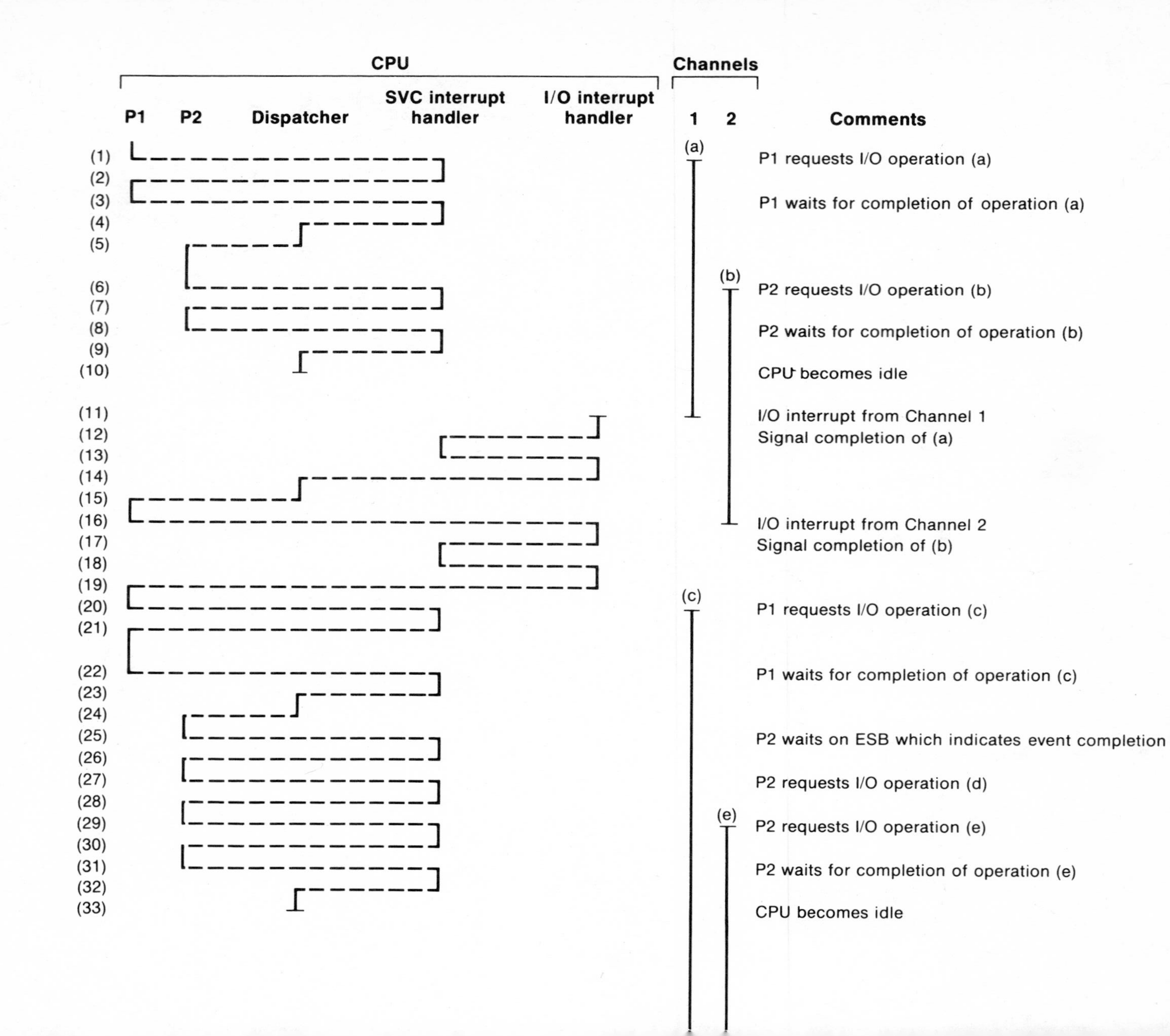

CPU
Channels
P1
P2
Dispatcher
SVC interrupt handler
I/O interrupt handler
1
2
Comments
(1)
(2)
(3)
(4)
(5)
(6)
(7)
(8)
(9)
(10)
(11)
(12)
(13)
(14)
(15)
(16)
(17)
(18)
(19)
(20)
(21)
(22)
(23)
(24)
(25)
(26)
(27)
(28)
(29)
(30)
(31)
(32)
(33)
(a)
(b)
(c)
(e)
P1 requests I/O operation (a)
P1 waits for completion of operation (a)
P2 requests I/O operation (b)
P2 waits for completion of operation (b)
CPU becomes idle
I/O interrupt from Channel 1
Signal completion of (a)
I/O interrupt from Channel 2
Signal completion of (b)
P1 requests I/O operation (c)
P1 waits for completion of operation (c)
P2 waits on ESB which indicates event completion
P2 requests I/O operation (d)
P2 requests I/O operation (e)
P2 waits for completion of operation (e)
CPU becomes idle

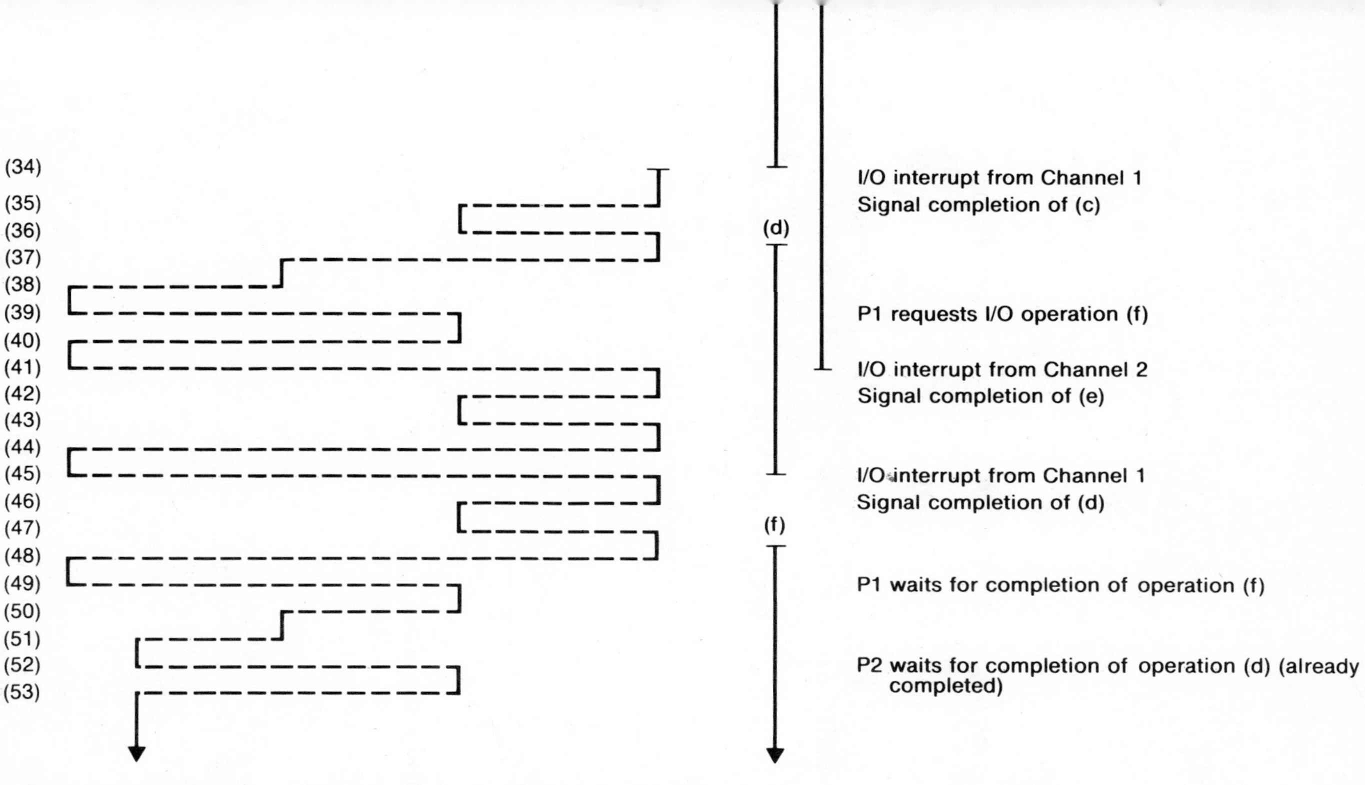

FIGURE 6.15 Example of I/O supervision and process-scheduling functions.

same processes that are outlined in Figs. 6.11 and 6.12. We assume the time-slice provided to each process by the dispatcher is relatively large so that timer interrupts do not ordinarily occur before the process must give up control for some other reason. The diagram in Fig. 6.15 shows the flow of activities being performed by the CPU, divided between the user processes and the parts of the operating system, and by the two I/O channels. The time scale runs from top to bottom on the diagram. Distances on this scale are not necessarily proportional to the actual lengths of time involved. Sequence numbers are provided to aid in the use of this example.

At the beginning of the example, processes P1 and P2 have both been initiated, and P1 has been dispatched first. Both I/O channels are idle. At sequence number (1), P1 makes its first I/O request by executing an SVC instruction. This causes an interrupt, which transfers control to the SVC-interrupt handler. For ease of reference, this I/O operation is designated in the diagram by (a). The I/O request specifies channel 1, which is currently free. Therefore, the SVC-interrupt handler starts the channel program and returns control to process P1 (sequence (2)).

At (3), P1 issues a WAIT request for operation (a) by executing another SVC instruction (SVC 0). Control is once again transferred to the SVC-interrupt handler. The ESB specified in this WAIT request indicates that the associated event has not yet occurred. Therefore, process P1 is placed in the blocked state, and the dispatcher is invoked (4). The dispatcher then switches control to process P2 (5). At sequence number (6), P2 issues its first I/O request, which is for device 2 on channel 2. Since channel 2 is free, the I/O operation is started, and control returns to P2. At (8), P2 must wait for the completion of its I/O request; since this event has not yet occurred, P2 becomes blocked. At (9), the dispatcher is invoked as before. This time, however, both processes are blocked, so the dispatcher places the CPU into an idle state (10). Note that both I/O channels are still active.

The CPU remains idle until (11), when channel 1 completes its I/O operation. We assume in this example that all operations are completed normally. The channel generates an I/O interrupt, which switches the CPU from its idle state to the I/O-interrupt handler. After determining that the operation was completed normally, the I/O-interrupt handler issues a SIGNAL request (SVC 1) for the associated ESB. This switches control to the SVC-interrupt handler (12). Process P1, which is waiting on this ESB, is placed in the ready state. The SVC handler then returns control to the I/O-interrupt handler (13). The dispatcher is invoked at sequence (14), and switches control to process P1 at (15). A similar series of operations occurs when channel 2 completes its operation (16); this causes process P2 to be made ready. However, since the CPU was not idle at the time of the interrupt, control does not pass immediately to P2. The I/O-interrupt handler restores control to the interrupted process P1. P2 does not receive control until P1 issues its next WAIT request at sequence (22).

You should follow carefully through the other steps in this example to be sure you understand the flow of control in response to the various interrupts. In doing this, you may find it useful to refer to the algorithms in Figs. 6.8, 6.9, 6.13, and 6.14. Note in particular the many different types of overlap between the CPU execution and the I/O operations of the different processes. The ability to provide such flexible sequencing of tasks is one of the most important advantages of an interrupt-driven operating system.

6.2.4 Management of Real Memory

Any operating system that supports more than one user at a time must provide a mechanism for dividing central memory among the concurrent processes. Many multiprogramming systems divide memory into *partitions,* with each process being assigned to a different partition. These partitions may be predefined in size and position (*fixed partitions*), or they may be allocated dynamically according to the requirements of the jobs being executed (*variable partitions*).

In this section we illustrate and discuss several different forms of partitioned memory management. Our examples use the sequence of jobs described in Fig. 6.16. We assume that the *level of multiprogramming* (i.e., the number of concurrent jobs) is limited only by the number of jobs that can fit into central memory.

Figure 6.17 illustrates the allocation of memory in fixed partitions. The total amount of memory available on the computer is assumed to

FIGURE 6.16 User jobs for memory allocation examples.

Job	Length (hexadecimal)
1	A000
2	14000
3	A800
4	4000
5	E000
6	B000
7	C000
8	D000

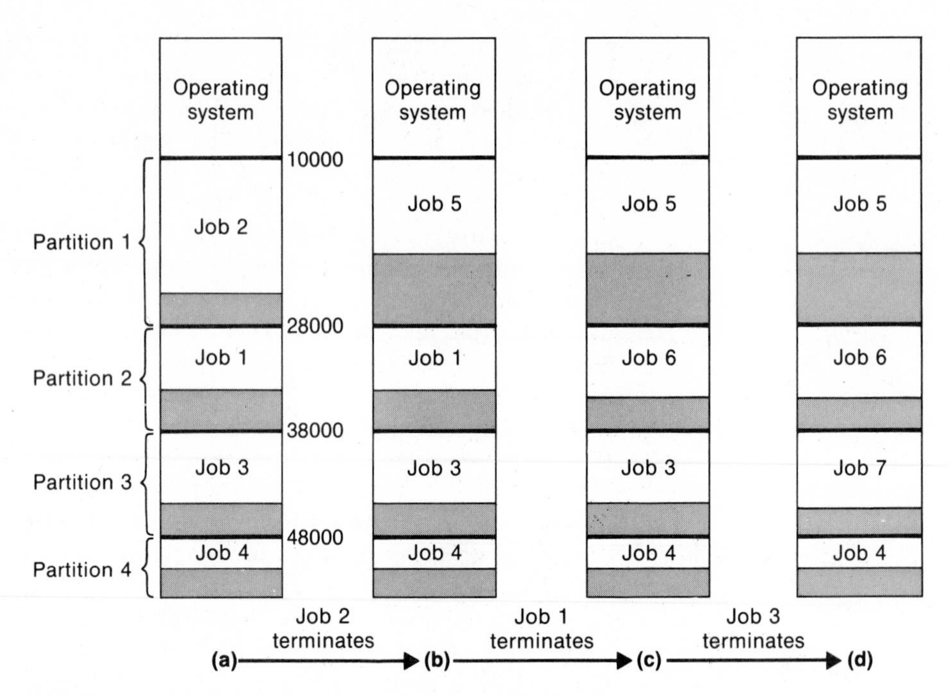

FIGURE 6.17 Memory allocation for jobs from Fig. 6.16 using fixed partitions.

be 50000 bytes; the operating system occupies the first 10000 bytes. Note that these sizes, and all other sizes and addresses used in this section, are given in hexadecimal. The memory that is not occupied by the operating system is divided into four partitions. Partition 1 begins at address 10000, immediately following the operating system, and is 18000 bytes in length. The other partitions follow in sequence: Partitions 2 and 3 are each 10000 bytes in length, and Partition 4 is 8000 bytes in length.

A simple allocation scheme using fixed partitions loads each incoming job into the smallest free partition in which it will fit. When the partition is larger than the size needed by the job, the excess memory within the partition is unused. With all four partitions initially empty, the system begins by loading Job 1 into Partition 2. Job 2 is then loaded into Partition 1, the only partition which is large enough to contain it. Jobs 3 and 4 are loaded into Partitions 3 and 4. All partitions are then occupied, so no more jobs can be loaded. The resulting memory allocations are shown in Fig. 6.17(a). The shaded areas of the diagram indicate unused memory locations.

Once it is loaded into a partition, a job remains until its execution is completed. After the job terminates, its partition becomes available for reuse. In Fig. 6.17(b), Job 2 has terminated and Job 5 has been loaded into Partition 1. The other parts of the figure show succeeding steps in the process as other jobs terminate.

Note that the partitions themselves remain fixed in size and position regardless of the sizes of the jobs that occupy them. The initial selection of the partition sizes is very important in a fixed partition scheme. There must be enough large partitions so that large jobs can be run without too much delay. If there are too many large partitions, however, a great deal of memory may be wasted when small jobs are run. The fixed partition technique is most effective when the sizes of jobs tend to cluster around certain common values, and when the distribution of job sizes does not change frequently. This makes it possible to make effective use of the available memory by tailoring a set of partitions to the expected population of job sizes.

Figure 6.18 illustrates the running of the same set of jobs using variable memory partitions. A new partition is created for each job to be loaded. This newly created partition is of exactly the size required to contain the job. When a job terminates, the memory assigned to its partition is released, and this memory then becomes available for use in allocating other partitions.

FIGURE 6.18 Memory allocation for jobs from Fig. 6.16 using variable partitions.

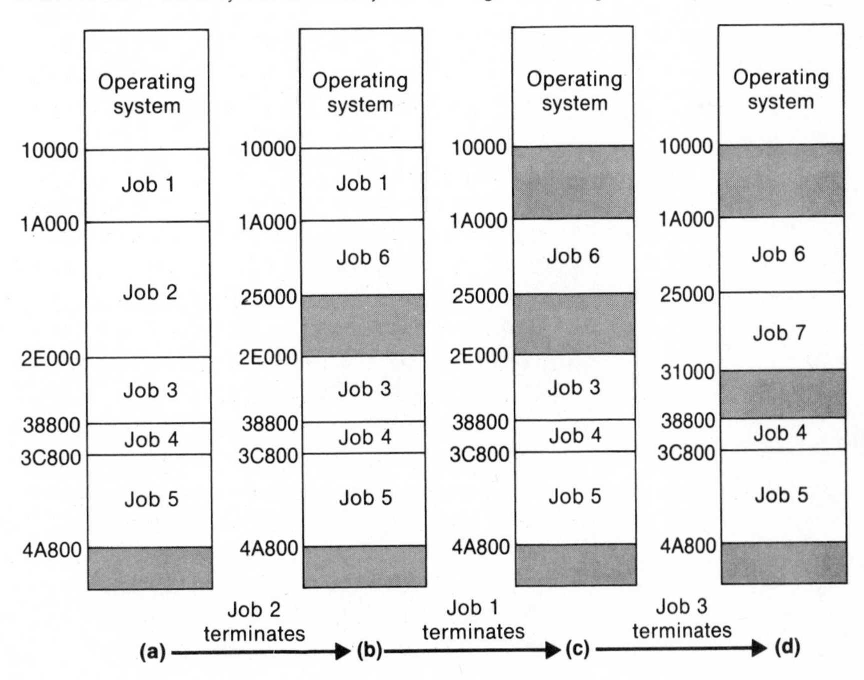

Initially, all memory except that assigned to the operating system is unallocated because there are no predefined partitions. When Job 1 is loaded, a partition is created for it. We assume this partition is allocated immediately after the operating system. Job 2 is then assigned a partition immediately following Job 1, and so on. Figure 6.18(a) shows the situation after the first five jobs are loaded. The free memory that remains following Job 5 is not large enough to load any other job.

When Job 2 terminates, its partition is released, and a new partition is allocated for Job 6. As shown in Fig. 6.18(b), this new partition occupies part of the memory previously assigned to Job 2. The rest of Job 2's former partition remains free. There are now two separate free areas of memory; however, neither of these is large enough to load another job. Figure 6.18(c) and (d) shows how the releasing and allocating of memory continue as other jobs terminate.

When variable partitions are used, it is not necessary to select partition sizes in advance. However, the operating system must do more work in keeping track of which areas of memory are allocated and which areas are free. Usually the system does this by maintaining a linked list of free memory areas. This list is scanned when a new partition is to be allocated. The partition is placed either in the first free area in which it will fit (*first-fit* allocation), or in the smallest free area in which it will fit (*best-fit* allocation). When a partition is released, its assigned memory is returned to the free list and combined with any adjacent free areas. A detailed discussion and comparison of such dynamic storage allocation algorithms can be found in Standish (1980).

Regardless of the partitioning technique that is used, it is necessary for the operating system and the hardware to provide *memory protection.* When a job is running in one partition, it must be prevented from modifying memory locations in any other partition or in the operating system. Some systems allow the reading of data anywhere in memory, but permit writing only within a job's own partition. Other systems restrict both reading and writing to the job's partition.

Some type of hardware support is necessary for effective memory protection. One simple scheme provides a pair of *bounds registers* that contain the beginning and ending addresses of a job's partition. These registers are not directly accessible to user programs; they can be used only when the CPU is in supervisor mode. The operating system sets the bounds registers when a partition is assigned to a user job. The values in these registers are automatically saved and restored during context switching operations such as those caused by an interrupt or an LPS instruction. Thus the bounds registers always contain the beginning and ending addresses of the partition assigned to the currently executing process. For every memory reference, the hardware automatically checks the referenced address against the bounds registers. If the address is outside the current job's partition, the memory reference is not performed and a program interrupt is generated.

A different type of memory protection scheme is used on SIC/XE. Each 800-byte (hexadecimal) block of memory has associated with it a 4-bit *storage protection key*. These keys can be set by the operating system using the privileged instruction SSK (Set Storage Key). Each user process has assigned to it a 4-bit *process identifier,* which is stored in the ID field of the status word SW. When a partition is assigned to a job, the operating system sets the storage keys for all blocks of memory within the partition to the value of the process identifier for that job. For each memory reference by a user program, the hardware automatically compares the process identifier from SW to the protection key for the block of memory being addressed. If the values of these two fields are not the same, the memory reference is not performed and a program interrupt is generated. However, this test is not performed when the CPU is in supervisor mode; the operating system is allowed to reference any location in memory.

One problem common to all general-purpose dynamic storage allocation techniques is *memory fragmentation*. Fragmentation occurs when the available free memory is split into several separate blocks, with each block being too small to be of use. Consider, for example, Fig. 6.18(c). There is more than enough total free memory to contain Job 7; however, this job cannot be loaded because no single free block is large enough.

Figure 6.19 illustrates one possible solution to this problem: *relocatable partitions*. After each job terminates, the remaining partitions are moved as far as possible toward one end of memory. This movement gathers all the available free memory together into one contiguous block that is more useful for allocating new partitions. As Fig. 6.19 shows, this technique can result in more efficient utilization of memory than is obtained with nonrelocatable partitions. However, the copying of jobs from one location in memory to another may require a substantial amount of time. This disadvantage can often outweigh the benefits of improved memory utilization.

The use of relocatable partitions also creates problems with program relocation. Consider, for example, the program P3 outlined in Fig. 6.20(a). The STA instruction specifies extended instruction format, so the actual address 08108 is contained in the assembled instruction. Using the methods described in Chapters 2 and 3, this address field would be flagged by the assembler. The value of the address field

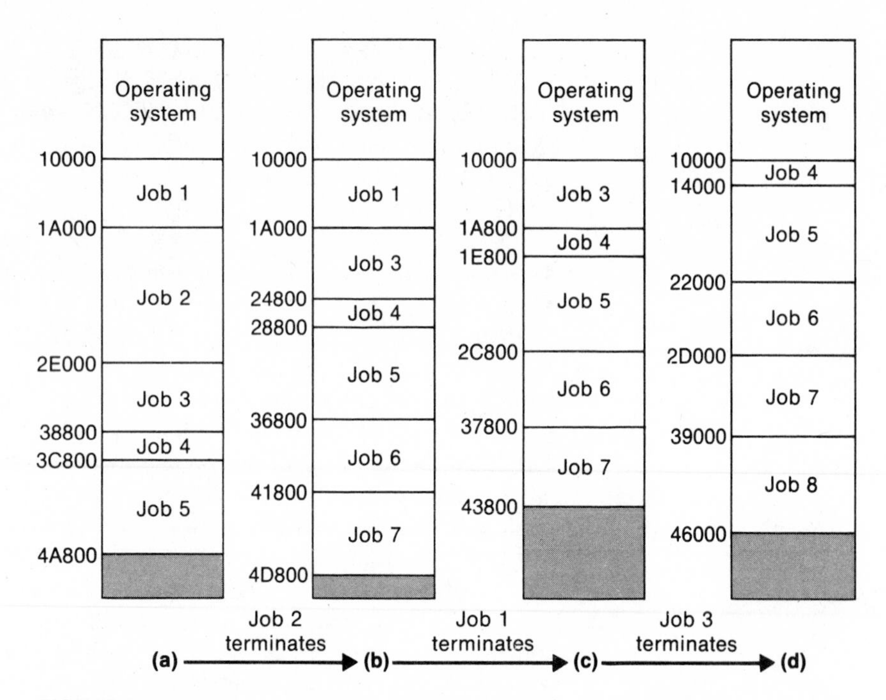

FIGURE 6.19 Memory allocation for jobs from Fig. 6.16 using relocatable partitions.

would be modified by the loader to reflect the actual address at which the program is loaded. For example, if P3 is initially loaded at actual address 2E000, the address field in the STA instruction would be changed to 36108, as illustrated in Fig. 6.20(b).

Now suppose the partition containing P3 is moved from address 2E000, as in Fig. 6.19(a), to address 1A000, as in Fig. 6.19(b). The address in the STA instruction would then be incorrect. In fact, it would refer to a memory location that is part of the partition assigned to a different job. A similar problem would occur if P3 loaded the address of some part of itself into a base register, or created a data structure that made use of address pointers. When a program is initially loaded, all relocatable values are defined by the assembler, which makes program relocation relatively simple. During execution, however, the program may use registers and memory locations in arbitrary ways. In general, it is not possible for the operating system to determine which values represent addresses and which represent other types of data. Thus relocating a program during execution is much more difficult than relocating it when it is first loaded.

In practice, the implementation of relocatable partitions requires some hardware support. One common method for achieving this is

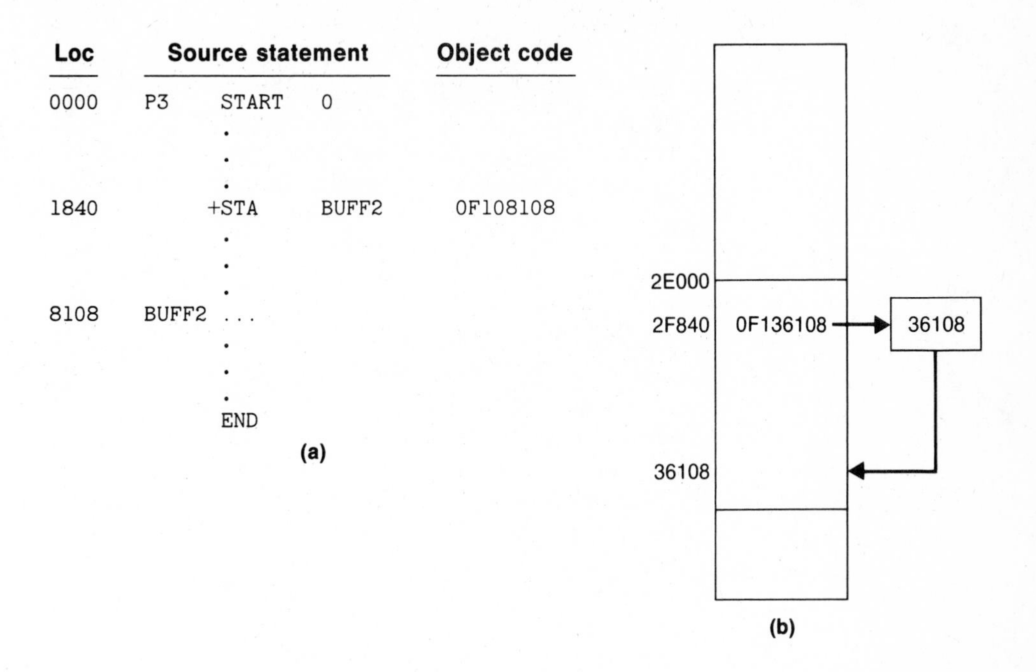

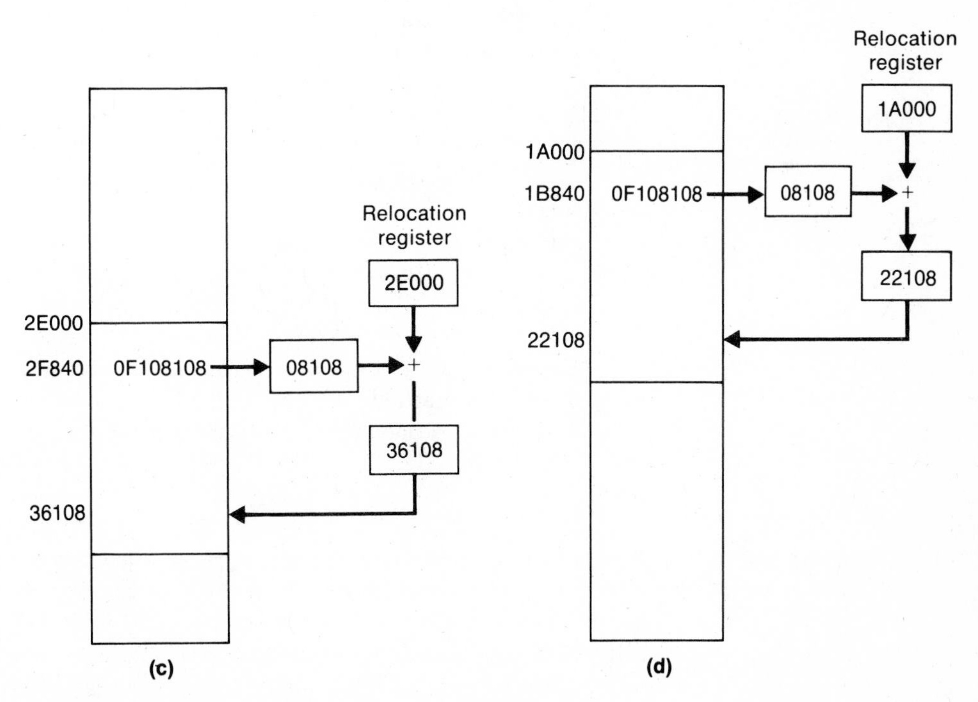

FIGURE 6.20 Use of relocation register in address calculation.

shown in Fig. 6.20(c). There is a special *relocation register* that is set by the operating system to contain the beginning address of the program currently being executed. This register is automatically saved and restored during context switching operations, and its value is modified by the operating system when a program is moved to a new location. The value in the relocation register is automatically added to the address for every memory reference made by the user program. For example, the STA instruction in Fig. 6.20(c) refers to address 08108; however, because of the relocation register, the address actually referenced is 36108. If program P3 were moved to address 1A000, as shown in Fig. 6.20(d), the operating system would change the value in the relocation register for P3 to 1A000. Thus the address actually referenced by the STA instruction would be 22108.

It is important to understand that the relocation register is under the control of the operating system; it is not directly available to the user program. Thus this register is quite different from the programmer-defined base registers found on SIC/XE, System/370, and many other computers. The relocation register is *automatically* involved each time a program refers to any location in memory, so it provides exactly the same effect as if each program were really loaded at actual address 00000. Indeed, on many computers there is no direct way for a user program to determine where it is actually located in memory. Thus this type of relocation applies to addresses given by pointers in data structures and values in base registers as well as to addresses in instructions. Note also that this automatic relocation performed by the hardware completely eliminates the need for program relocation by the loader.

In this section we described methods for allocating program partitions of predetermined size. Some operating systems allow user programs to dynamically request additional memory during execution. The additional memory assigned need not necessarily be contiguous with the original partition. Such dynamic storage allocation is usually performed with methods similar to those used in managing storage for data structures. A good discussion of such techniques can be found in Standish (1980).

6.2.5 Management of Virtual Memory

A *virtual* resource is one that appears to a user program to have characteristics that are different from those of the actual implementation of the resource. Consider, for example, Fig. 6.21. User programs are allowed to use a large contiguous *virtual memory,* sometimes called a *virtual address space.* This virtual memory may even be larger than the total amount of real memory available on the computer. The vir-

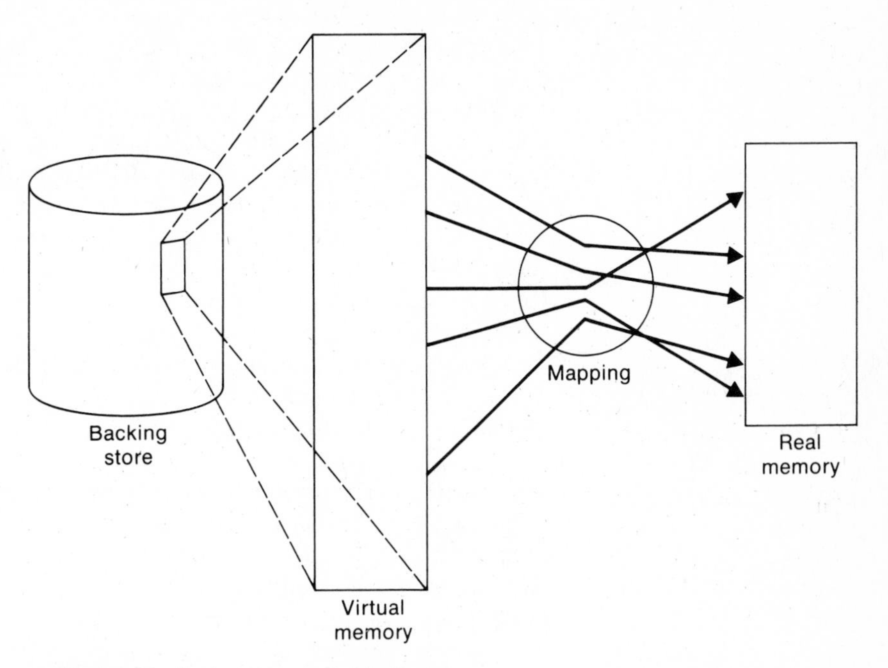

FIGURE 6.21 Basic concept of virtual memory.

tual memory used by a program is stored on some external device (the *backing store*). Portions of this virtual memory are mapped into real memory as they are needed by the program. The backing store and the virtual-to-real mapping are completely invisible to the user program. The program is written exactly as though the virtual memory really existed.

In this section we describe *demand paging,* which is one common method for implementing virtual memory. References to discussions of other types of virtual memories can be found at the end of the section.

In a demand-paging system, the virtual memory of a process is divided into *pages* of some fixed length. The real memory of the computer is divided into *page frames* of the same length as the pages. Any page from any process can potentially be loaded into any page frame in real memory. The mapping of pages onto page frames is described by a *page map table* (PMT); there is one PMT for each process in the system. The page map table is used by the hardware to convert addresses in a program's virtual memory into the corresponding addresses in real memory. This conversion process is similar to the use of the relocation register described in the last section. However, there is one PMT entry for each page instead of one relocation register for the entire program.

This conversion of virtual addresses to real addresses is known as *dynamic address translation*.

These concepts are illustrated by the program outlined in Fig. 6.22. The program is divided into pages that are 1000 bytes (hexadecimal) in length. Virtual addresses 0000 through 0FFF are in Page 0; addresses 1000 through 1FFF are in Page 1, and so on. When the execution of the program is begun, the operating system loads Page 0, the page containing the first executable instruction, into some page frame in real memory. Other pages are loaded into memory as they are needed.

The processes of dynamic address translation and page loading are illustrated in Fig. 6.23. In Fig. 6.23(a), Page 0 of the program has been loaded into page frame 1D (i.e., real memory addresses 1D000–1DFFF). Consider first the JEQ instruction that is located at virtual address 0103. The operand address for this instruction is virtual address 0420. We used an instruction format that provides direct addressing to make this initial example easier to follow. The operand address 0420 is located within Page 0, at offset 420 from the beginning of the page. The page map table indicates that Page 0 of this program is loaded in page frame 1D (that is, beginning at address 1D000). Thus the real address calculated by the dynamic address translation is 1D420.

Next let us consider the LDA instruction at virtual address 0420. The operand for this instruction is at virtual address 6FFA (Page 6, offset FFA). However, Page 6 has not yet been loaded into real memory, so the dynamic address translation hardware is not able to compute a real address. Instead, it generates a special type of program interrupt called a *page fault* (see Fig. 6.23b). The interrupt-handling routine, which we discuss later, responds to this interrupt by loading the required page into some page frame. Let us assume page frame 29 is chosen. The instruction that caused the interrupt is then reexecuted. This time, as shown in Fig. 6.23(c), the dynamic address translation is successful.

The other pages of the program are loaded on demand in a similar way. Assume that Fig. 6.22 shows all the Jump instructions in the program as well as all instructions whose operands are located in another page. When control passes from the last instruction in Page 0 to the first instruction in Page 1, the instruction-fetch operation causes a page fault, which results in the loading of Page 1. The STA instruction at virtual address 1840 causes Page 8 to be loaded. Page 2 is then loaded as a result of the instruction-fetch cycle, just as Page 1 was. Now consider the two Jump instructions at addresses 2020 and 2024. If the first of these jumps is executed (i.e., if the less-than condition is

```
          Loc       Source statement          Object code

        ( 000000    P3     START   0
        | .
        | .
Page 0  { 000103           +JEQ    SKIP1       33100420
        | .
        | .
        | 000420    SKIP1  +LDA    BUFF1       03106FFA
        ( .
        ( .
        | .
        | .
Page 1  { .
        | 001840    SKIP2  +STA    BUFF2       0F108108
        | .
        | .
        ( .
        ( .
        | .
        | .
Page 2  { 002020           +JLT    SKIP3       3B104A00
        | 002024            J      SKIP2       3F281C
        | .
        | .
        ( .
        ( .
        | .
        | .
Page 3  { .
        | .
        | .
        | .
        ( .
        ( .
        | .
        | 004A00    SKIP3   LDX    #8          050008
Page 4  { .
        | .
        | 004A20           +STA    BUFF1,X     0F906FFA
        | .
        ( .
        ( .
        | .
        | .
Page 5  { .
        | .
        | .
        | .
        ( .
```

FIGURE 6.22 Program for illustration of demand paging.

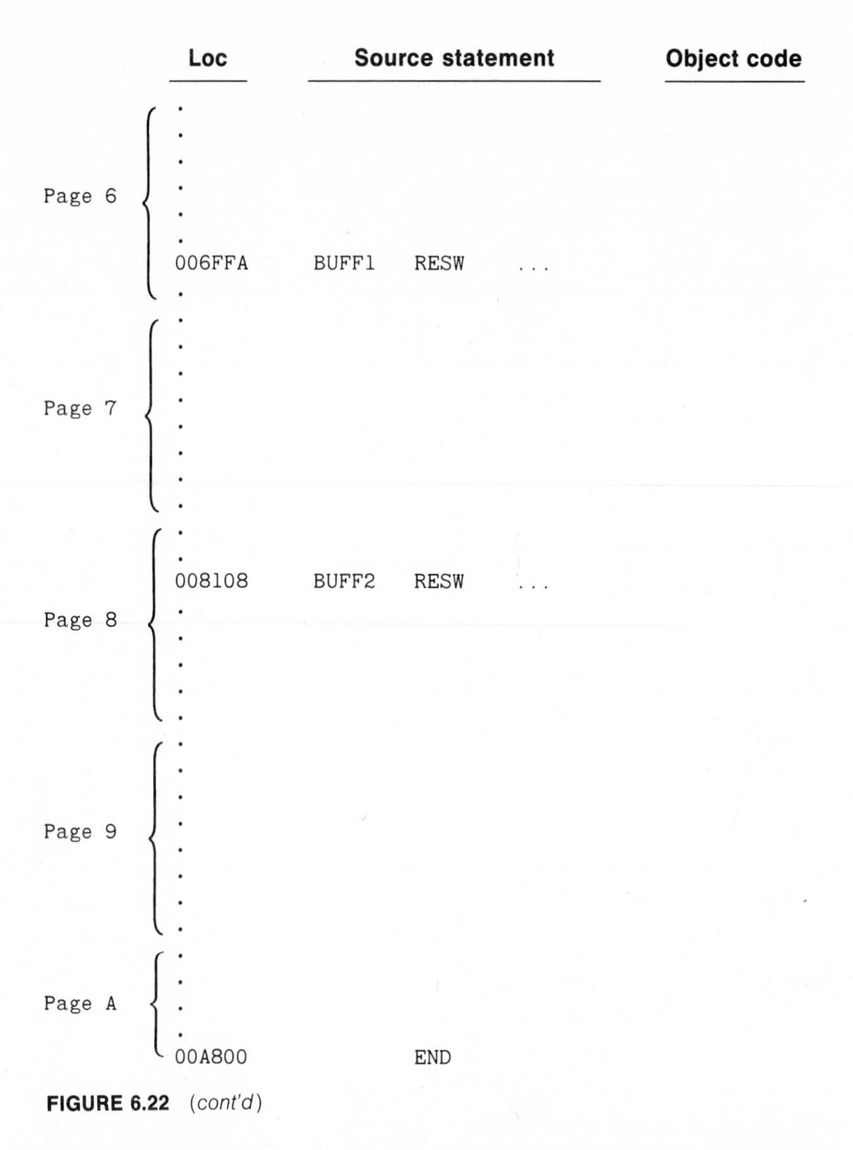

FIGURE 6.22 *(cont'd)*

true), it causes Page 4 to be loaded; otherwise, Page 4 remains unloaded. In this latter case, the unconditional jump at 2024 transfers control back to a location in Page 1 (which has already been loaded). After control passes to Page 4, the STA instruction at 4A20 is executed. This instruction specifies an operand address of 6FFA, with indexed addressing. We assume the value 8 remains in the index register. The resulting target address is 7002; as a result of this instruction, Page 7 is loaded.

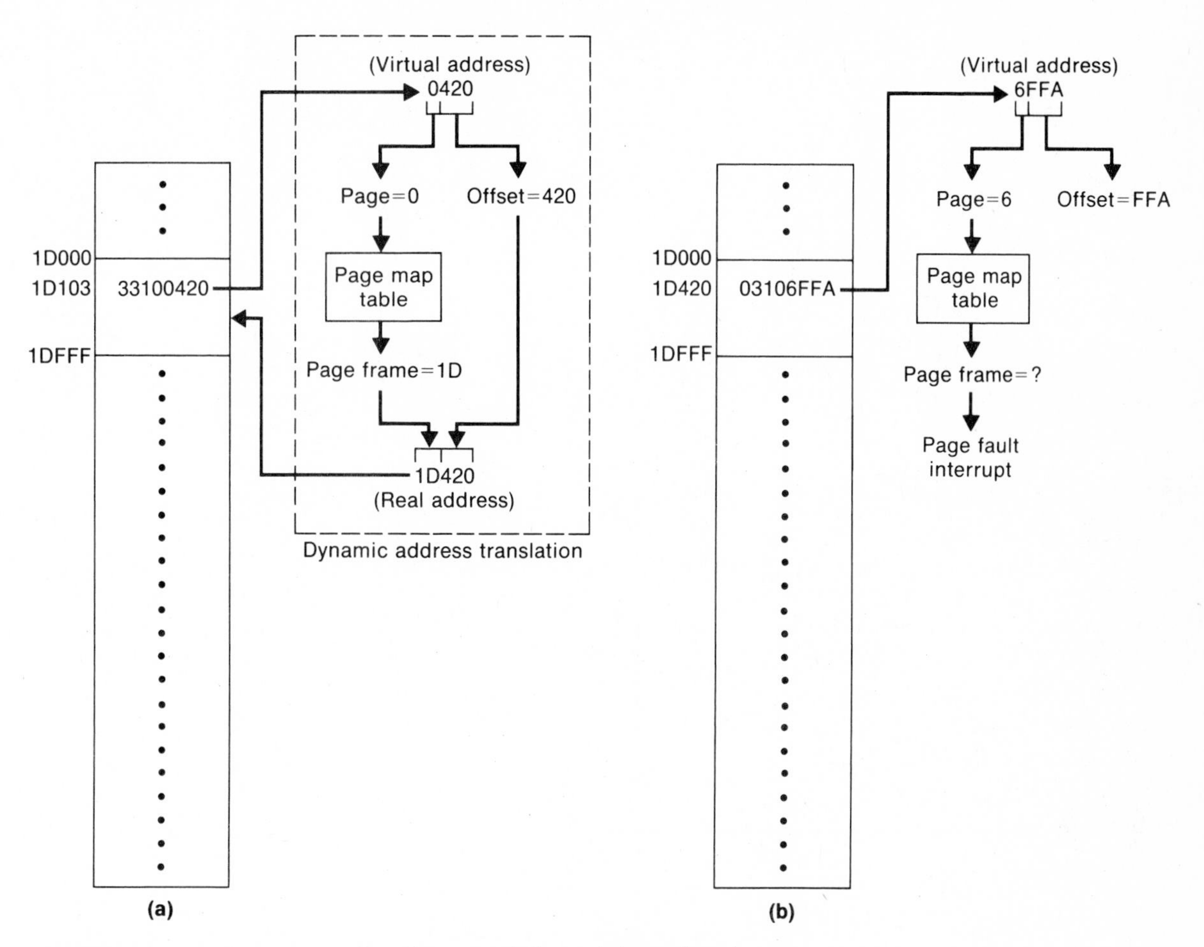

FIGURE 6.23 Examples of dynamic address translation and demand paging.

Figure 6.24 shows the situation after the sequence of events just described. The page map table for P3 reflects the fact that Pages 0, 1, 2, 4, 6, 7, and 8 are currently loaded, and gives the corresponding page-frame numbers. There is a similar page table for every other program in the system. Note that the page map table also indicates which pages have been modified since they were loaded (in this case, Pages 7 and 8). This information is used by the page-fault interrupt-handling routine when it is necessary to remove a page already in memory.

Figure 6.25 summarizes the address-translation and demand-paging functions illustrated in the previous discussion. Figure 6.25(a) describes the dynamic address translation algorithm used. Recall that this algorithm is implemented directly by the hardware of the ma-

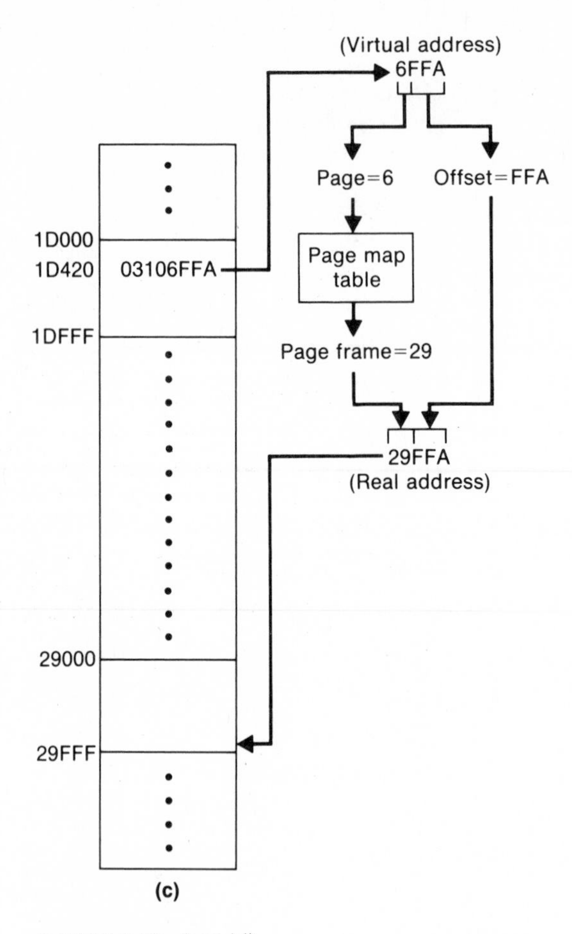

(c)

FIGURE 6.23 *(cont'd)*

chine. If the dynamic address translation cannot be completed because the required page is not in memory, a page fault interrupt occurs.

Figure 6.25(b) describes the interrupt-handling routine that is invoked for a page fault. The operating system maintains a table describing the status of all page frames. The first step in processing a page fault interrupt is to search this table for an empty page frame. If an empty page frame is found, the required page can be loaded immediately. Otherwise, a page currently in memory must be removed to make room for the page to be loaded. If the page being removed has been modified since it was last loaded, then the updated version must be rewritten on the backing store. If the page has not been modified, the image in memory can simply be discarded.

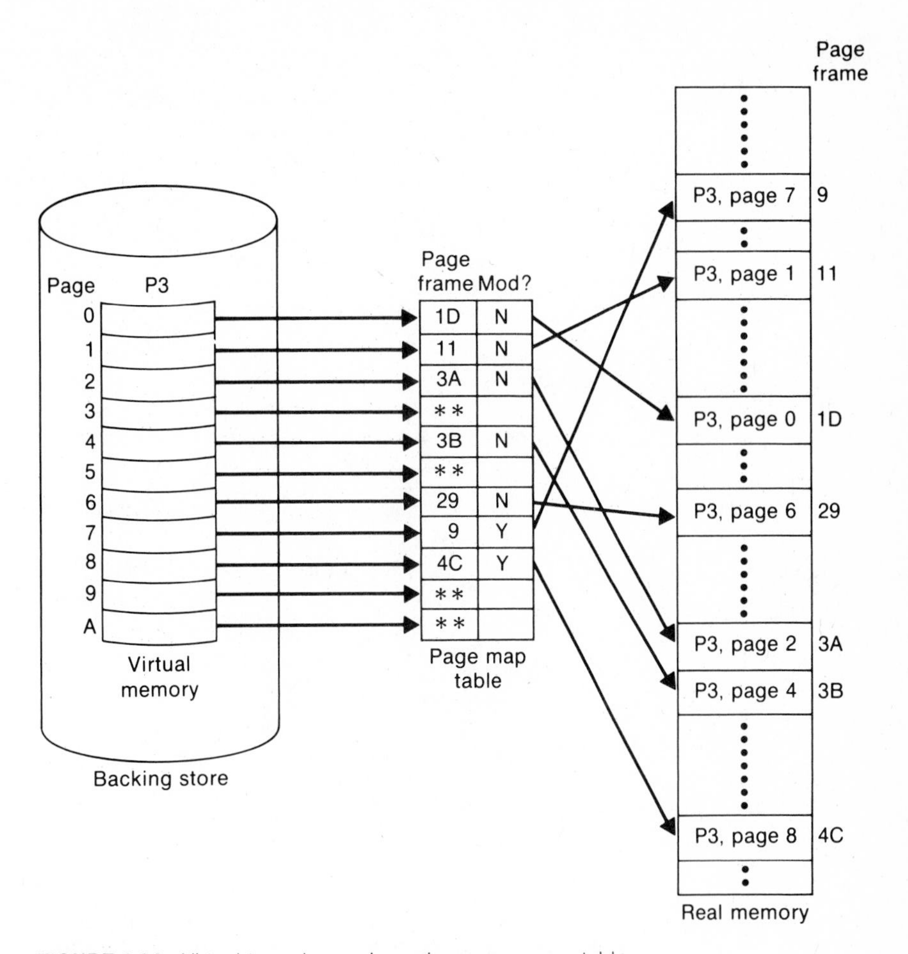

FIGURE 6.24 Virtual-to-real mapping using a page map table.

The page-fault interrupt-handling routine described in Fig. 6.25(b) requires the performance of at least one physical I/O operation. Thus this routine takes much longer to execute than any of the other interrupt handlers we have discussed. It is usually not desirable to inhibit interrupts for long periods of time. For example, it is clearly advantageous to allow other programs to use the CPU while paging I/O is being performed. The management of this multiprogramming requires that the interrupt system be available. Therefore, the page-fault interrupt handler first determines what actions need to be performed and saves the status information from the interrupted process. The interrupt handler than enables the interrupt system during the remainder of the processing.

```
procedure DAT          {implemented in hardware}
   decompose virtual address into (page number, offset)
   find entry in PMT for this page
   if the page is currently in memory then
      begin
         combine (page frame address, offset) to form real address
         if this is a "store" instruction then
            mark PMT entry to indicate page modified
      end {if page is in memory}
   else
      generate page fault interrupt
```

(a)

```
procedure PAGEFAULT  {implemented as part of the operating system}
   save process status from interrupt work area
   mark process as Blocked
   if there is an empty page frame then
      begin
         select an empty page frame
         mark the selected page frame table entry as committed
         enable all interrupts using LPS
      end {if empty page frame}
   else
      begin
         select page to be removed
         mark the selected page frame table entry as committed
         update PMT to reflect the removal of the page
         enable all interrupts using LPS
         if the selected page has been modified then
            begin
               issue I/O request to rewrite page to backing store
               wait for completion of the write operation
            end {if modified}
      end {if no empty page frame}
   issue I/O request to read page into the selected page frame
   wait for completion of the read operation
   update PMT and page frame table
   mark process as Ready
   restore status of user process that caused the page fault
```

(b)

FIGURE 6.25 Algorithms for dynamic address translation and page-fault interrupt processing.

More specifically, the interrupt handler selects a page frame to receive the required page and marks this frame as *committed* so that it will not be selected again because of a subsequent page fault. If a page is to be removed, the PMT for the process that owns that page is updated to reflect its removal. This prevents that process from attempting to reference the page while it is being removed. The interrupt handler saves the status information from the program-interrupt work area, placing it in some location associated with the process so that this information will not be destroyed by a subsequent program interrupt. Then the interrupt handler turns on the interrupt system by loading a status word that is the same as the current SW, except that all MASK bits are set to 1. The remainder of the interrupt-processing routine functions in much the same way as a user process: it makes SVC requests to initiate I/O operations and to wait for the results. After the completion of the paging operation, the interrupt handler uses the saved status information to return control to the instruction that caused the page fault. The dynamic address translation for this instruction will then be repeated.

The algorithm description in Fig. 6.25(b) leaves several important questions unanswered. The most obvious of these is which page to select for removal. Some systems keep records of when each page in memory was last referenced and replace the page that has been unused for the longest time. This is called the *least recently used* (LRU) method. Since the overhead for this kind of record keeping can be high, simpler approximations to LRU are often used. Other systems attempt to determine the set of pages that are frequently used by the process in question (the so-called *working set* of the process). These systems attempt to replace pages in such a way that each process always has its working set in memory. Discussions and evaluations of various page replacement strategies can be found in Deitel (1984).

Another unanswered question concerns the implementation of the page tables themselves. One possible solution is to implement these tables as arrays in central memory. A register is set by the operating system to point to the beginning of the PMT for the currently executing process. This method can be very inefficient because it requires an extra memory access for each address translation. Some systems, however, use such a technique in combination with a high-speed buffer to improve average access time. Another possibility is to implement the page map tables in a special high-speed associative memory. This is very efficient, but may be too expensive for systems with large real memories. Further discussions of these and other PMT implementation techniques can be found in Deitel (1984).

Demand-paging systems avoid most of the wasted memory due to fragmentation that is often associated with partitioning schemes. They also save memory in other ways. For example, parts of a program that are not used during a particular execution need not be loaded. However, demand-paging systems are vulnerable to other serious problems. For example, suppose that referencing a word in central memory requires 1 microsecond, and that fetching a page from the backing store requires an average of 10 milliseconds (10,000 microseconds). Suppose also that on the average, considering all jobs in the system, only 1 out of 100 virtual memory references causes a page fault. Even with this apparently low page-fault rate, the system will not perform well. For every 100 memory references (requiring 100 microseconds), the system will spend 10,000 microseconds fetching pages from the backing store. Thus the computing system will spend approximately 99 percent of its time swapping pages, and only 1 percent of its time doing useful work. This total collapse of service because of a high paging rate is known as *thrashing*.

To avoid thrashing in the situation just described, it is necessary for the page-fault rate to be much lower (perhaps on the order of one fault for every 10,000 memory references). At first glance, this might seem to make demand paging useless. It appears that all of a program's pages would need to be in memory to achieve acceptable performance. However, this is not necessarily the case. Because of a property called *locality of reference,* which can be observed in most real programs, memory references are not randomly distributed through a program's virtual address space. Instead, memory references tend to be clustered together in the address space, as illustrated in Fig. 6.26(a). This clustering is due to common program characteristics such as sequential instruction execution, compact coding of loops, sequential processing of data structures, and so on.

Because of locality of reference, it is possible to achieve an acceptably low page-fault rate without keeping all of a program's address space in real memory. Figure 6.26(b) shows the page fault rate for a hypothetical program as a function of the number of the program's pages kept in memory. The scaling of this curve varies markedly from one program to another; however, the general shape is typical of many real programs. Often, as in this example, there is a critical point W. If fewer than W pages are in memory, thrashing will occur. If W pages or more are in memory, performance will be satisfactory. This critical point W is the size of the program's working set of pages that was mentioned earlier. For further discussion of the issues of working-set size and thrashing, see Deitel (1984) and Lorin (1981).

Demand paging provides yet another example of delayed binding: the association of a virtual-memory address with a real-memory address is not made until the memory reference is performed. This delayed binding requires more overhead (for dynamic address translation, page fetching, etc.). However, it can provide more convenience for the programmer and more effective use of real memory. You may want to compare these observations with those made in the previous examples of delayed binding (Sections 3.4.2, 5.3.1, and 5.4.2).

In this section we described an implementation of virtual memory

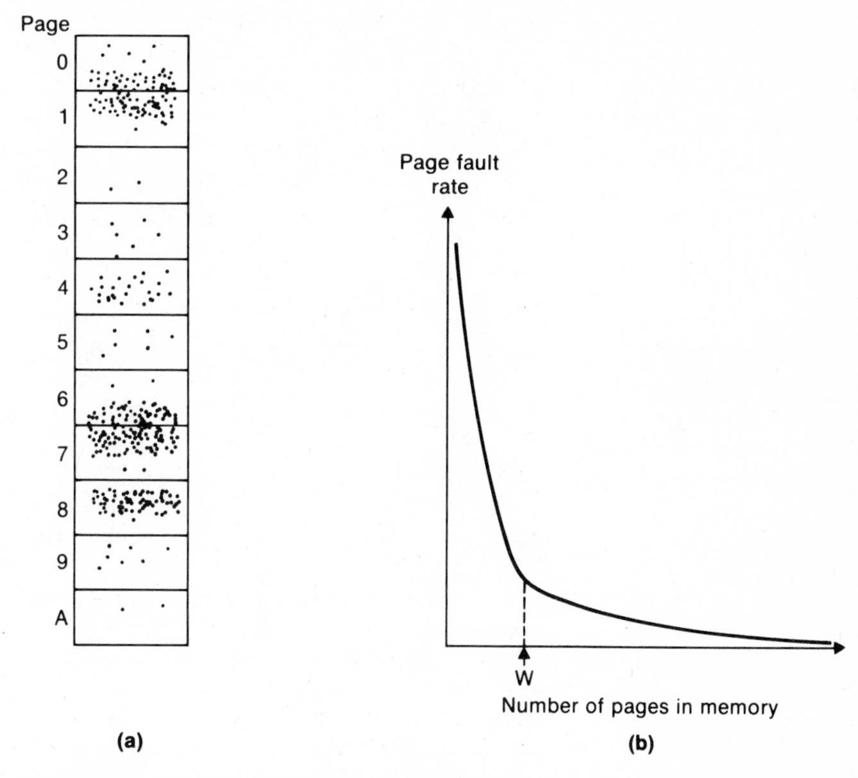

FIGURE 6.26 (a) Localized memory references. (Each dot represents one memory location referenced in a given period of time during the execution of the program.) (b) Effect of localized references on page-fault rate.

using demand paging. A different type of virtual memory can be implemented using a technique called *segmentation*. In a segmented virtual-memory system, an address consists of a segment number and an offset within the segment being addressed. The concepts of mapping and dynamic address translation are similar to those we have discussed. However, segments may be of any length (as opposed to pages, which are of a fixed length for the entire system). Also, segments usually correspond to logical program units such as procedures or data areas (as opposed to pages, which are arbitrary divisions of the address space). This makes it possible to associate protection attributes such as *read only* or *execute only* with certain segments. It is also possible for segments to be shared between different user jobs. Segmentation is often combined with demand paging. This combination requires a two-level mapping and address-translation procedure. For further information about segmentation and its implementation, see Deitel (1984), Lorin (1981), and Madnick (1974).

6.3 MACHINE-INDEPENDENT OPERATING SYSTEM FEATURES

In this section we briefly describe several common functions of operating systems that are not directly related to the architecture of the machine on which the system runs. These features tend to be implemented at a higher level—that is, further removed from the machine level—than the features we have discussed so far. For this reason, such topics are not as fundamental to the basic purpose of an operating system as are the hardware-support topics discussed in the last section. Because of space limitations, these subjects are not discussed in as much detail as were the machine-dependent features. References are provided for readers who want to learn more about the topics introduced here.

In Section 6.2.3 we described a technique that can be used to manage I/O operations. Section 6.3.1 examines a related topic at a higher level: the management and processing of logical files. Similarly, Section 6.3.2 discusses the problem of job scheduling, which selects user jobs as candidates for the lower-level process scheduling discussed previously.

Section 6.3.3 discusses the general subject of resource allocation by an operating system and describes some of the problems that may occur. Finally, Section 6.3.4 provides a brief introduction to the important topics of protection and operating system security.

6.3.1 File Processing

In this section we introduce some of the functions performed by a typical operating system in managing and processing files. On most systems, it is possible for user programs to request I/O by using the mechanisms described in Section 6.2.3: constructing channel programs and making SVC requests to execute them. However, this is generally inconvenient. The programmer is required to be familiar with the details of the channel command codes and formats. The user program must know the correct channel number and device number. In the case of a direct-access device such as a disk, the program also must know the actual address of the desired record on the device. In addition, the program must take care of details such as waiting for I/O completion, and the buffering and blocking functions described later in this section.

The file-management function of an operating system is an intermediate stage between the user program and the I/O supervisor. This function is illustrated in Fig. 6.27. The user program makes requests, such as "Read the next record from file F," at a logical level, using file names, keys, etc. The file-management routine, which is sometimes

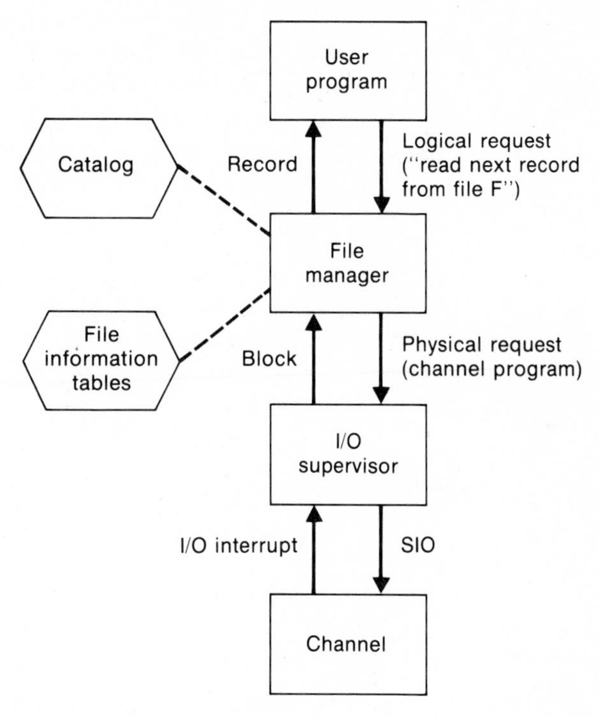

FIGURE 6.27 I/O using a file manager routine.

called an *access method,* translates these logical requests into physical I/O requests (i.e., channel programs), and passes the requests to the I/O supervisor. The I/O supervisor then functions as described in Section 6.2.3 to direct the operation of the I/O channels.

To convert the program's logical requests into channel programs, the file manager must have information about the location and structure of the file. It obtains such information from data structures we call the *catalog* and the *file information tables.* The actual terms used for these structures, as well as their formats and contents, vary considerably from one operating system to another. The catalog relates logical file names to their physical locations and may give some basic information about the files. The file information table for a file gives additional information such as file organization, record length and format, and indexing technique, if any. To begin the processing of a file, the file manager searches the catalog and locates the appropriate file information table. The file manager may also create buffer areas to receive the blocks being read or written. This initialization procedure is known as *opening* the file. After the processing of the file is completed, the buffers and any other work areas and pointers are deleted. This procedure is called *closing* the file.

One of the most important functions of the file manager is the automatic performance of *blocking* and *buffering* operations on files being read or written. Figure 6.28 illustrates these operations on a sequential input file. We assume the user program starts reading records at the beginning of the file and reads each record in sequence until the end. The file logically consists of records that are 1024 bytes long; however, the file is physically written in 8192-byte blocks, with each block containing 8 logical records. This sort of blocking of records is commonly done with certain types of storage devices to save processing time and storage space. For a detailed discussion, see Loomis (1983).

FIGURE 6.28 Blocking and buffering of a sequential file.

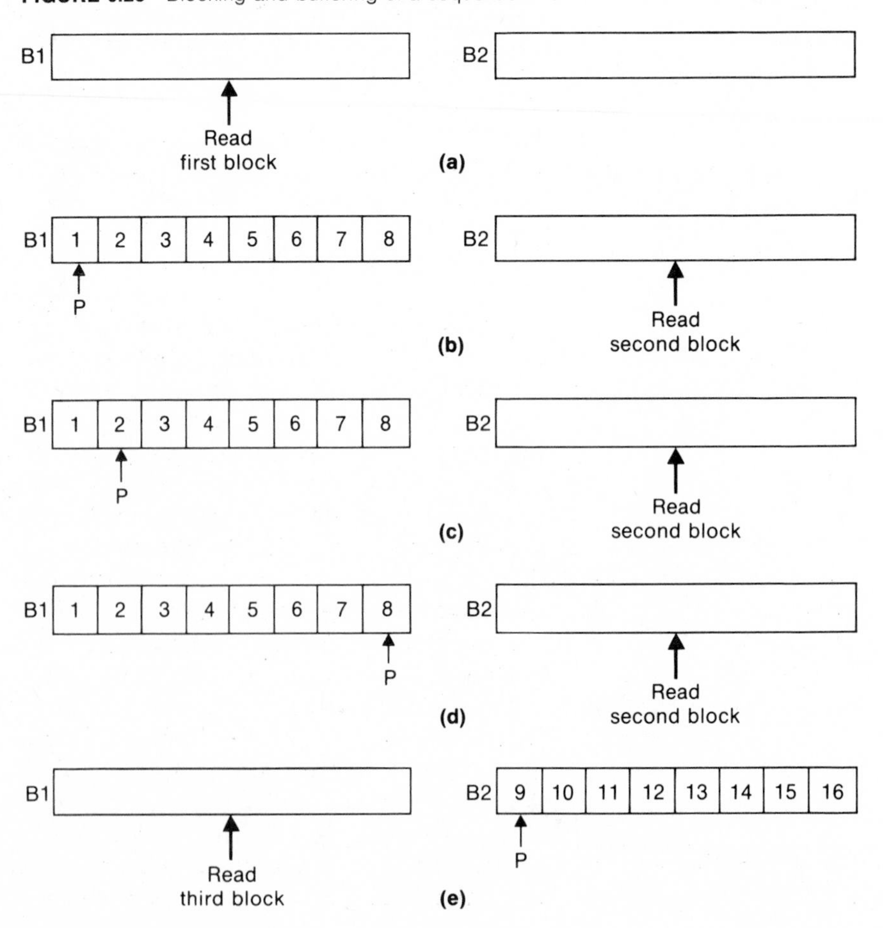

Figure 6.28(a) shows the situation after the file has been opened and the user program has made its first read-record request. The file manager has issued an I/O request to read the first block of the file into buffer B1. The file manager must wait for the completion of this I/O operation before it can return the requested record to the user. In Fig. 6.28(b), the first block has been read. This block, containing logical records number 1 through 8, is present in buffer B1. The file manager can now return the requested record to the user program. In this case, the requested record is returned by setting a pointer P to the first logical record. The file manager also issues a second physical I/O request to read the second block of the file into buffer B2.

The next time the user program makes a read-record request, it is not necessary to wait for any physical I/O activity. The file manager simply advances the pointer P to logical record 2, and returns to the user. This operation is illustrated in Fig. 6.28(c). Note that the physical I/O operation that reads the second block into buffer B2 is still in progress. The same process continues for the rest of the logical records in the first block (see Fig. 6.28d).

If the user program makes its 9th read-record request before the completion of the I/O operation for block 2, the file manager must again cause the program to wait. After the second block has been read, the pointer P is switched to the first record in buffer B2. The file manager then issues another I/O request to read the third block of the file into buffer B1, and the process continues as just described. Note that the use of two buffer areas allows overlap of the internal processing of one block with the reading of the next. This technique, often called *double buffering,* is widely used for input and output of sequential files.

The user program in the previous example simply makes a series of read-record requests. It is unaware of the buffering operations and of the details of the physical I/O requests being performed. Compare this with the program in Fig. 6.11, which performs a similar buffering function by dealing directly with the I/O supervisor. Clearly, the use of the file manager makes the user program much simpler and easier to write, and therefore less error prone. It also avoids the duplication of similar code in a large number of programs.

File-management routines also perform many other functions, such as the allocation of space on external storage devices and the implementation of rules governing file access and use. For further discussions of such topics, see Deitel (1984) and Madnick (1974).

6.3.2 Job Scheduling

Job scheduling is the task of selecting the next user job to begin execution. In a single-job system, the job scheduler completely specifies the order of job execution. In a multiprogramming system, the job scheduler specifies the order in which jobs enter the set of tasks that are being executed concurrently.

Figure 6.29(a) illustrates a typical two-level scheduling scheme for a multiprogramming system. Jobs submitted to the system become part of an *input queue;* a *job scheduler* selects jobs from this workload. The jobs selected become *active,* which means they begin to participate in the process-scheduling operation described in Section 6.2.2. This two-level procedure is used to limit the *multiprogramming level,* which is the number of user jobs sharing the CPU and the other system resources. Such a limitation is necessary in a multiprogramming system to maintain efficient operation. If the system attempts to run too many jobs concurrently, the overhead of resource management becomes too large, and the amount of resources available to each job becomes too small. As a result, system performance is degraded.

In the scheme just described, the job scheduler is used as a tool to maintain a desirable level of multiprogramming. However, this ideal multiprogramming level may vary according to the jobs being executed. Consider, for example, a system that uses demand-paged memory management. The number of user jobs that can share the real memory is essentially unlimited. Each job can potentially be executed with as little as one page in memory. However, thrashing occurs when a job does not have a certain critical number of pages in memory, and the performance of the overall system suffers. Unfortunately, the num-

FIGURE 6.29 (a) Two-level scheduling system and (b) three-level scheduling system.

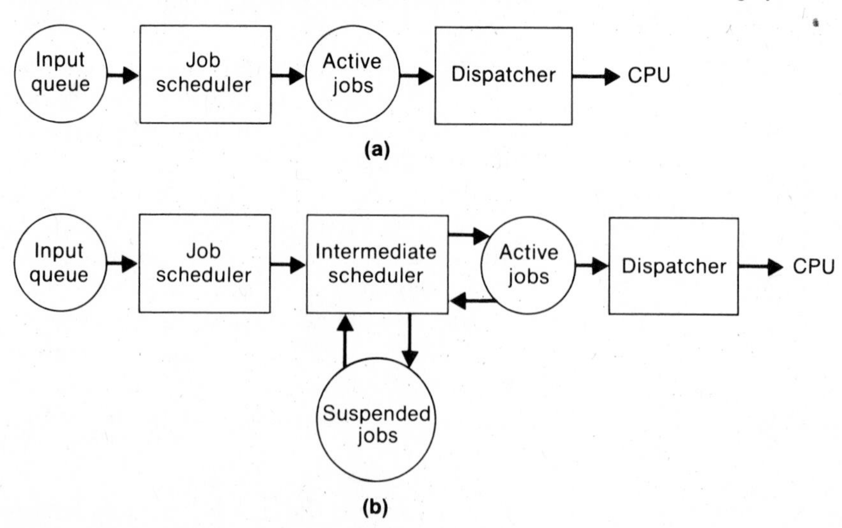

ber of pages a job requires to prevent thrashing is difficult to predict. In addition, this critical number of pages may change considerably during the execution of the program, so the desired level of multiprogramming may change during the operation of the system.

The multiprogramming level can be increased easily enough by simply invoking the job scheduler, assuming that the input queue is not empty. It is more difficult to decrease the level of multiprogramming, which would be done, for example, to stop a system from thrashing. Figure 6.29(b) shows a three-level scheduling procedure that is commonly used to accomplish this decrease. The job scheduler and the process scheduler (i.e., the dispatcher) operate as before. However, there is also an *intermediate-level scheduler* that monitors system performance and adjusts the multiprogramming level as needed. If the multiprogramming level is too high, the intermediate scheduler lowers it by *suspending* or *rolling out* one or more active jobs. If the multiprogramming level is too low, the intermediate scheduler resumes the execution of a suspended job or calls on the job scheduler for a new job to be made active. Such intermediate schedulers can also be used to adjust the dispatching priority of active jobs, based on observation of the jobs during execution.

The overall scheduling system is usually based on a system of priorities that are designed to help meet desired goals. For example, one goal might be to achieve the maximum system *throughput*—that is, to perform the most computing work in the shortest time. Clearly, achieving this goal is based on making effective use of overall system resources. Another common goal is to achieve the lowest average *turnaround time,* which is the time between the submission of a job by a user and the completion of that job. A related goal for a timesharing system is to minimize expected *response time,* which is the length of time between pressing the ENTER key on a terminal and the acceptance of this request by the system.

There are many other possible scheduling goals for a computing system. For example, we might want to provide a guaranteed level of service by limiting the maximum possible turnaround time or response time. Another alternative is to be equitable by attempting to provide the same level of service for all. On the other hand, it may be desirable to give certain jobs priority for external reasons such as meeting deadlines or providing good service to important or influential users. On some systems it is even possible for users to get higher priority by paying higher rates for service, in which case the overall scheduling goal of the system might be to make the most money.

The first two goals mentioned above—high throughput and low average turnaround time or response time—are commonly accepted as desirable system characteristics. Unfortunately, these two goals are often incompatible. Consider, for example, a timesharing system with a large number of terminals. We might choose to provide better response time by switching control more rapidly among the active user terminals. This could be accomplished by giving each process a shorter time-slice when it is dispatched. However, the use of shorter time-slices would mean a higher frequency of context switching operations, and would require the operating system to make more frequent decisions about the allocation of the CPU and other resources. This means the operating system overhead would be higher and correspondingly less time would be available for user jobs, so the overall system throughput would decline.

On the other hand, consider a batch processing system that runs one job at a time. The execution of two jobs on such a system is illustrated in Fig. 6.30(a). Note the periods of CPU idle time, represented by gaps in the horizontal lines for Job 1 and Job 2. If both jobs are submitted at time 0, then the turnaround time for Job 1 (T_1) is 2 minutes, and the turnaround time for Job 2 (T_2) is 5 minutes. The average turnaround time, T_{avg}, is 3.5 minutes.

Now consider a multiprogramming system that runs two jobs con-

FIGURE 6.30 Comparison of turnaround time and throughput for (a) a single-job system and (b) a multiprogramming system.

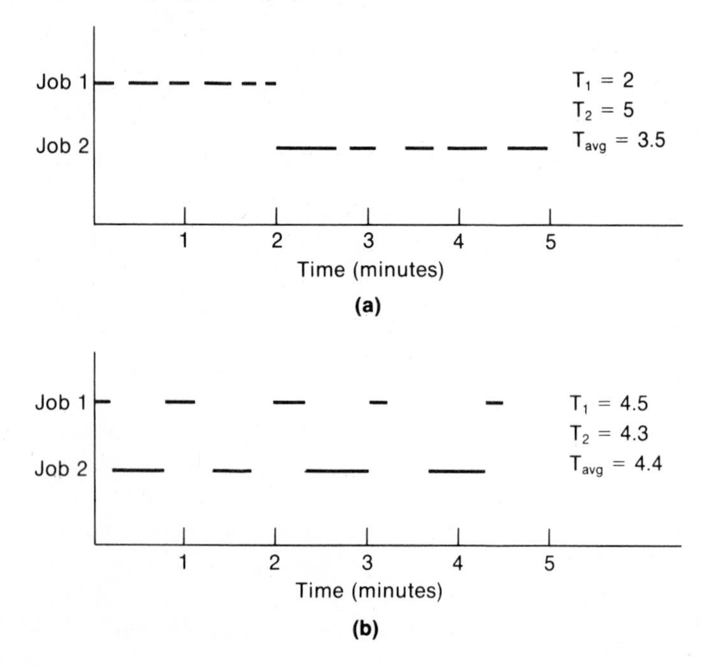

currently, as illustrated in Fig. 6.30(b). Note that the two concurrent jobs share the CPU so that there is less overall idle time; this is the same phenomenon we studied in Fig. 6.15. Because there is less idle time, the two jobs are completed in less total time: 4.5 minutes instead of 5 minutes. This means the system throughput has been improved: we have done the same amount of work in less time. However, the average turnaround time has become worse: 4.4 minutes instead of 3.5 minutes.

Two common job-scheduling policies are *first come–first served* (FCFS) and *shortest job first* (SJF). FCFS tends to treat all jobs equally, so it minimizes the range of turnaround times. SJF provides a lower average turnaround time because it runs short jobs much more quickly; however, long jobs may be forced to wait a long time for service. For examples of these characteristics and discussions of other scheduling policies, see Deitel (1984), Lorin (1981), and Madnick (1974).

6.3.3 Resource Allocation

In Section 6.2 we discussed how an operating system might control the use of resources such as central memory, I/O channels, and the CPU. Such resources are needed by all user jobs, and their allocation is handled automatically by the system. In this section, we describe a more general resource-allocation function provided by many operating systems. Such a facility can be used to control the allocation of user-defined resources such as files and data structures.

The need for a general resource-allocation function is illustrated by the two programs in Fig. 6.31(a). Both these programs utilize a sequential stack that is defined by some other program. The external variable STACK indicates the base address of the stack; TOP contains the relative location of the item currently on top of the stack. We assume that external references to the variables STACK and TOP are handled by linking methods like those discussed in Chapter 3. Program P1 adds items to the stack by incrementing the previous value of TOP by 3, storing a new item from register A on the top of the stack, and then saving the new value of TOP (lines 24–27). Program P2 removes items by loading the value from the top of the stack into register A, and then subtracting 3 from the value of TOP (lines 37–40). For simplicity, we have not shown the code needed to handle stack overflow and underflow.

If processes P1 and P2 are executed concurrently, they may or may not work properly. For example, suppose the present value of TOP is 12. If P1 executes its instructions numbered 24–27, it will add a new item in bytes 15–17 of the stack, and the new value of TOP will be 15.

```
 1        P1        START     0
 2                  EXTREF    STACK,TOP
 3                  LDS       #3               REGISTER S = CONSTANT 3
                     .
                     .
                     .
24                  +LDX      TOP              GET POINTER TO TOP OF STACK
25                   ADDR     S,X              INCREMENT POINTER
26                  +STA      STACK,X          ADD NEW ITEM TO STACK
27                  +STX      TOP              STORE NEW TOP POINTER
                     .
                     .
                     .
48                   END
```

```
 1        P2        START     0
 2                  EXTREF    STACK,TOP
 3                  LDS       #3               REGISTER S = CONSTANT 3
                     .
                     .
                     .
37                  +LDX      TOP              GET POINTER TO TOP OF STACK
38                  +LDA      STACK,X          GET TOP ITEM FROM STACK
39                   SUBR     S,X              DECREMENT POINTER
40                  +STX      TOP              STORE NEW TOP POINTER
                     .
                     .
                     .
75                   END
```

(a)

FIGURE 6.31 Control of resources using operating system service requests.

If P2 then executes its instructions 37–40, it will remove the item just added by P1, resetting the value of TOP to 12. This represents a correct functioning of P1 and P2: the two processes perform their intended operations on the stack without interfering with each other. Another correct sequence would occur if P2 executed lines 37–40, and then P1 executed lines 24–27.

On the other hand, suppose that P1 has just executed line 24 when its current time-slice expires. The resulting timer interrupt causes all register values to be saved; the saved value for register X is 12. Suppose now that the dispatcher transfers control to P2, which executes lines 37–40. These instructions will cause P2 to remove the item from bytes 12–14 of the stack because the value of TOP has not yet been updated by P1; P2 will then set TOP to 9. When P1 regains control of the CPU, its register X will still contain the value 12. Thus lines 25–27 will add the new item in bytes 15–17 of the stack, setting TOP to 15.

```
 1      P1      START    0
 2              EXTREF   STACK,TOP
 3              LDS      #3            REGISTER S = CONSTANT 3
                  .
                  .
                  .
22              LDT      =SNAME        SET POINTER TO RESOURCE NAME
23              SVC      3             REQUEST RESOURCE
24             +LDX      TOP           GET POINTER TO TOP OF STACK
25              ADDR     S,X           INCREMENT POINTER
26             +STA      STACK,X       ADD NEW ITEM TO STACK
27             +STX      TOP           STORE NEW TOP POINTER
28              SVC      4             RELEASE RESOURCE
                  .
                  .
                  .
47      SNAME   BYTE     C'STACK1'
48              END
```

```
 1      P2      START    0
 2              EXTREF   STACK,TOP
 3              LDS      #3            REGISTER S = CONSTANT 3
                  .
                  .
                  .
35              LDT      =STKNM        SET POINTER TO RESOURCE NAME
36              SVC      3             REQUEST RESOURCE
37             +LDX      TOP           GET POINTER TO TOP OF STACK
38             +LDA      STACK,X       GET TOP ITEM FROM STACK
39              SUBR     S,X           DECREMENT POINTER
40             +STX      TOP           STORE NEW TOP POINTER
41              SVC      4             RELEASE RESOURCE
                  .
                  .
                  .
74      STKNM   BYTE     C'STACK1'
75              END
```

(b)

FIGURE 6.31 *(cont'd)*

The sequence of events just described has resulted in an incorrect operation of P1 and P2. The item that was removed by P2 is still logically a part of the stack, and the stack appears to contain one more item than it should. Several other sequences of execution also yield incorrect results. Similar problems may occur whenever two concurrent processes attempt to update the same file or data structure.

Problems of this sort can be prevented by granting P1 or P2 exclusive control of the stack during the time it takes to perform the updating operations. Figure 6.31(b) illustrates a common type of solution using operating system service requests. P1 requests exclusive control of the stack by executing an SVC 3 instruction. The resource being requested is specified by register T, which points to the (user-defined) logical name that has been assigned to the stack. After adding the new item to the stack and updating TOP, P1 releases control of the stack by executing an SVC 4. P2 performs a similar sequence of request and release operations.

The operating system responds to a request for control of a resource by checking whether or not that resource is currently assigned to some other process. If the requested resource is free, the operating system returns control to the requesting process. If the resource is busy, the system places the requesting process in the blocked state until the resource becomes available. For example, suppose the resource STACK1 is currently free. If P1 requests this resource (line 23), the system will return control directly to P1. As in the previous discussion, suppose the time-slice for P1 expires after line 24 has just been executed. Control of the CPU then passes to P2; however, STACK1 remains assigned to P1. Thus P2 will be placed in the blocked state when it requests STACK1 at line 36. Eventually P1 will regain control and complete its operation on the stack. When P1 releases control of STACK1 (line 28), P2 will be assigned control of this resource and moved from the blocked state to the ready state. It can then continue with its updating operation when it next receives control of the CPU. You should trace through this sequence of events carefully to see how the problems previously discussed are prevented by this scheme.

Unfortunately, the use of request and release operations can lead to other types of problems. Consider, for example, the programs in Fig. 6.32. P3 first requests control of resource RES1; later, it requests resource RES2. P4 utilizes the same two resources; however, it requests RES2 before RES1.

Suppose P3 requests, and receives, control of RES1, and that its time-slice expires before it can request RES2. P4 may then be dispatched. Suppose P4 requests, and receives, control of RES2. This sequence of events creates a situation in which neither P3 nor P4 can complete its execution. Eventually, P4 will reach its line 4 and request control of RES1; it will then be placed into the blocked state because RES1 is assigned to P3. Similarly, P3 will eventually reach its line 4 and request control of RES2; P3 will then be blocked because RES2 is assigned to P4. Neither process can acquire the resource it needs to continue, so neither process will ever release the resource needed by the other.

This situation is an example of a *deadlock:* a set of processes each of which is permanently blocked because of resources held by the oth-

```
1     P3      START   0
                :
2             LDT     =R1        REQUEST RES1
3             SVC      3
                :
4             LDT     =R2        REQUEST RES2
5             SVC      3
                :
6             LDT     =R2        RELEASE RES2
7             SVC      4
                :
8             LDT     =R1        RELEASE RES1
9             SVC      4
                :
10    R1      BYTE     C'RES1 '
11    R2      BYTE     C'RES2 '
12            END
```

```
1     P4      START   0
                :
2             LDT     =R2        REQUEST RES2
3             SVC      3
                :
4             LDT     =R1        REQUEST RES1
5             SVC      3
                :
6             LDT     =R1        RELEASE RES1
7             SVC      4
                :
8             LDT     =R2        RELEASE RES2
9             SVC      4
                :
10    R1      BYTE     C'RES1 '
11    R2      BYTE     C'RES2 '
12            END
```

FIGURE 6.32 Resource requests leading to potential deadlock.

ers. Once a deadlock occurs, the only solution is to release some of the resources currently being held; this usually means cancelling one or more of the jobs involved. There are a number of methods that can prevent deadlocks from occurring. For example, the system could require that a process request all its resources at the same time, or that it request them in a particular order (such as RES1 before RES2). Unfortunately, such methods may require that resources be tied up for longer than is really necessary, which can degrade the overall operation of the system. Discussions of methods for detecting and preventing deadlocks can be found in Deitel (1984).

The problems we have discussed in this section are examples of the more general problems of *mutual exclusion* and *process synchronization*. Discussions of these problems, and techniques for their solution, can be found in Deitel (1984) and Holt (1978).

6.3.4 Protection

An operating system that serves a number of different individuals or groups should provide some mechanism for protecting each user from unauthorized actions by the others. For example, a user should be able to create files that cannot be read or modified by others. The overall problem of security and protection is a complex one. In this section we briefly introduce the basic protection functions that are performed by a typical operating system. These controls must be combined with physical security, administrative procedures, and other types of controls to provide effective protection. Further information concerning protection, security, and privacy issues can be found in Deitel (1984), Denning (1982), and Fernandez (1981).

Most multi-user operating systems provide some type of *access control* or *authorization* mechanism, which is often based on an *access matrix* similar to the one shown in Fig. 6.33. In this example, the three users are designated as u_1, u_2, and u_3. Each entry in the access matrix specifies the rights of a user with respect to some object such as a file or program. Thus user u_2 may read file f_1, read or write file f_2, and execute program p_2, but u_2 is not allowed any type of access to file f_3 or program p_1. Access rights to a newly created object are usually specified by the creator of that object. On some systems, users having certain access rights are allowed to pass those rights on to others. Users

FIGURE 6.33 Example of an access matrix.

	Files			Programs	
Users	f_1	f_2	f_3	p_1	p_2
u_1	R		R	E	E
u_2	R	R,W			E
u_3	R,W		R		R,E

who want to access the controlled objects must do so via operating system service requests. The system will deny access to any user who is not properly authorized.

In most realistic situations, the access matrix is quite sparse. Most users are allowed access to relatively few of the total number of objects in the system. For this reason, the information concerning access rights is often stored as an *authorization list* (i.e., a list of authorized users) for each object, or as a *capability list* (i.e., a list of objects that can be accessed) for each user. For further information about implementing the access-matrix model, see Denning (1982).

The effectiveness of a technique such as the one just described obviously depends upon the proper identification of users. It must be possible for the operating system to verify the identity of any user so that it may apply the correct access-control rules. One of the most common methods for user identification is a system of *passwords*. On most systems, the user must supply a secret password to enter a job into the system or log on using a timesharing terminal. Sometimes this scheme is extended so that passwords must be provided to access certain files, programs, etc. On some systems, there is a master table of passwords that is used by the operating system to verify passwords submitted by users; however, this table is a potentially vulnerable point in the overall system security. To avoid this problem, many operating systems store the table in an encoded form. Details about this and other similar techniques can be found in Denning (1982).

A system of user identification and authorization does not always solve the overall security problem, because information must sometimes leave the secure environment. Consider, for example, Fig. 6.34(a). The user at the timesharing terminal is properly identified and authorized to access a certain file F. We assume the computer system provides adequate access controls and is contained in a physically secure environment. We also assume the terminal itself is physically protected. However, the communication link between the computer system and the terminal may be difficult or impossible to protect. (Consider, for example, the problem of preventing wiretapping when the terminal is connected to the computer via public telephone lines.) This means that any information from file F that is transmitted to the terminal is vulnerable during the transmission process. Similarly, a wiretapper could observe passwords being sent by the user and thereby gain access to protected objects.

The usual solution to this type of security problem is *data encryption* (see Fig. 6.34b). Information to be sent over a nonsecure communication link is *encrypted* (encoded) while still in the secure environment of the sender. The transmitted information is *decrypted* (decoded) after

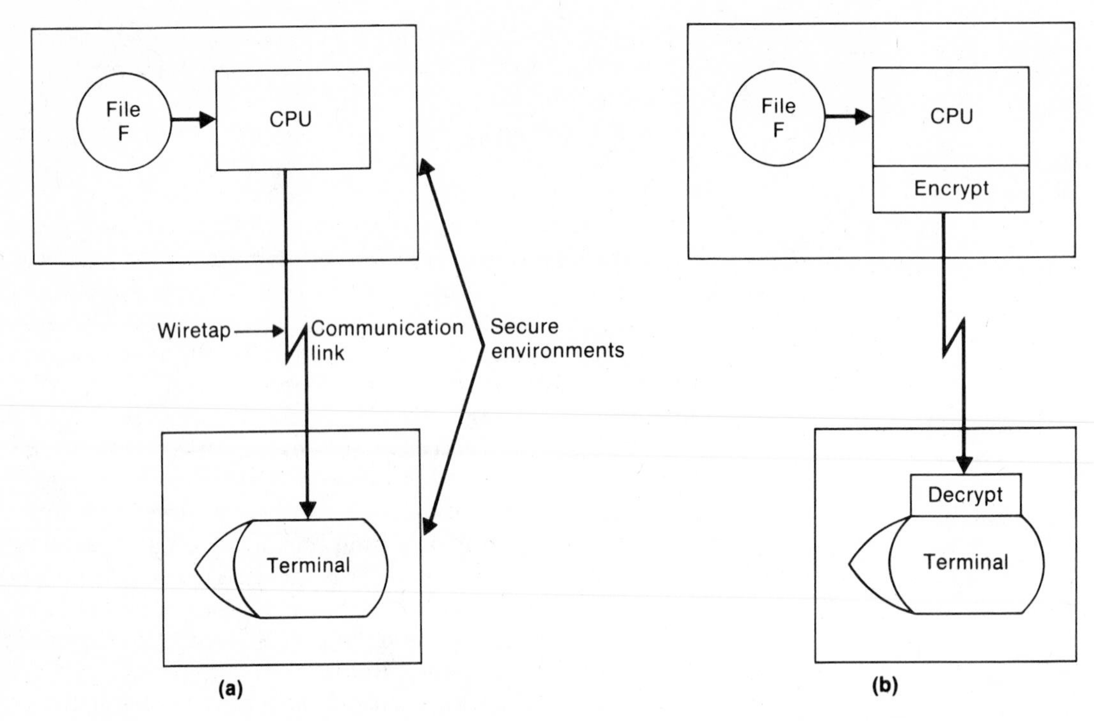

FIGURE 6.34 Use of encryption to protect data during transmission.

entering the secure environment of the receiver. Wiretapping is still possible; however, the eavesdropper would be unable to interpret the encrypted information being transmitted. The encryption and decryption operations can be performed efficiently by either hardware or software. There are a large number of different encryption techniques available. For a comprehensive discussion of such methods, see Denning (1982).

Of course, the effectiveness of any protection system depends entirely on the correctness and protection of the security system itself. In the system we discussed, the access-control information must be protected against unauthorized modifications. There must also be a mechanism to ensure that users cannot access the protected objects without going through the security system. Hardware features such as user/supervisor modes, privileged instructions, and memory protection mechanisms are often useful in dealing with issues of this kind. It is also important to be sure that the part of the operating system that applies the protection rules (the *security kernel*) performs its task correctly. Further discussions of these issues can be found in Deitel (1984), which also contains a survey of common flaws in operating system security mechanisms and an interesting case study on penetrating the security of a system.

6.4 OPERATING SYSTEM DESIGN OPTIONS

In this section we briefly describe some important concepts related to the design and structure of an operating system. Section 6.4.1 introduces the notion of hierarchical operating system structure, which has been used in the design of many real systems. Section 6.4.2 describes how an operating system may provide multiple *virtual machines*. Such a system gives each user the impression of running on a dedicated piece of hardware. Section 6.4.3 introduces the topic of operating systems for multiprocessors and discusses some options for the division of tasks between the processors.

6.4.1 Hierarchical Structure

Many real operating systems are designed and implemented using a *hierarchical* structure. Figure 6.35 shows an example of such an operating system structure. The basic principle, illustrated in Fig. 6.35(a), is a generalization of the extended-machine concept shown in Fig. 6.1. Each layer, or *level,* of the structure can use the functions provided by lower levels just as if they were part of the real machine. Thus Level 0, often called the *kernel* of the operating system, deals directly with the underlying hardware; Level 1 deals with the interface provided by Level 0, and so on. User programs deal with the highest-level interface (in this case, Level 3), which represents the user interface discussed in Section 6.1.

Figure 6.35(b) shows the operating system functions assigned to each level of our sample structure. The placement of the functions is governed by the relationships between the operations that must be performed. In general, functions at one level are allowed to refer only to functions provided by the same or lower levels; that is, there should be no outward calls. In our example, the file-management routines (Level 3) must use the memory manager (Level 2) to allocate buffers, and the I/O supervisor (Level 1) to read and write data blocks. If demand-paged memory management is being used, the memory manager must also call on the I/O supervisor to transfer pages between the real memory and the backing store. All levels of the system use the process-scheduling and resource-management functions provided by Level 0.

There are many advantages to such a hierarchical structure. Operating system routines at a given level can use the relatively simple functions and interfaces provided by lower levels. It is not necessary for the programmer to understand how these lower-level functions are

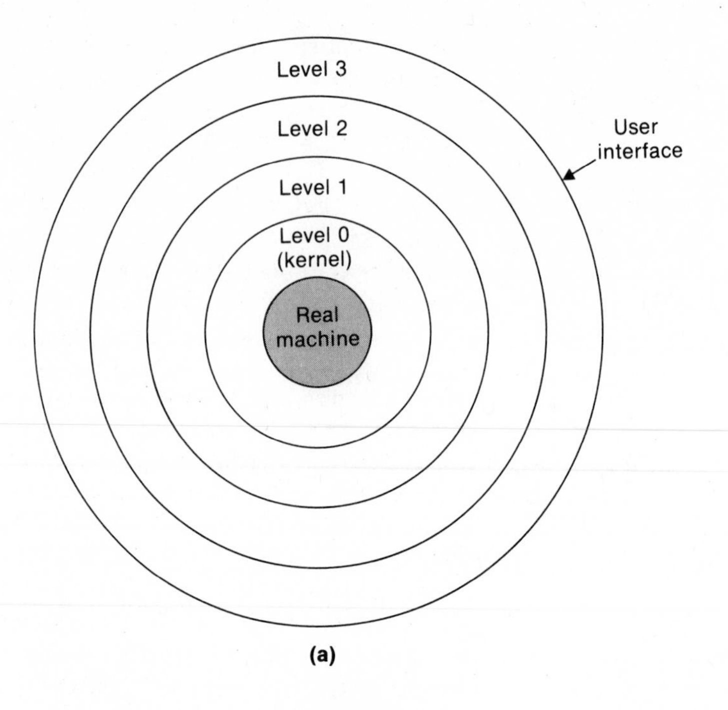

(a)

Level	Functions
3	File management
2	Memory management
1	I/O supervision
0	Dispatching, resource management

(b)

FIGURE 6.35 Example of a hierarchical operating system structure.

actually implemented. The operating system can be implemented and tested one level at a time, beginning with Level 0. This greatly reduces the complexity of each part of the system and makes the tasks of implementation and debugging much simpler.

The placement of functions shown in Fig. 6.35(b) is typical; however, there are many variations between different systems. Consider, for example, the interrupt-handling routines. Many systems place all *first-level interrupt handlers* (FLIH) in the kernel (Level 0). After initial interrupt processing, an FLIH can transfer control to a routine at some higher level; this is an exception to the no-outward-calls rule. Thus, for example, the FLIH for a page-fault interrupt might save status information, enable other interrupts, and then transfer to a routine at Level 2. See Fig. 6.25(b) for an example of such processing.

In some operating systems, the placement of functions in a hierarchy is made more difficult by special situations. For discussions of such situations, and other examples of hierarchical structures, see Lorin (1981) and Peterson (1983).

Hierarchical systems also differ in the rules for passing control from one level to another. In a *strict hierarchy,* each level may refer only to the level immediately beneath it. Thus Level 3 could communicate directly only with Level 2. If it were necessary for the file-management routine in our example to call the I/O supervisor, this request would have to be passed on from Level 2 to Level 1. This approach has the advantage of simplicity of use: each level has only one interface with which it must be concerned. However, such a restriction can lead to inefficiency because it increases the number of calls that must be performed to reach the inner level. In a *transparent hierarchy,* each level may communicate directly with the interface of any lower level. Thus, for example, a user program could invoke file-management routines at Level 3, or it could call the I/O-supervisor functions of Level 1 directly.

Further discussions of hierarchical operating system structures can be found in Peterson (1983), Lorin (1981), and Madnick (1974).

6.4.2 Virtual Machines

The hierarchical structure concepts discussed in the previous section can be extended to provide users, including operating systems, with the illusion of running on separate *virtual machines*. This virtual-machine approach makes it possible to run different operating systems concurrently on the same real machine. Thus we can think of virtual machines as an extension of the concept of multiprogramming down to the lowest level of the operating system.

Consider, for example, Fig. 6.36. Operating system OS1 is a multiprogramming system supporting three concurrent users. Operating system OS2 is a single-job system. OS3 is an operating system that is currently being tested. There is also a user program (User5) that is designed to run in supervisor mode as a stand-alone program, not under the control of any operating system.

All three operating systems, plus the stand-alone user, are actually running on the same real machine. However, they do not deal directly with the real machine. Instead, they deal with a *virtual machine monitor* (VMM), which provides each user with the illusion of running on a separate machine. Thus it is possible to test new operating systems and to allow users with special needs to run in supervisor

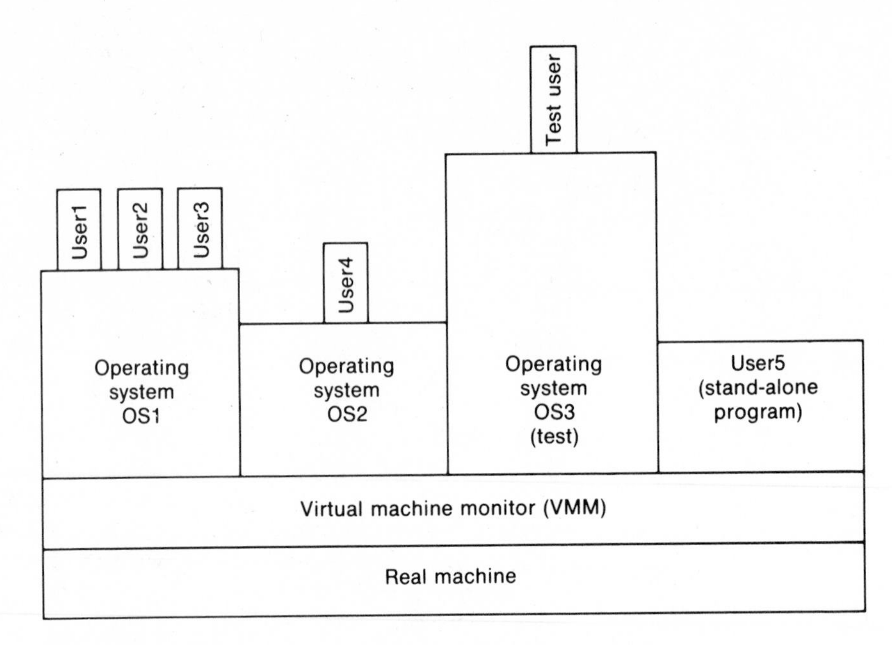

FIGURE 6.36 Multiple users of a virtual-machine operating system.

mode, while at the same time continuing to serve ordinary users in the usual way. If the operating system being tested, or the stand-alone user, causes a system to crash, this crash affects only its own virtual machine. The other users of the real machine can continue their operation without being disturbed.

Figure 6.37 illustrates how this illusion can be accomplished. The lowest-level routines of the operating system deal with the virtual machine monitor instead of with the real machine. The VMM, which is completely invisible to the operating system and the user program, provides resources, services, and functions that are the same as those available on the underlying real machine.

Each direct user of a virtual machine, such as OS1 or User5 in Fig. 6.36, actually runs in user mode, not supervisor mode, on the real machine. When such a user attempts to execute a privileged instruction such as SIO, STI, or LPS, a program interrupt occurs. This interrupt transfers control to the virtual machine monitor. The VMM simulates (with respect to the virtual machine) the effect of the privileged operation that was being attempted, and then returns control to the user of the virtual machine. Similarly, an interrupt on the real machine also activates the virtual machine monitor. The VMM determines which virtual machine should be affected by the interrupt and makes the appropriate changes in the status of that virtual machine.

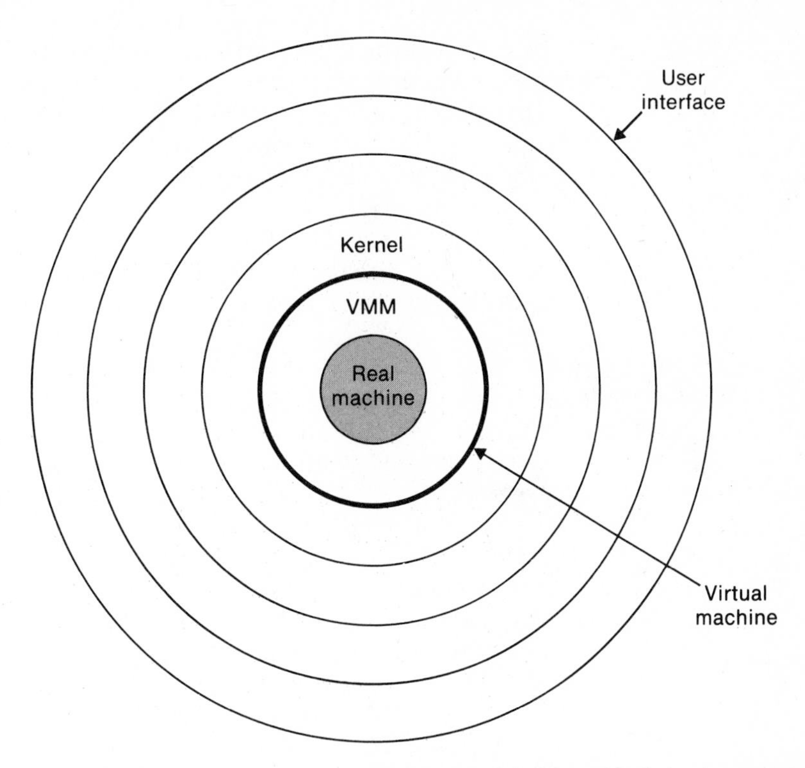

FIGURE 6.37 Virtual machine as an extension of the hierarchical structure concept.

The virtual machine monitor is actually a complete, but simple, operating system for the real machine. The other operating systems and stand-alone users of virtual machines are the "users" of the real operating system (the VMM). Thus the virtual machine monitor must provide all of the essential machine-dependent functions we have discussed. The VMM saves status information for each virtual machine and switches the real CPU between the various virtual machines; this is the same as the process-scheduling function discussed in Section 6.2. It also provides a separate virtual memory and a set of virtual I/O channels for each virtual machine, using techniques similar to those we have already discussed.

The most obvious advantages of the virtual-machine approach are flexibility and convenience. Different operating systems can be run concurrently to serve the needs of different types of users. Operating systems and stand-alone programs can be tested while still making the machine available to ordinary users. The use of separate virtual machines can also provide a higher degree of protection since each virtual machine has no access to the resources of any other. The disadvantage, of course, is the higher system overhead required to simulate virtual-machine operation. For example, if an operating system running on a virtual machine uses virtual-memory techniques itself, it may be necessary to have two separate levels of dynamic address translation. The efficiency of a virtual-machine operating system depends heavily on how many operations must be simulated by the VMM, and how many can be performed directly on the real machine.

Further discussions of virtual machine operating systems can be found in Deitel (1984) and Peterson (1983).

6.4.3 Multiprocessor Systems

Our previous discussions of operating systems have concerned machines with one central processing unit (CPU). In this section we briefly describe some design alternatives for multiprocessor operating systems.

An operating system for a machine with multiple processors must perform the same basic tasks we have discussed for single-processor systems. Of course, some of these functions may need to be modified. For example, the process scheduler may have more than one CPU to assign to user jobs, so more than one process might be in the running state at the same time. There are also a number of less obvious issues, many of which are related to the overall organization of the system.

Figure 6.38 illustrates the simplest form of multiprocessor organization. Each processor has its own memory, I/O devices, and other resources. In addition, each processor runs its own operating system. The processors are interconnected by communication links. The only way for one processor to access the resources belonging to another processor is by sending a request to the other processor via a communication link. This arrangement, which is often called *loosely coupled multiprocessing,* is quite similar to a network of single-processor systems. Such systems are often used when there is a specialization of function among the various processors. For example, some time-sharing systems have a front-end processor that manages the details of communication with the user terminals. The main processor does all the actual computational work and communicates with the front-end processor whenever terminal I/O is required. The operating system used by each processor in a loosely coupled multiprocessing system is quite similar to those we have discussed for single-processor systems. The only really new function that must be performed is the management of messages sent across the communication links.

Loosely coupled multiprocessor systems are relatively simple because each resource is dedicated to a single processor. On the other hand, some multiprocessor systems allow direct sharing of resources

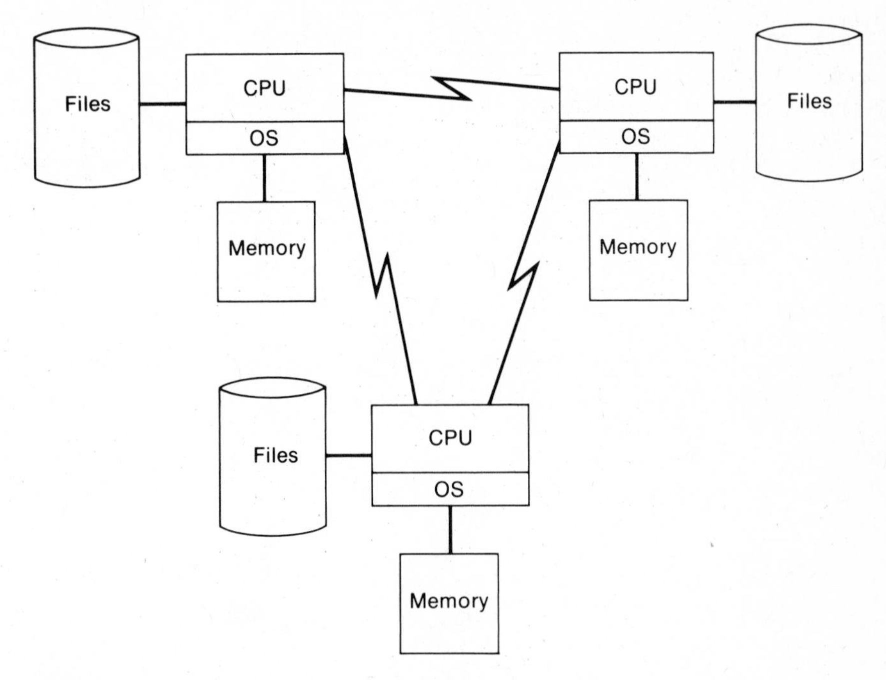

FIGURE 6.38 Loosely coupled multiprocessing system.

between processors. This approach, called *tightly coupled multiprocessing,* makes the task of an operating system somewhat more difficult.

Figure 6.39 illustrates two variations of tightly coupled multiprocessing. Figure 6.39(a) shows a system that uses *master–slave* processing. With this approach, one master processor performs all resource management and other operating system functions. This processor completely controls the activities of the slave processors, which execute user jobs. The processors communicate with each other either via direct communication links or through work areas in the shared memory. In a master–slave multiprocessing system, the processors can share system resources such as memory and data files. However, the routines and data structures that make up the operating system itself are not shared; they are used only by the master processor. Thus this type of operating system is also rather similar to the single-processor systems we have discussed.

The most significant problem with master–slave multiprocessing systems is the unbalanced use of resources. For example, the master processor might be overloaded by requests for operating system services, which could cause the slave processors to remain idle much of the time. In addition, any hardware failure in the master processor

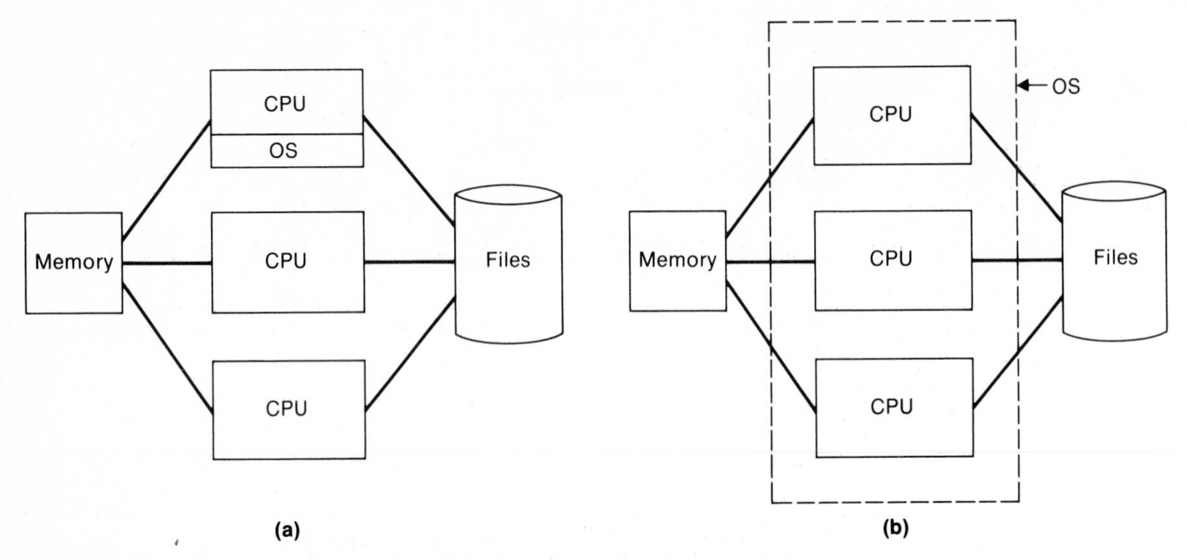

FIGURE 6.39 Tightly coupled multiprocessing systems with (a) master–slave processing and (b) symmetric processing.

causes the entire system to stop functioning. Such problems can be avoided by allowing any processor to perform any function required by the operating system or by a user program. This approach, called *symmetric* processing, is illustrated in Fig. 6.39(b). Since all processors have the ability to perform the same sets of functions, the potential bottlenecks of a master–slave system are avoided. In addition, the failure of any one processor will not necessarily cause the entire system to fail. The other processors can continue to perform all the required functions.

In a symmetric multiprocessing system, different parts of the operating system can be executed simultaneously by different processors. Because of this, such a system may be significantly more complicated and more difficult to design than the other types of operating systems we have discussed. For example, all processors must have access to all the data structures used by the operating system. However, the processors deal with these data structures independently of one another, and problems can arise if two processors attempt to update the same structure at the same time (see Section 6.3.3). Symmetric multiprocessing systems must provide some mechanism for controlling access to critical operating system tables and data structures. The request and release operations described in Section 6.3.3 are not sufficient to handle this problem because two different processors might perform request operations simultaneously. The solution usually requires a special hardware feature that allows one processor to seize control of a critical resource, locking out all other processors in a single step. Discussions of such mechanisms can be found in Peterson (1983) and Lorin (1981).

Further information concerning operating systems for multiprocessors can be found in Deitel (1984), Lorin (1981), and Madnick (1974).

6.5 IMPLEMENTATION EXAMPLES

In this section we present brief descriptions of several real operating systems. These systems have been chosen to illustrate some of the variety of design and purpose in such software. As in our previous examples, we do not attempt a complete high-level description of any system. Instead, we focus on some of the more interesting or unusual features provided and give references for readers who want more information.

Our first two examples are operating systems that are not tied to any one machine. Section 6.5.1 discusses the UCSD Pascal system, which is a portable operating system designed for single users on small computers. Section 6.5.2 describes UNIX, a more sophisticated operating system that has also been implemented on a variety of computers.

The other examples are systems that were developed for particular families of computers. Section 6.5.3 describes NOS™, which runs on the CDC CYBER series of computers; this is an example of a multiprocessor operating system. Section 6.5.4 discusses the VAX/VMS™ system, which incorporates several interesting techniques for managing virtual memory. Section 6.5.5 describes IBM's VM/370, an operating system that can provide multiple virtual machines.

6.5.1 UCSD Pascal System

The UCSD Pascal system is an operating system designed for small computers. Almost all the system is written in Pascal and runs on a pseudo-machine, or P-machine (see Sections 5.4.3 and 5.5.2). Thus the operating system can be run on any machine for which there exists a P-machine interpreter. The UCSD Pascal operating system itself is relatively simple. Because it is designed to support a single user on a small computer, sophisticated techniques are not necessary.

One of the most interesting parts of the UCSD Pascal system is the user interface it provides. This interface can be visualized as a tree structure. At each level of the tree, the system displays a *prompt line,* also often called a *menu.* This prompt line specifies the possible commands that can be entered by the user. Entering one of these commands selects a branch and causes the user to descend one level in the tree structure. A new prompt line is then displayed, providing a selection of commands at this lower level. At the lowest level, the system prompts the user for any additional information required and executes the command.

For example, when the system is first loaded it displays the prompt line

Command: E(dit, R(un, F(ile, C(omp, L(ink, X(ecute, A(ssem, D(ebug.

Typing E causes the user to descend one level in the tree to enter the Editor. A new prompt line for the Editor is then displayed. This prompt line might contain options such as

Edit: A(djst, C(py, D(lete, F(ind, I(nsrt, J(mp, R(eplace, Q(uit, X(chng, Z(ap.

(The details of the actual prompt lines vary slightly from one version of the system to another.) Typing I causes the system to prompt the user for the text to be inserted. Typing Q causes the user to leave the Editor and return to the root of the tree structure (the original prompt line).

The commands available at the outermost level provide easy access to the most important system functions. The Editor allows the user to enter text and to perform editing operations on any specified text file or on a system *workfile*. The Filer provides commands for maintaining files on disk. These commands include, for example, listing directories, transferring files from one location to another, and creating new workfiles. The Compiler, Assembler, and Linker are system programs with functions like those described earlier in this text.

Entering R(un at the outermost command level directs the system to execute the program in the current workfile. Entering X(ecute causes the system to prompt for a file name and to execute the program in the specified file. D(ebug executes the program in the current workfile, providing debugging facilities. If an error occurs or a user-defined breakpoint is reached, the system Debugger is called.

A number of utility programs are also provided by the system. These include, for example, programs to tailor the system to a particular hardware configuration, to create bootstrap routines, and to perform disk-formatting operations not provided by the Filer. Such utility programs are entered from the outermost command level via the X(ecute command.

The UCSD Pascal operating system provides only a basic set of service functions for user programs. Dynamic allocation of memory is supported to meet the needs of Pascal programs, and the system has the ability to rearrange the contents of main memory when necessary. Only a minimal amount of concurrency support is provided. Most system activities, including the performing of I/O operations, are synchronous. This means, for example, that a READ operation causes the user program to wait until the physical I/O operation is completed; no other tasks can be executed during this period.

Further information about the UCSD Pascal system can be found in SofTech Microsystems (1980 and 1983).

6.5.2 UNIX*

The UNIX operating system was originally developed at Bell Laboratories for the DEC PDP-11 family of computers. It has since been implemented on many other machines, from large mainframes to microprocessors. There are currently several versions of UNIX. In this section, we briefly describe some of the more interesting UNIX features that are common to most of these versions.

The user interface of a UNIX system is provided by a program called the *shell.* The shell is a command-line interpreter that reads lines typed by the user and interprets them as requests to execute other programs. Ordinarily, the shell responds to each command by calling the appropriate program. When the called program terminates, the shell resumes its own execution and prompts the user for the next command. As an alternative, the shell can activate the called program and return immediately for another command without waiting for the program to complete. The shell can also perform parameter substitution, execute commands conditionally based on character-string comparisons, and perform transfers of control within the command sequence.

The standard shell described above is designed to allow a user to access all the facilities of the UNIX system. Sometimes it is desirable to provide a different user interface, which is easily done by the use of a nonstandard shell. A user's entry in the password file can contain the name of a program to be used instead of the standard shell. This program can interpret the user's commands to provide an interface that is tailored to the requirements of the specific user. For example, the password-file entries for users of a secretarial editing system might specify that a text-editing program is to be used instead of the standard shell. Thus these users could begin work immediately upon logging in. The shell can also help provide protection in the system: users of a particular command shell can be prevented from invoking UNIX programs that are not intended for their use.

On the UNIX system, processes are allowed to create other processes that compete independently for system resources. The system call *fork* causes a process to split into two independently executing

* Adapted from D. M. Ritchie and K. Thompson, "The UNIX Time-Sharing System." *Communications of the ACM,* Vol. 17, No. 7. Copyright 1974, Association for Computing Machinery, Inc.

processes, one of which is designated as the *parent,* and the other as the *child.* UNIX provides process-synchronization tools by which a parent process can wait for the completion of a child process. Certain status information is available from the child process upon its termination.

UNIX processes can communicate directly with each other via logical interprocess channels called *pipes.* Ordinary read and write system calls are used to pass messages between the processes. Neither process need know that a pipe, rather than an ordinary file, is involved. The synchronization of the processes and buffering of messages are handled automatically by UNIX. Several processes can be connected by pipes in a linear fashion to form a *pipeline* in which the output from one process becomes the input for the next.

UNIX provides a hierarchical file system. Users have tree-structured directories to their files, and the system maintains several directories for its own use. The same file may appear in several different directories, possibly under different names. All such directory entries have equal status. That is, a file does not belong to any one particular directory. Files may also exist without being included in any directory.

The most unusual feature of the UNIX file system is the existence of *special files.* Each I/O device supported by UNIX is associated with at least one such file. Special files are read and written like ordinary disk files; however, requests to read or write the special file result in activation of the associated device. This makes file I/O and device I/O as similar as possible. For example, devices are subject to the same protection mechanisms as ordinary files.

Further information and references concerning UNIX systems can be found in Bourne (1983), Deitel (1984), and Ritchie (1978).

6.5.3 NOS

NOS is an operating system designed for use on CDC CYBER and 6000 series computers. These machines are multiprocessors, having one or two *central processing units* (CPU) and up to 20 *peripheral processors* (PP). The CPU and the PPs share access to central memory; thus, NOS is an example of a tightly coupled multiprocessing system. However, the CPU cannot directly access any of the I/O channels or devices. This access must be provided by the peripheral processors. PPs have the ability to start and stop the CPU using special hardware instructions.

Under NOS, the CPU is used almost exclusively for the execution of user programs. The PPs perform all I/O and most operating system tasks; however, certain system work that can be done more efficiently in the CPU is processed there. Each PP is assigned a block of eight words in central memory through which communication with the system is conducted.

A system monitor, which is composed of a PP routine MTR and a CPU routine CPUMTR, is in complete supervisory control of the machine. MTR continuously scans for service requests by user jobs. The mechanism used to accomplish this communication will be described later. CPUMTR assigns tasks, one at a time, to each PP. When an assigned task is completed, the PP signals the system. CPUMTR waits for this signal before assigning another task to the PP.

Each user job is allocated a certain area of central memory. The machines on which NOS runs have hardware relocation registers like those described in Section 6.2.4, so it is possible to move jobs from one location to another during execution. NOS takes advantage of this capability in two ways. First, jobs are allowed to increase or decrease the amount of central memory assigned to them via operating system service requests. NOS rearranges the contents of central memory as needed to accommodate these requests. Second, NOS may temporarily suspend execution of a job to make central memory available for a higher-priority process. This is known as *rolling out* a job. When the job is rolled back in to continue execution, it need not be returned to the same area of central memory.

The first 64 words of the area of memory assigned to a job are reserved for communication with the operating system. To make an operating system service request, the user program places a request code in a fixed location within this communication area. The PP routine MTR continuously scans the communication areas assigned to user jobs, checking for the presence of a request code. When a request is detected, MTR activates CPUMTR to assign the requested task to a PP.

The mechanism used by NOS to synchronize program execution with the performance of service functions differs somewhat from the method described in Sections 6.2.2 and 6.2.3. A user program making an operating system service call may request that it be temporarily placed into a blocked state. When the service function has been completed, the program is once again activated. In NOS, this activation is termed *automatic recall.* The automatic recall feature is often used, for example, to wait for the completion of a requested I/O operation.

If a program wants to continue its processing while a service is being performed, it requests that service without specifying automatic recall. When the program must wait for the completion of the requested service, it makes a special *periodic recall* request. This request places the program into a blocked state for a certain interval of time. After this time interval has elapsed, the program is reactivated; it then checks to determine whether or not it can proceed. If the service has not yet been completed, the program again requests periodic recall. A similar technique can be used to make several service requests, such as unrelated I/O operations, that can be performed simultaneously.

Further information about NOS can be found in CDC (1981 and 1983).

6.5.4 VAX/VMS

VAX/VMS is an operating system designed for use on VAX computers. It is a multipurpose system that supports batch, timesharing, and real-time applications. Virtual memory is an integral part of a VAX system. The management of this virtual memory is one of the most interesting features of the VAX/VMS operating system.

The VAX/VMS virtual memory consists of 2^{32} bytes; this is divided into a *system address space* and a *process address space,* each of which has 2^{31} bytes. The system address space is shared by all processes; however, each user process has its own process address space. The process address space is further divided into a *program region* and a *control region.* The program region contains the program currently being executed by the process. The control region contains the user and system stacks, and other information maintained by the operating system.

Virtual memory is divided into 512-byte pages. There is a *system page table* that contains one entry for each page of system address space. This entry indicates either the page frame in which the associated page is loaded, or the address of the page in the backing store. Two special-purpose registers specify the starting address and the length of this system page table. Process address space is described by two page tables: one for the program region and one for the control region. These tables reside in the system address space, and may themselves be paged. The memory management routines also maintain a table describing the status of all page frames in memory and the status and location of all virtual pages in the system.

When a page fault occurs, a system routine called the *pager* is activated. The pager selects an available page frame and brings the required virtual page into memory. There is a limit on the number of pages that each process can have loaded at any one time. If this limit has been reached, the pager selects a page to be removed. The page limit for a process is dynamically adjusted, depending upon the page-fault rate for that process.

Modified pages selected for removal are not immediately written to the backing store. Instead, they are placed on a list of pages called the *modified page list.* The pager attempts to write these pages in clusters (that is, several pages in one write operation) to minimize the overhead for paging I/O. If a page on this list is demanded before it is actually written, the page can be retrieved without performing any physical

I/O. Similarly, unmodified pages selected for removal are placed on a *free page list.* If a page on this list is required before the page frame has been reused, it can be recalled without any physical I/O.

Another operating system routine, called the *swapper,* is used to move the working sets of entire processes in and out of real memory. The purpose of the swapper is to keep the highest-priority executable processes in memory so they can be scheduled for execution. Thus the function of the swapper is similar to that of the intermediate level scheduler mentioned in Section 6.3.2.

The VAX/VMS process-scheduling function is based on a priority system with 32 levels of priority. The lower 16 levels of priority are reserved for normal processes; the higher 16 priorities are reserved for real-time processes. Each process is defined to the software by a *process control block* and a process header. The process control block contains clusters of *event flags* that can be used by the processes to communicate and to coordinate their operation. VAX/VMS also provides *locking* mechanisms for synchronizing the use of resources by cooperating processes. There are six different choices of lock mode, each of which provides a different level of sharing.

Support for I/O operations is provided at two levels: *record management services* (RMS) and *I/O system services.* RMS procedures provide device-independent, file-structured access to I/O devices. Functions such as the blocking of records may be performed either by RMS or by the user. I/O system services provide both device-dependent and device-independent services. Users with sufficient privilege can perform direct I/O operations to define their own file structures on private disk or tape devices. Both types of I/O support use the same lower-level system routines to queue and execute physical I/O operations.

Further information about VAX/VMS can be found in Deitel (1984) and DEC (1982).

6.5.5 VM/370*

VM/370 is an operating system for the IBM 370 that provides multiple virtual machines. Every user appears to have a complete replica of a System/370, including I/O devices.

VM/370 has two main components: the *control program* (CP) and the *conversational monitor system* (CMS). CP, which is the resource manager of the system, performs the same functions as the virtual machine monitors discussed in Section 6.4.2. CP also provides commands to assist in the control and debugging of an operating system in

* Adapted from "VM/370—A Study of Multiplicity and Usefulness," by L. H. Seawright and R. A. MacKinnon, *IBM Systems Journal,* Vol. 18, No. 1, copyright 1979 International Business Machines Corporation.

a virtual machine. These commands allow such operations as the displaying of contents of virtual memory and the setting of breakpoints based on instruction address. In essence, these are the same debugging functions a system programmer might perform at the console of a real CPU.

All the operating systems that are normally used on 370 systems can be executed in virtual machines under CP. This includes, of course, CP itself. Running CP in a virtual machine, under the control of another copy of CP, provides a virtual VM/370 environment. This is particularly useful in debugging parts of CP or introducing new versions of CP into a production system.

The conversational monitor system (CMS) provides interactive support for a single user of VM/370 at a single terminal. Access to multiple terminals is accomplished by CP's ability to support multiple virtual machines, each running CMS. CMS is designed specifically for VM/370 and depends directly on CP for its execution. Unlike the other 370 operating systems, CMS cannot operate independently on a real machine. However, CMS performs more efficiently under CP than do these other operating systems.

CMS can be said to present a user-friendly interface to its users. This interface consists of a series of user-oriented commands rather than a series of job control statements as in other 370 operating systems. The user is required to know relatively little about the operating system beyond the activities directly required to accomplish the desired function.

Total system performance is always a concern when considering the use of a virtual-machine operating system. The overhead imposed by the activities of the virtual machine monitor normally decreases system throughput and increases response time. VM/370 has a performance measurement facility that can provide resource-utilization information while CP is running, as well as collecting measurement data for later analysis.

There are also a number of options that can be used to improve the performance of the system. For example, specified pages can be locked into real storage, and devices and channels can be dedicated to a particular virtual machine. A single virtual machine can be run in virtual-equals-real mode, in which case the virtual machine's memory is not demand-paged by CP.

Other software developments are more specialized and involve changes to operating systems that use the virtual-machine environment. These changes allow the operating system in a virtual machine to recognize that it is running under VM/370 and to communicate directly with CP. This means that the system no longer has to perform operations that are redundant when it is running in a virtual-machine environment. These changes, collectively termed *handshaking,* result in greater operating efficiency and improved virtual-machine performance.

Certain models of the System/370 and related processors provide special features designed for use with VM/370. These features can handle in hardware some of the most frequently executed CP functions. This approach can also result in considerably improved virtual-machine performance.

Further information and references concerning VM/370 can be found in Deitel (1984) and Seawright (1979).

EXERCISES

Section 6.2

1. In Section 6.2.1, we assumed that the occurrence of an interrupt inhibited all other interrupts of equal or lower priority. Would the scheme described in the text work if we simply inhibited all other interrupts of the same class?

2. Suppose the processing of a certain type of interrupt is unusually complex. It might not be desirable to leave other interrupts inhibited for the length of time required to complete the interrupt processing. Suggest a method of interrupt handling that would allow all interrupts to be enabled during most of the interrupt processing.

3. What are the advantages of having several different classes of interrupts, instead of just one class with flag bits to indicate the interrupt type?

4. Suppose there is a limit on the total amount of CPU time a job is allowed to use. This limit can vary from one job to another. What part of the operating system would be responsible for enforcing this time limit, and how might such a function be accomplished?

5. On SIC/XE, setting the IDLE bit of SW to 1 places the CPU into an idle status. Is such a hardware feature necessary on a computer that supports multiprogramming?

6. Suppose you wanted to implement a multiprogramming batch operating system on a computer that has no hardware interval timer. What problems might arise? Can you think of a way to solve these problems using some other hardware or software mechanism?

7. How would your answer to Exercise 6 change if the operating system also supported real-time processing?

8. Consider a multiprogramming operating system. Suppose there is only one user job currently ready to use the CPU. With the methods we have described, this user job would periodically be interrupted by the expiration

of its time-slice. The status of the job would be saved; the dispatcher would then immediately restore this status and dispatch the job for another time interval. How might an operating system avoid this unnecessary overhead while still being able to service other jobs when they become ready?

9. Instead of maintaining a single list of all jobs with an indication of their status, some systems keep separate lists (i.e., a ready list, a blocked list, etc.). What are the advantages and disadvantages of this approach?

10. In the example shown in Fig. 6.15, we assumed that no timer interrupts occurred. Suppose the time-slice assigned to process P1 runs out between sequence numbers (2) and (3). Redraw the diagram, through the equivalent of sequence number (10), showing a possible series of events that might occur after the timer interrupt.

11. Redraw the diagram in Fig. 6.15, assuming that process P2 is given higher dispatching priority than process P1 and that preemptive process scheduling is used.

12. Suppose two jobs are being multiprogrammed together. Job A uses a great deal of CPU time and performs relatively little I/O. Job B performs many I/O operations, but requires very little CPU time. Which of these two jobs should be given higher dispatching priority to improve the overall system performance?

13. Suppose the I/O supervisor is able to select I/O requests from channel queues based on a priority system. Which of the two jobs in Exercise 12 should be given higher I/O priority in this way?

14. Suppose a certain I/O device can be reached via either one of two I/O channels; however, the device can be used by only one channel at a time. How would the I/O supervisor routines described in the text need to be changed to accommodate this situation?

15. How would you select the number and size of partitions for a system using fixed-partition memory management?

16. What are the advantages and disadvantages of the bounds-register approach to memory protection, as compared to the protection-key approach?

17. In a multiprogramming system, frequently an I/O operation is being performed for one job while another job is in control of the CPU. How could memory protection be provided for such an I/O operation? For example, how could one job be prevented from reading data into another job's partition?

18. Suppose a certain machine includes flag bits that indicate the type of each item stored in memory or in a register (for example, integer, character, floating-point number, instruction, or address pointer). How could this information be used to implement relocatable partitions on such a machine?

19. Is memory-protection hardware necessary on a machine that uses demand-paged memory management?

20. Is a relocating loader necessary on a machine that uses demand-paged memory management?

21. Some demand-paging systems select pages to be removed from memory based in part upon whether the page has been modified. That is, the system prefers to replace a page that has not been modified since it was last loaded, instead of a page that has been modified. What are the advantages and disadvantages of such an approach?

22. What methods might a programmer use to improve the locality of reference of a program? What programming techniques and data structures would probably lead to poor locality of reference?

23. What would the diagram in Fig. 6.26(b) look like for a program with no locality of reference (i.e., a program in which each memory reference is independent of the previous references)? What would the diagram look like for a program with perfect locality of reference (i.e., one with each reference made to the same page as the previous reference)?

24. In a multiprogramming system, other jobs can perform useful work while one job is waiting for an I/O operation to complete. Why, then, does thrashing by one program in a virtual memory system create a problem for other programs in the same system?

25. Outline algorithms for the four interrupt handlers on a SIC/XE multiprogramming operating system that uses demand paging.

26. Why was the timer interrupt assigned a lower priority than the SVC interrupt on the SIC/XE machine (see Section 6.2.1)?

27. Why was the I/O interrupt assigned a lower priority than the SVC interrupt on the SIC/XE machine?

Section 6.3

1. Give an algorithm for a file-manager routine that performs the blocking and buffering operations illustrated in Fig. 6.28.

2. When might it be advantageous to use more than two buffers for a sequential file?

3. Draw a state-transition diagram similar to Fig. 6.7 for the three-level scheduling procedure illustrated in Fig. 6.29(b).

4. Is it possible for a job to have a shorter turnaround time under a multiprogramming operating system than under a single-job system for the same machine?

5. Describe algorithms and data structures for operating system routines that implement the request and release functions described in Section 6.3.3.

6. Suppose there is an event status block defined for each resource in the system in such a way that "the-event-has-occurred" is logically equivalent to "the-resource-is-free." Could the request and release functions be imple-

mented using the WAIT and SIGNAL operations described in Section 6.2.2?

7. Consider the two programs in Fig. 6.31. Instead of using request and release operations, we could simply inhibit timer interrupts between lines 24 and 27 of P1, and between lines 37 and 40 of P2. (The inhibiting and enabling of these interrupts could be done via operating system service calls.) Would this be a practical solution to the problem discussed in the text?

8. How might the operating system detect that a deadlock has occurred?

CHAPTER 7

OTHER SYSTEM SOFTWARE

In this chapter we present brief overviews of several different types of system software. The purpose of this chapter is to introduce the basic concepts and terms related to these pieces of software. Because of space limitations, we do not attempt to give detailed discussions of implementation. References are provided in each section for readers who want to study these topics further.

Section 7.1 describes the purpose and functions of a generalized database management system (DBMS), and discusses the relationship of the DBMS to the operating system. Section 7.2 gives a brief description of interactive text-editing systems, discussing the general approaches used in such editors.

Section 7.3 introduces the topic of interactive program-debugging systems. Although the desirability of such systems has been recognized for some time, there are relatively few actual debugging systems in practical use. Therefore, Section 7.3 presents a discussion of the functions and characteristics that are likely to be important in the future development of such software.

7.1 DATABASE MANAGEMENT SYSTEMS (DBMS)

This section describes the basic purpose and functions of a generalized database management system (DBMS). Our main focus in this section is the user's view of a DBMS. We also discuss how the DBMS functions are related to other types of software in the system. Comprehensive discussions of database management system theory and implementation can be found in Wiederhold (1983), Date (1981), and Ullman (1980), among others.

Section 7.1.1 discusses the problems that led to the development of database management systems, and presents the basic concept of a DBMS. Section 7.1.2 describes in general terms how such systems appear to the user. Section 7.1.3 discusses the relationship of a DBMS to other system software, particularly the operating system.

7.1.1 Basic Concept of a DBMS

The development of database management systems resulted in large part from two major problems with conventional file-processing systems: *data redundancy* and *data dependence*. Consider for example, Fig. 7.1, which shows a simplified set of file-processing systems for a hypothetical university. This example includes a registration system to keep track of the enrollment of students in courses, a financial aid system to control payments of scholarship funds to students, and a payroll system for the entire university, including both faculty and student employees.

Each file-processing system consists of a set of programs and a set of data files. The format, organization, and content of these data files are specified by the person who initially defines the system. The files for each system contain only the information needed by that system; there is little or no coordination between files belonging to different systems.

FIGURE 7.1 Separate file-processing systems.

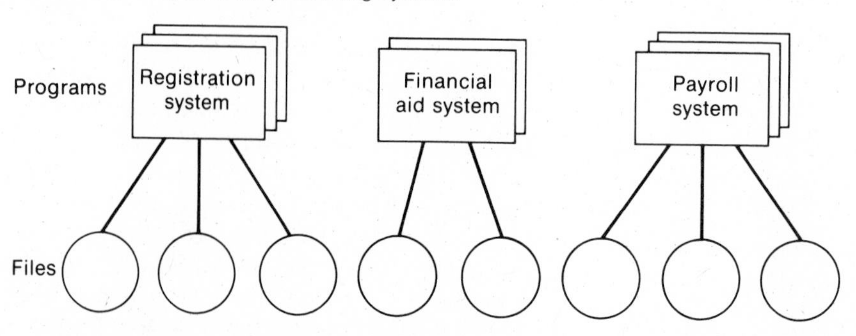

The use of separate file-processing systems like those in Fig. 7.1 often leads to a large amount of *data redundancy,* which is the duplication of data items in different files. For example, the number of courses in which a student is enrolled might be stored in one of the files for the registration system, and also in a file for the financial aid system. If a student were an employee of the university and a scholarship recipient, the student's name and address might appear in three different places.

The most obvious disadvantage of data redundancy is the additional storage space required. However, there are more serious problems associated with such duplication. A piece of information that is stored in several different places must be updated several times when its value changes. This multiple updating requires more computing time and more I/O operations. Even more serious is the possibility of inconsistent data because of program errors or differences in the schedules for updating the files. For example, an entry in a registration file might reflect the fact that a student had withdrawn from a certain course; however, this information might not be incorporated into the files for the financial aid system.

One possible solution to the problem of data redundancy is illustrated in Fig. 7.2(a). The information from all of the file systems is gathered into a single integrated *database* for the entire university. This database contains only one copy of each logical data item, so redundancy is eliminated. The database itself consists of a set of files. Different applications that require the same data item may share the file that contains the needed information.

Although the approach just described solves the problem of data redundancy, it may cause other difficulties. Application programs that deal directly with physical files are *data dependent,* which means they depend on characteristics such as record format and file organization and sequence. Whenever these file characteristics are changed, the application programs must also be modified.

Suppose, for example, that a student-advising system is added to the set of applications. The new system may require certain information that is not already present in the database, so one or more new files may need to be created. On the other hand, some of the information required is already present. The advising system may need to refer to existing files for such items as each student's major code and current enrollment.

Unfortunately, this existing information may not be present in the form required by the new system. Suppose, for example, that the advising system must provide interactive access to enrollment information for all students who have a given advisor. This would require that the

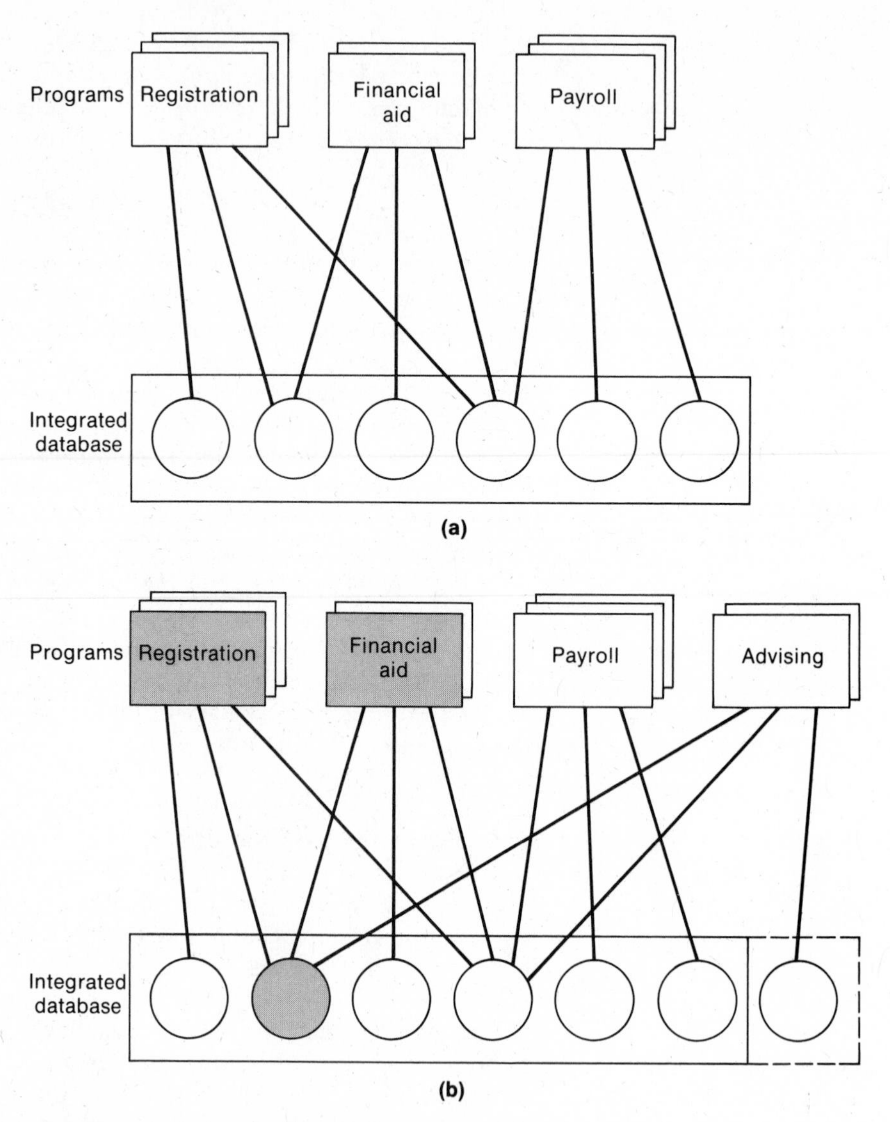

FIGURE 7.2 Data-dependent programs using an integrated database. (Shading indicates changes required by the addition of a new system.)

advisor's name be added to the enrollment information for a student and that the enrollment data be indexed by advisor. The content and organization of some existing files would need to be changed, and, because of data dependence, all other programs using these files would also have to be modified.

This situation is illustrated in Fig. 7.2(b). The advising system uses three database files: one new file and two existing ones. It was necessary to modify the format and structure of one of the existing files used by the new system (shown by shading in the figure). All other programs that use this file must therefore be changed. In this case, the changes involve programs in the existing registration and financial aid systems.

Problems like the one just described can be avoided by making application programs independent of details such as file organization and record format. Figure 7.3(a) shows how this can be accomplished. The user programs do not deal directly with the files of the database. Instead, they access the data by making requests to a *database management system* (DBMS). Application programs request data at a logical level, without regard for how the data is actually stored in files. For example, a program can request current enrollment information for a particular student. The DBMS determines which physical files are involved, and how these files are to be accessed, by referring to a stored *data mapping description*. It then reads the required records from the files of the database and converts the information into the form requested by the application program. This process is discussed in more detail in Sections 7.1.2 and 7.1.3.

The *data independence* provided by this approach means that file structures can be changed without affecting the application programs. Consider, for example, Fig. 7.3(b). As before, a new advising system has been added; a new file has been added to the database; one existing file has been modified (indicated by the shading in the figure). The data mapping description has been modified to reflect these changes, but the application programs themselves remain unchanged. The same logical request from an application program may now result in a different set of operations on the files of the database. The application programs, however, are unaware of this difference because they are concerned only with logical items of information, not with how this information is stored in files.

The data independence provided by a database management system is also important for other reasons. The techniques used for physical storage of the database can be changed whenever it is desirable to do so. For example, some or all of the database can be moved to a different type of storage device. Files can be reorganized, sorted in different sequences, or indexed by a different set of keys. Decisions of this sort are usually made by a *database administrator,* who attempts to organize and store the data in a way that leads to the most efficient overall use of the database. All these changes can be made without

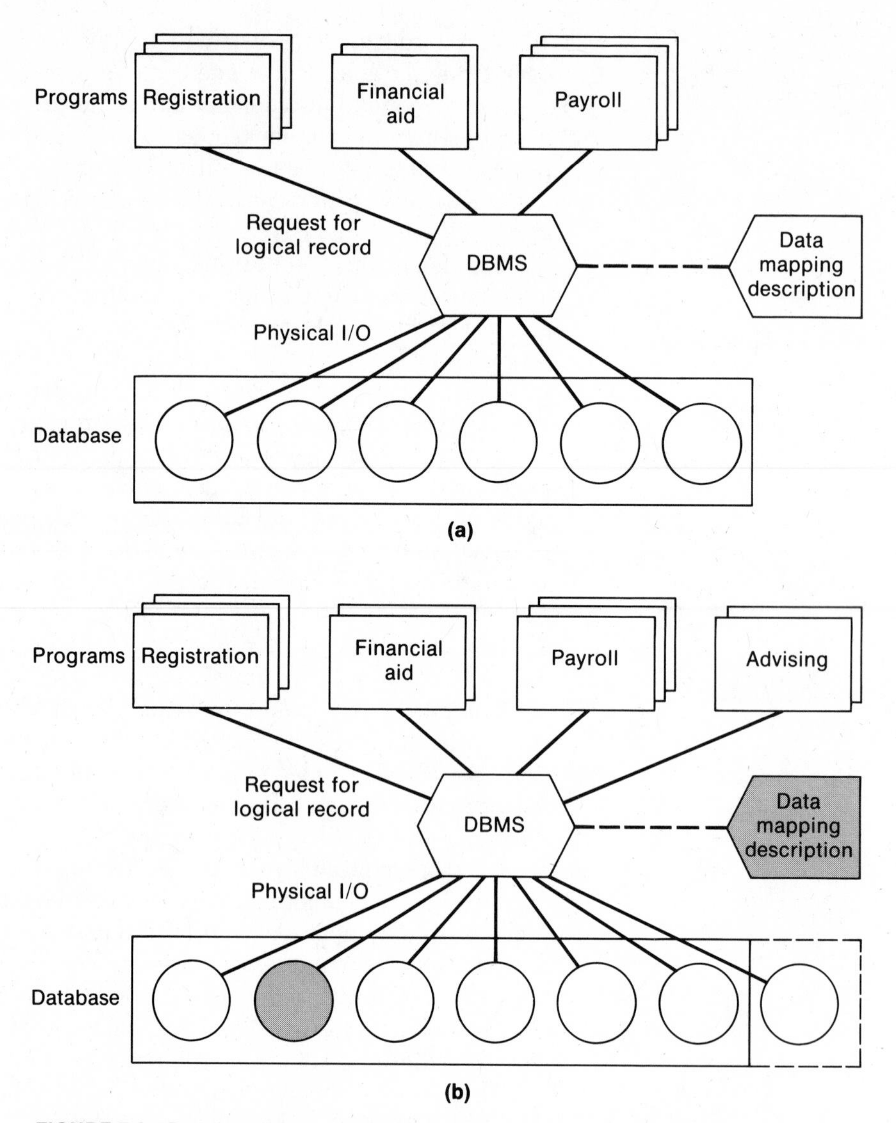

FIGURE 7.3 Data-independent programs using a database management system. (Shading indicates changes required by the addition of a new system.)

affecting any of the application programs. Indeed, the programs are in general not able to detect that such changes have been made.

Because the data mapping description must be consulted for each reference to the database, using a DBMS involves more system overhead than running data-dependent programs. However, the benefits of data independence and reduced data redundancy usually outweigh the additional overhead required. This is particularly true when the content and structure of the database are subject to periodic changes because of new applications.

7.1.2 Levels of Data Description

As we discussed in the previous section, the use of a database management system makes application programs independent of the way data is physically stored. With conventional file systems, the programmer is concerned with descriptions of the organization, sequencing, and indexing techniques of files to be processed. However, such details are not usually known to a programmer using a DBMS. Even if they are known, the programmer cannot rely on this knowledge in writing a program because the physical storage of the database may be changed at any time. Thus the application programmer's view of the data must be at a *logical* level, independent of file structure and other such questions of physical storage.

The information stored by a DBMS can be viewed in a number of different ways, depending upon the needs of the user. The most general such view is an overall logical database description called the *schema*. Figure 7.4 shows an example of such a schema. This example contains a number of logical database records, such as STUDENT and COURSE. Some of these records are connected by lines that indicate possible relationships between records of the given types. For example, the enrolled-in relationship specifies which students are enrolled in which courses. This example is intended to represent a database for a hypothetical university. The data items shown in each logical record

FIGURE 7.4 Schema for sample university database.

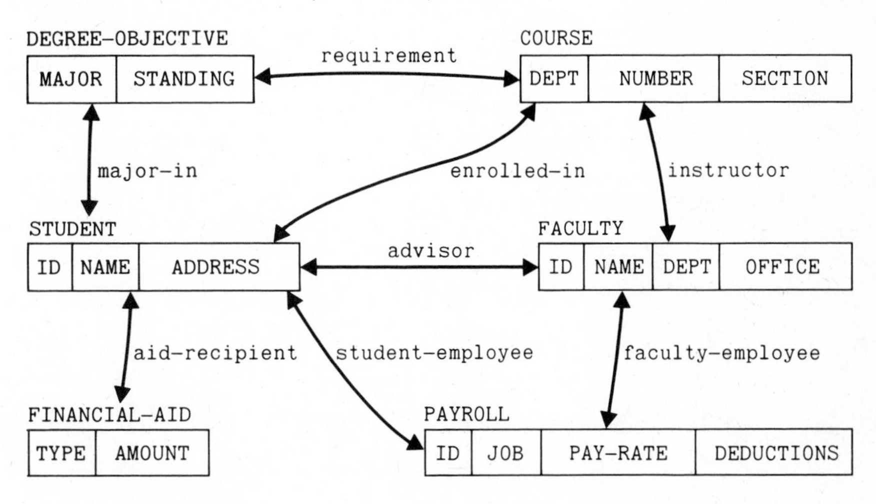

illustrate the kinds of information that might be contained in such records. For a real database, there would probably be more logical record types, and each record would contain many more data items.

Database management systems differ considerably in the kinds of records and relationships that can be included in the schema. For example, some systems require that the logical database representation be expressed as a hierarchy or a tree structure, while others allow more general types of interconnections between records. Some systems do not allow explicit connections between records at all. These connections are expressed implicitly through the values of corresponding data items. A discussion of such *data models* is beyond the scope of this book. For further information and references, see Date (1981).

The schema gives a complete description of the logical structure and content of the database. However, most application programmers are concerned with only a small fraction of this information; a particular program usually deals with only a few types of records and relationships. The description of the data required by an application program is given by a *subschema*. There are usually many different subschemas corresponding to one schema. Each subschema gives a view of the database suited to the needs of the programs that use that subschema.

Figure 7.5 shows three different subschemas that correspond to various parts of the schema in Fig. 7.4. Subschema (a) is one that

FIGURE 7.5 Three possible subschemas corresponding to the schema of Fig. 7.4.

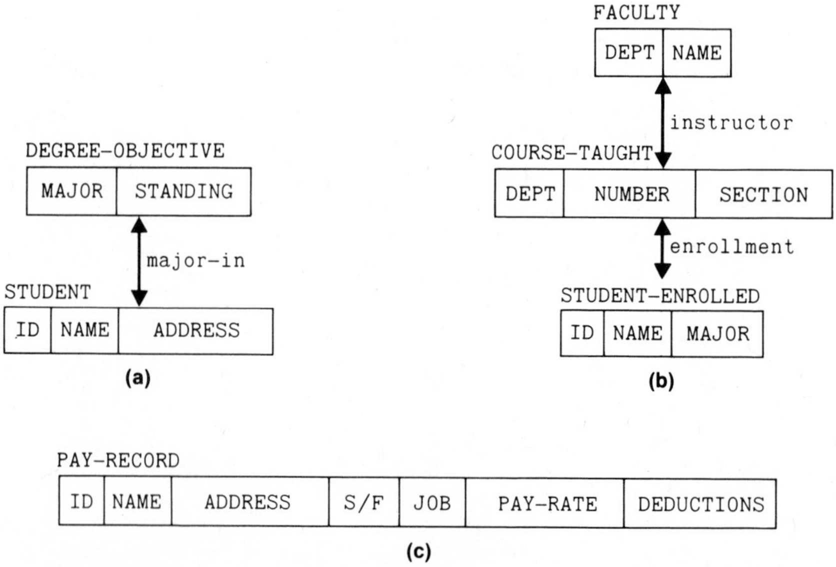

DEGREE-OBJECTIVE
MAJOR | STANDING
major-in
STUDENT
ID | NAME | ADDRESS
(a)

FACULTY
DEPT | NAME
instructor
COURSE-TAUGHT
DEPT | NUMBER | SECTION
enrollment
STUDENT-ENROLLED
ID | NAME | MAJOR
(b)

PAY-RECORD
ID | NAME | ADDRESS | S/F | JOB | PAY-RATE | DEDUCTIONS
(c)

might be used by a program that produces a listing of students by major. To a program using subschema (a), the database appears to consist of a number of DEGREE-OBJECTIVE records, with each such record linked to a set of STUDENT records (one for each student with the given major). Subschema (b) might be used by a program that prints class rolls for each faculty member. This view of the database is a three-level structure. A record for each faculty member is linked to a set of records representing the courses he or she is teaching; each of these course records is similarly linked to a set of records representing the students enrolled in the course. Subschema (c) is designed for use by a program that processes the payroll for the university. Each logical record in this subschema contains the information needed to issue a paycheck to an employee.

These three subschemas provide quite different views of the database. A subschema must be consistent with the schema—that is, it must be possible to derive the information in the subschema from the schema. Subschema (a) is simply a subset of the schema: the record names, data items, and relationships are the same as those contained in the corresponding part of the schema. This need not be true, however, for all subschemas.

In subschema (b), the application program uses record names that are different from those contained in the schema. It is also possible to use different names for individual data items. The FACULTY record in subschema (b) contains some, but not all, of the information from the FACULTY record of the schema. The COURSE-TAUGHT subschema record contains the same information that is present in the COURSE schema record. The STUDENT-ENROLLED record in the subschema contains information from the STUDENT record in the schema. However, the STUDENT-ENROLLED record also contains information about the student's major, which is contained in the DEGREE-OBJECTIVE record that is logically connected to the STUDENT record by the major-in relationship.

Subschema (c) consists of a single logical record type PAY-RECORD, whose data may come from three different schema records. Information concerning rate of pay and deductions is taken from the PAYROLL schema record. Information such as employee name and address is obtained from either the STUDENT record or the FACULTY record, whichever is appropriate, by using the student-employee and faculty-employee relationships. The data field in PAY-RECORD that is designated S/F indicates whether the record pertains to a student employee or to a faculty member.

A subschema provides an application program with a view of the database that is suited to the needs of the particular program. The

DBMS takes care of converting information from the database into the form specified by the subschema (see Section 7.1.3). As a result, the application program is simpler and easier to write because the programmer does not have to be concerned with data items and relationships that are not relevant to the application. The subschema is also an aid in providing data security because a program has no way of referring to data items not described in its subschema.

We have now discussed three different levels of data description in database management systems: the subschemas, the schema, and the data mapping description. A DBMS supplies languages, called *data description languages,* for defining the database at each of these levels. The subschemas are used by application programmers and are written in a *subschema description language* designed to be convenient for the programmer. Often subschema description languages are extensions of the data description capabilities in the programming language to be used. However, the subschemas are created and maintained by the database administrator. In defining a subschema, the database administrator must be sure that the view of data given in the subschema is derivable from the schema, and that it contains only those data items the application program is authorized to use.

The schema itself, and the physical data mapping description, are normally used only by the database administrator. On many systems, the *schema description language* is closely related to the subschema description language. It is also possible to use a more generalized language, because the schema is not used directly by application programmers. The *physical data description language* is influenced by the types of logical structures supported by the schema, and also by the types of files and storage devices supported by the DBMS.

Further discussions and examples of data description languages can be found in Date (1981).

7.1.3 Use of a DBMS

In the two preceding sections we introduced basic concepts and terminology related to database management systems. In this section we complete the picture by discussing how a user interacts with a DBMS, and how the DBMS is related to other pieces of system software.

The two principal methods for user interaction with a DBMS are illustrated in Fig. 7.6. The user can write a source program in the normal way, using a general-purpose programming language such as COBOL, PL/I, or assembler language. However, instead of writing I/O statements of the form provided by the programming language, the programmer writes commands in a *data manipulation language* (DML) defined for use with the DBMS. These commands are often designed so

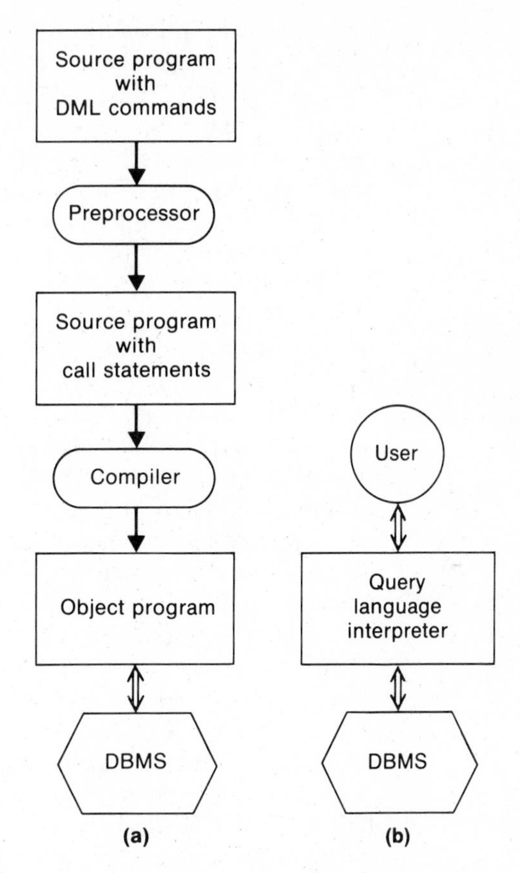

FIGURE 7.6 Interaction with a DBMS using (a) a data manipulation language and (b) a query language.

that the DML appears to be merely an extension of the programming language. As shown in Fig. 7.6(a), a preprocessor may be used to convert the DML commands into programming language statements that call DBMS routines. The modified source program is then compiled in the usual way. Another approach is to modify the compiler to handle the DML statements directly. Some data manipulation languages are defined as a set of CALL statements using the programming language itself, which avoids the need for preprocessing or compiler modification.

The other approach to DBMS interaction, illustrated in Fig. 7.6(b), does not require the user to write programs to access the database. Instead, users enter commands in a special *query language* defined by the DBMS. These commands are processed by a query-language interpreter, which calls DBMS routines to perform the requested operations.

Each of these approaches to user interaction has its own advantages. With a query language, it is possible to obtain results much more quickly because there is no need to write and debug programs. Query languages can also be used effectively by nonprogrammers, or by individuals who program only occasionally. Most query languages, however, have built-in limitations. For example, it may be difficult or impossible to perform a function for which the language was not designed. On the other hand, a DML allows the programmer to use all the flexibility and power of a general-purpose programming language; however, this approach requires much more effort from the user. Most modern database management systems provide both a query language and a data manipulation language so that a user can choose the form of interaction that best meets his or her needs. Further discussions and examples of DMLs and query languages can be found in Date (1981).

The sequence of operations performed by a DBMS in processing a request is essentially the same regardless of whether a DML or a query language is being used. These actions are illustrated in Fig. 7.7. The sequence of events begins when the DBMS is entered via a call from application program A (step 1 in the figure). If a query language is being used, program A is the query-language interpreter. We assume this call is a request to read data from the database. The sequences of events for other types of database operations are similar.

The request from program A is stated in terms of the subschema being used by A. For example, a program using the subschema in Fig. 7.5(c) might request the PAY-RECORD for a specified employee. To process such a request, the DBMS must first examine the subschema definition being used (step 2). The DBMS must also consider the relationship between the subschema and the schema (step 3) to interpret the request in terms of the overall logical database structure. Thus, for example, the DBMS would detect that it needed to read the schema PAYROLL record for the specified employee (see Fig. 7.4) to supply program A with its expected PAY-RECORD. In addition to this PAYROLL record, the DBMS would also need to examine the student-employee and faculty-employee relationships for the PAYROLL record in question, and read the corresponding STUDENT or FACULTY record.

After determining the logical database records that must be read (in terms of the schema), the DBMS examines the data mapping description (step 4). This operation gives the information needed to locate the required records in the files of the database. At this point, the DBMS has converted a logical request for a subschema record into

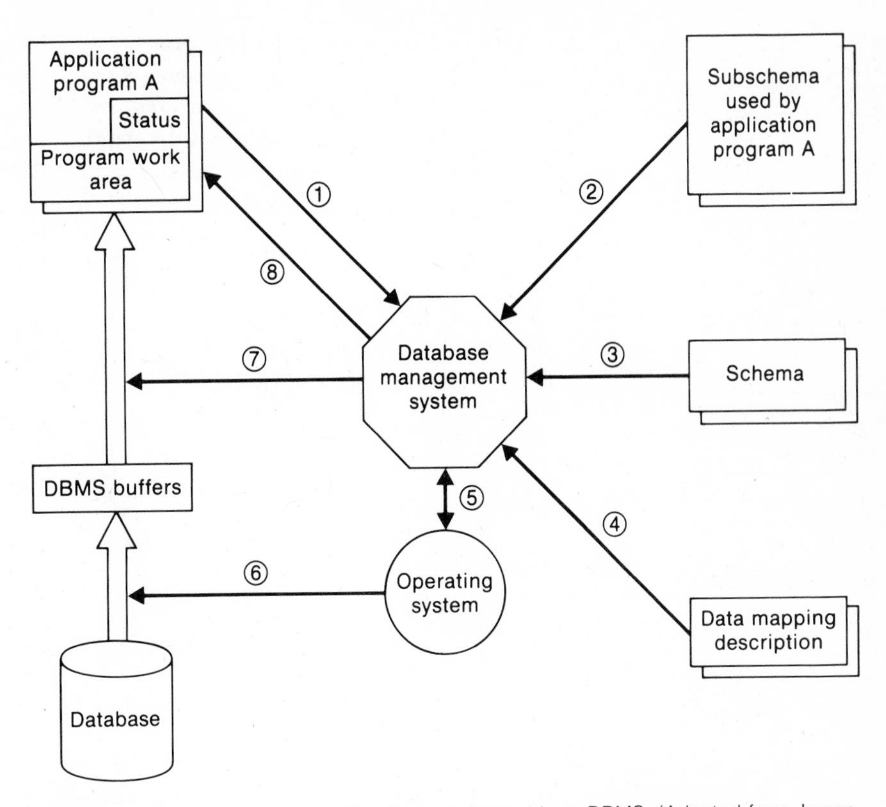

FIGURE 7.7 Typical sequence of actions performed by a DBMS. (Adapted from James Martin, *Computer Data-Base Organization*, 2nd ed., © 1977, p. 83. Reprinted by permission of Prentice-Hall Inc., Englewood Cliffs, N.J.)

physical requests to read data from one or more files. These requests for file I/O are passed to the operating system (step 5) using the types of service calls discussed in Chapter 6. The operating system then issues channel and device commands to perform the necessary physical I/O operations (step 6). These I/O operations read the required records from the database into a DBMS buffer area.

After the physical I/O operations have been completed, all the data requested by the application program is present in central memory. However, this information must still be converted into the form expected by the program. The DBMS accomplishes this conversion (step 7) by again comparing the schema and the subschema. In the example we are discussing, the DBMS would extract data from the PAYROLL record and the associated STUDENT or FACULTY record, and construct the PAY-RECORD requested by program A. The PAY-RECORD would then be placed into a work area supplied by the application program; this completes the processing of the program's request for data. Finally, the DBMS returns control to the application program and makes available to the program a variety of status information, including any possible error indications.

Further details concerning the topics discussed in this section can be found in Wiederhold (1983) and Date (1981).

7.2 TEXT EDITORS*

The interactive text editor has become an important part of almost any computing environment. No longer are editors thought of as tools only for programmers or for secretaries transcribing from marked-up copy generated by authors. It is now increasingly recognized that a text editor should be considered the primary interface to the computer for all types of "knowledge workers" as they compose, organize, study, and manipulate computer-based information.

In this section we briefly discuss interactive text-editing systems from the points of view of both the user and the system. Section 7.2.1 gives a general overview of the editing process. Section 7.2.2 expands upon this introduction by discussing various types of user interfaces and input/output devices. Section 7.3.3 describes the structure of a typical text editor and discusses a number of system-related issues.

7.2.1 Overview of the Editing Process

An *interactive editor* is a computer program that allows a user to create and revise a target *document.* The term document includes objects such as computer programs, text, equations, tables, diagrams, line art, and photographs—anything that one might find on a printed page. In this discussion, we restrict our attention to *text editors,* in which the primary elements being edited are character strings of the target text.

The document-editing process is an interactive user–computer dialogue designed to accomplish four tasks.

1. Select the part of the target document to be viewed and manipulated
2. Determine how to format this view on-line and how to display it
3. Specify and execute operations that modify the target document
4. Update the view appropriately.

* Adapted from Norman Meyrowitz and Andries van Dam, "Interactive Editing Systems: Part I and Part II." *ACM Computing Surveys,* September 1982. Copyright 1982, Association for Computing Machinery, Inc. These publications also contain much more detailed discussions of the editing process, descriptions of a large number of actual editors, and a comprehensive bibliography.

Selection of the part of the document to be viewed and edited involves first *traveling* through the document to locate the area of interest. This search is accomplished with operations such as *next screenful, bottom,* and *find pattern.* Traveling specifies where the area of interest is; the selection of what is to be viewed and manipulated there is controlled by *filtering.* Filtering extracts the relevant subset of the target document at the point of interest, such as the next screenful of text or the next statement. *Formatting* then determines how the result of the filtering will be seen as a visible representation (the *view*) on a display screen or a hard-copy device.

In the actual *editing* phase, the target document is created or altered with a set of operations such as *insert, delete, replace, move,* and *copy.* The editing functions are often specialized to operate on *elements* meaningful to the type of editor. For example, a manuscript-oriented editor might operate on elements such as single characters, words, lines, sentences, and paragraphs; a program-oriented editor might operate on elements such as identifiers, keywords, and statements.

In a simple scenario, then, the user might travel to the end of the document. A screenful of text would be filtered, this segment would be formatted, and the view would be displayed on an output device. The user could then, for example, delete the first three words of this view.

7.2.2 User Interface

The user of an interactive editor is presented with a *conceptual model* of the editing system. This model is an abstract framework on which the editor and the world on which it operates are based. The conceptual model, in essence, provides an easily understood abstraction of the target document and its elements, with a set of guidelines describing the effects of operations on these elements. Some of the early *line editors* simulated the world of the keypunch. These editors allowed operations on numbered sequences of 80-character card-image lines, either within a single line or on an integral number of lines. Some more modern *screen editors* define a world in which a document is represented as a quarter-plane of text lines, unbounded both down and to the right. Operations manipulate portions of this quarter-plane without regard to line boundaries. The user sees through a cutout only a rectangular subset of this plane on a multiline display terminal. The cutout can be moved left or right, and up or down, to display other portions of the document.

Besides the conceptual model, the user interface is concerned with the *input devices,* the *output devices,* and the *interaction language* of the system. Brief discussions and examples of these aspects of the user interface are presented in the remainder of this section.

Input devices are used to enter elements of the text being edited, to enter commands, and to designate editable elements. These devices, as used with editors, can be divided into three categories: text devices, button devices, and locator devices. *Text* or *string* devices are typically typewriter-like keyboards on which a user presses and releases keys, sending a unique code for each key. Virtually all current computer keyboards are of the QWERTY variety (named for the first six letters in the second row of the keyboard). Several alternative keyboard arrangements have been proposed, some of which offer significant advantages over the standard keyboard layout. None of these alternatives, however, seems likely to be widely accepted in the near future because of the retraining effort that would be required.

Button or *choice* devices generate an interrupt or set a system flag, usually causing invocation of an associated application-program action. Such devices typically include a set of special function keys on an alphanumeric keyboard or on the display itself. Alternatively, buttons can be simulated in software by displaying text strings or symbols on the screen. The user chooses a string or symbol instead of pressing a button.

Locator devices are two-dimensional analog-to-digital converters that position a cursor symbol on the screen by observing the user's movement of the device. Such devices include *joysticks, touch screen panels, data tablets,* and *mice.* The latter two are the most common locator devices for editing applications. The data tablet is a flat, rectangular, electromagnetically sensitive panel. Either a ballpoint-pen-like *stylus* or a *puck,* a small device that fits in the palm of the hand, is moved over the surface. The tablet returns to a system program the coordinates of the position on the data tablet at which the stylus or puck is currently located. The program can then map these data-tablet coordinates to screen coordinates and move the cursor to the corresponding screen position. The mouse is another hand-held device similar to the puck. As it is moved on any flat surface, the motion of the mouse causes relative changes in position to be sampled by a system program. These mouse coordinates are again mapped to screen coordinates to move a cursor on the screen.

A locator device combined with a button device allows the user to specify either a particular point on the screen at which text should be inserted or deleted, or the start and endpoints of a string of characters to be operated upon. In fact, the mouse and puck usually have built-in buttons for the user to signal a selection. When the cursor has been positioned over an element, the user presses a button to indicate the selection. The system associates the cursor position with the element that it covers and performs the desired action.

Text devices with arrow (cursor) keys are often used to simulate locator devices. Each of these keys shows an arrow that points up, down, left, or right, Pressing an arrow key typically generates an appropriate character sequence; the program interprets this sequence and moves the cursor in the direction of the arrow on the key pressed.

Still in the research stage, *voice-input devices,* which translate spoken words to their textual equivalents, may prove to be the text devices of the future. While currently restricted to a small vocabulary (typically fewer than 1000 words), voice recognizers may soon be commercially viable for command recognition.

Formerly limited in range, output devices for editing are becoming more diverse. The output device serves to let the user view the elements being edited and the results of the editing operations. The first output devices were the (now obsolete) teletypewriters and other character-printing terminals that generated output on paper. Next, "glass teletypes" based on *cathode ray tube* (CRT) technology used the CRT screen essentially to simulate a hard-copy teletypewriter (although a few operations, such as backspacing, were performed more elegantly). Today's advanced CRT terminals use hardware assistance for such features as moving the cursor, inserting and deleting characters and lines, and scrolling lines and pages. The new *professional work stations,* based on personal computers with high-resolution displays, support multiple proportionally spaced character fonts to produce realistic facsimiles of hard-copy documents. Thus the user can see the document portrayed essentially as it will look when printed on paper.

The interaction language of a text editor is generally one of several common types. The *typing-oriented* or *text command-oriented* method is the oldest of the major editor interfaces. The user communicates with the editor by typing text strings both for command names and for operands. These strings are sent to the editor and are usually echoed to the output device.

Typed specification often requires the user to remember the exact form of all commands, or at least their abbreviations. If the command language is complex, the user must continually refer to a manual or an on-line *help* function for a description of less frequently used commands. In addition, the typing required can be time consuming, especially for inexperienced users. The *function-key* interface addresses these deficiencies. Here each command has associated with it a marked key on the user's keyboard. For example, the *insert character* command might have associated with it a key marked IC. Function-key command specification is typically coupled with cursor-key movement for specifying operands, which eliminates much typing.

For the common commands in a function-key editor, usually only a single key need be pressed. For less frequently invoked commands or options, an alternative textual syntax may be used. More commonly, however, special keys are used to shift the standard function-key interpretations, just as the SHIFT key on a typewriter shifts from lowercase to uppercase. As an alternative to shifting function keys, the standard alphanumeric keyboard is often *overloaded* to simulate function keys. For example, the user may press a *control* key simultaneously with a normal alphanumeric key to generate a new character that is interpreted like a function key.

Typing-oriented systems require familiarity with the system and language, as well as some expertise in typing. Function key-oriented systems often have either too few keys, requiring multiple-keystroke commands, or have too many unique keys, which results in an unwieldy keyboard. In either case, the function-key systems demand even more agility of the user than a standard keyboard does. The *menu-oriented* user interface is an attempt to address these problems. A *menu* is a multiple-choice set of text strings or *icons,* which are graphic symbols that represent objects or operations. The user can perform actions by selecting items from the menu. The editor prompts the user with a menu of only those actions that can be taken at the current state of the system.

One problem with a menu-oriented system can arise when there are many possible actions and several choices are required to complete an action. The display area for the menu is usually rather limited; therefore, the user might be presented with several consecutive menus in a hierarchy before the appropriate command and its options appear. Since this can be annoying and detrimental to the performance of an experienced user, some menu-oriented systems allow the user to turn off menu control and return to a typing or function-key interface. Other systems have the most-used functions on a main command menu and have secondary menus to handle the less frequently used functions. Still other systems display the menu only when the user specifically asks for it. For example, the user might press a button on a mouse to display a menu with the full choice of applicable commands (perhaps temporarily overlaying some existing information on the screen). The mouse could be used to select the appropriate command. The system would then execute the command and delete the menu. Interfaces like this, in which prompting and menu information are given to the user at little added cost and little degradation in response time, are becoming increasingly popular.

7.2.3 Editor Structure

Most text editors have a structure similar to that shown in Fig. 7.8 regardless of the particular features they offer and the computers on

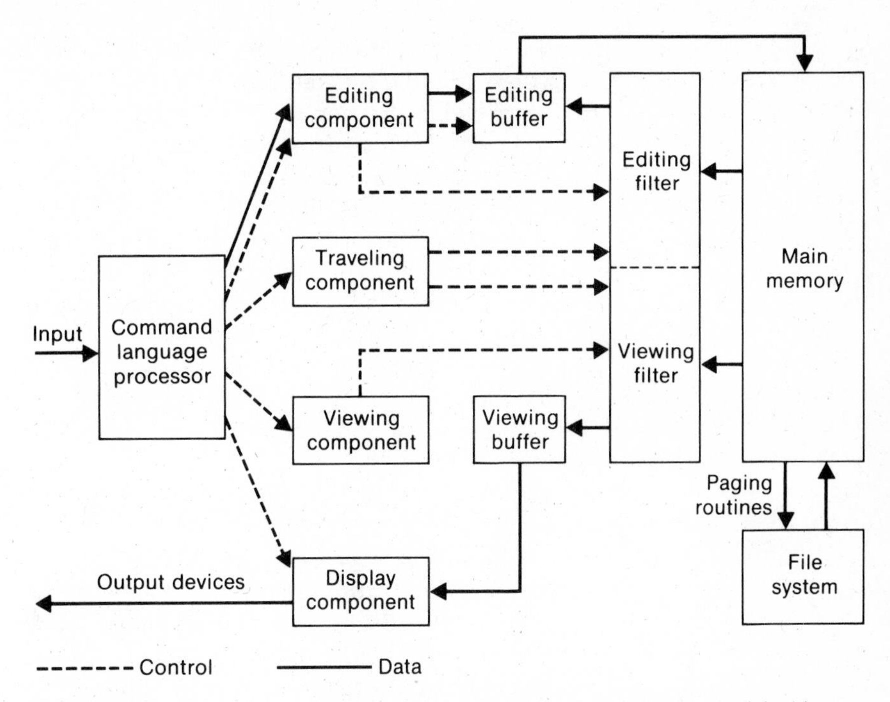

FIGURE 7.8 Typical editor structure. (Adapted from Norman Meyrowitz and Andries van Dam, "Interactive Editing Systems: Part I and Part II," *ACM Computing Surveys*, 1982. Copyright 1982, Association for Computing Machinery, Inc.)

which they are implemented. The *command language processor* accepts input from the user's input devices, and analyzes the tokens and syntactic structure of the commands. In this sense, the command language processor functions much like the lexical and syntactic phases of a compiler. Just as in a compiler, the command language processor may invoke semantic routines directly. In a text editor, these semantic routines perform functions such as editing and viewing. Alternatively, the command language processor may produce an intermediate representation of the desired editing operations. This intermediate representation is then decoded by an interpreter that invokes the appropriate semantic routines. The use of an intermediate representation allows the editor to provide a variety of user-interaction languages with a single set of semantic routines that are driven from a common intermediate representation.

The semantic routines involve traveling, editing, viewing, and display functions. Editing operations are always specified explicitly by the user, and display operations are specified implicitly by the other three categories of operations. However, the traveling and viewing operations may be invoked either explicitly by the user or implicitly by the editing operations. The relationship between these classes of operations may be considerably more complicated than the simple model described in Section 7.2.1. In particular, there need not be a simple one-to-one relationship between what is currently displayed on the screen and what can be edited. To illustrate this, we take a closer look at the components of Fig. 7.8 that are meant to be conceptual entities.

In editing a document, the start of the area to be edited is determined by the *current editing pointer* maintained by the *editing component,* which is the collection of modules dealing with editing tasks. The current editing pointer can be set or reset explicitly by the user with traveling commands, such as *next paragraph* and *next screen,* or implicitly by the system as a side effect of the previous editing operation, such as *delete paragraph.* The *traveling component* of the editor actually performs the setting of the current editing and viewing pointers, and thus determines the point at which the viewing and/or editing filtering begins.

When the user issues an editing command, the editing component invokes the *editing filter*. This component filters the document to generate a new *editing buffer* based on the current editing pointer as well as on the editing filter parameters. These parameters, which are specified both by the user and the system, provide such information as the range of text that can be affected by an operation. Filtering may simply consist of the selection of contiguous characters beginning at the current point. Alternatively, filtering may depend on more complex user specifications pertaining to the content and structure of the document. Such filtering might result in the gathering of portions of the document that are not necessarily contiguous. The semantic routines of the editing component then operate on the editing buffer, which is essentially a filtered subset of the document data structure. Note that this explanation is at a conceptual level—in a given editor, filtering and editing may be interleaved, with no explicit editing buffer being created.

Similarly, in viewing a document, the start of the area to be viewed is determined by the *current viewing pointer*. This pointer is maintained by the *viewing component* of the editor, which is a collection of modules responsible for determining the next view. The current viewing pointer can be set or reset explicitly by the user with a traveling command or implicitly by the system as a result of the previous editing operation. When the display needs to be updated, the viewing component invokes the *viewing filter*. This component filters the document to generate a new *viewing buffer* based on the current viewing pointer as well as on the viewing filter parameters. These parameters, which are specified both by the user and by the system, provide such information as the number of characters needed to fill the display, and how to select them from the document. In line editors, the viewing buffer may contain the current line; in screen editors, this buffer may contain a rectangular cutout of the quarter-plane of text. This viewing buffer is then passed to the *display component* of the editor, which produces a display by mapping the buffer to a rectangular subset of the screen, called a *window* or *viewport*.

The editing and viewing buffers, while independent, can be related in many ways. In the simplest case, they are identical: the user edits the material directly on the screen (see Fig. 7.9). On the other hand, the editing and viewing buffers may be completely disjoint. For example, the user of a certain editor might travel to line 75, and after viewing it, decide to change all occurrences of "ugly duckling" to "swan" in lines 1 through 50 of the file by using a *change* command such as

```
[1,50] c/ugly duckling/swan/
```

As part of this editing command, there is implicit travel to the first line of the file. Lines 1 through 50 are then filtered from the document to become the editing buffer, and successive substitutions take place in this editing buffer without corresponding updates of the view. If the pattern is found, the current editing and viewing pointers are moved to the last line on which it is found, and that line becomes the default

FIGURE 7.9 Simple relationship between editing and viewing buffers.

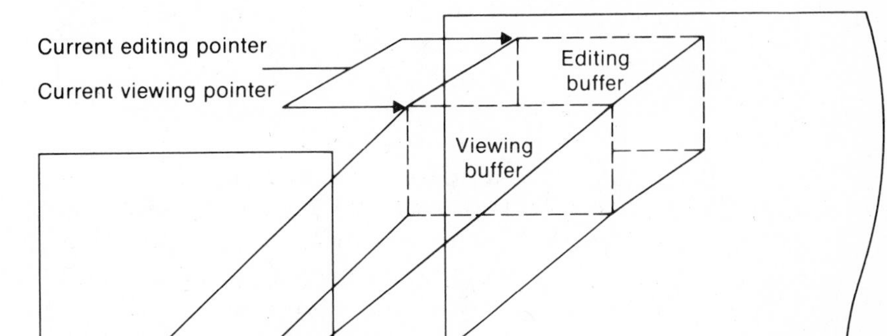

contents of both the editing and viewing buffers. If the pattern is not found, line 75 remains in the editing and viewing buffers.

The editing and viewing buffers can also partially overlap, or one may be completely contained in the other. For example, the user might specify a search to the end of the document, starting at a character position in the middle of the screen. In this case the editing filter creates an editing buffer that contains the document from the selected character to the end of the document. The viewing buffer contains the part of the document that is visible on the screen, only the last part of which is in the editing buffer.

Windows typically cover either the entire screen or a rectangular portion of it. Mapping viewing buffers to windows that cover only part of the screen is especially useful for editors on modern graphics-based work stations. Such systems can support multiple windows, simultaneously showing different portions of the same file or portions of different files. This approach allows the user to perform inter-file editing operations much more effectively than with a system having only a single window.

The mapping of the viewing buffer to a window is accomplished by two components of the system. First, the viewing component formulates an ideal view, often expressed in a device-independent intermediate representation. This view may be a very simple one consisting of a window's worth of text arranged so that lines are not broken in the middle of words. At the other extreme, the idealized view may be a facsimile of a page of fully formatted and typeset text with equations, tables, and figures. Second, the *display component* takes this idealized view from the viewing component and maps it to a physical output device in the most efficient manner possible.

Updating of a full-screen display connected over low-speed lines (1200 baud or less) is slow if every modification requires a full rewrite of the display surface. Much research is concerned with optimal screen-updating algorithms that are based on comparing the current version of the screen with the following screen. Such techniques make use of the innate capabilities of the terminal, such as *insert-character* and *delete-character* functions, transmitting only those characters needed to generate a correct display.

Device-independent output, like device-independent input, helps provide portability of the interaction language. Decoupling editing and viewing operations from display functions for output avoids the need to have a different version of the editor for each output device. Many editors make use of a *terminal-control database*. Instead of having explicit terminal-control sequences in the display routines, these editors simply call terminal-independent library routines such as *scroll down* or *read cursor position*. These library routines use the terminal-control database to look up the appropriate control sequences for a particular terminal. Consequently, adding a new terminal merely entails adding a database description of that terminal.

The components of the editor deal with a user document on two levels: in main memory and in the disk file system. Loading an entire document into main memory may be infeasible. However, if only part of a document is loaded, and if many user-specified operations require a disk read by the editor to locate the affected portion, editing might be unacceptably slow. In some systems, this problem is solved by mapping the entire file into virtual memory and letting the operating system perform efficient demand paging. An alternative is to provide *editor paging routines,* which read one or more logical portions of a document into memory as needed. Such portions are often termed *pages,* although there is usually no relationship between these pages and hard-copy document pages or virtual-memory pages. These pages remain resident in main memory until a user operation requires that another portion of the document be loaded.

Documents are often represented internally not as sequential strings of characters, but in an *editor data structure* that allows addition, deletion, and modification with a minimum of I/O and character movement. When stored on disk, the document may be represented in terms of this data structure or in an editor-independent general-purpose format, which might consist of character strings with imbedded control characters such as *linefeed* and *tab*.

Editors function in three basic types of computing environments: timesharing, stand-alone, and distributed. Each type of environment imposes some constraints on the design of an editor. The timesharing editor must function swiftly within the context of the load on the computer's processor, central memory, and I/O devices. The editor on a stand-alone system must have access to the functions that the timesharing editor obtains from its host operating system. These may be provided in part by a small local operating system, or they may be built into the editor itself if the stand-alone system is dedicated to editing. The editor operating in a distributed resource-sharing local network must, like a stand-alone editor, run independently on each user's machine and must, like a timesharing editor, contend for shared resources such as files.

Some timesharing editing systems take advantage of terminal-based hardware to perform editing tasks. These *intelligent terminals* have their own microprocessors and local buffer memories in which editing manipulations can be done. Small actions are not controlled by the CPU of the host processor, but are handled by the local terminal itself. For example, the editor might send a full screen of material from the host processor to the terminal. The user would then be free to add and delete characters and lines, using the terminal buffer and terminal-control commands. After the buffer has been edited in this way, its updated contents would be transmitted back to the host processor.

The advantage of this scheme is that the host need not be concerned with each minor change or keystroke; however, this is also the major disadvantage. With a nonintelligent terminal, the CPU sees every character as it is typed and can react immediately to perform error checking, to prompt, to update the data structure, etc. With an intelligent terminal, the lack of constant CPU intervention often means that the functionality provided to the user is more limited. Also, local work on the intelligent terminal may be lost in the event of a system crash. On the other hand, systems that allow each character to interrupt the CPU may not use the full hardware editing capabilities of the terminal because the CPU needs to see every keystroke and provide character-by-character feedback.

7.3 INTERACTIVE DEBUGGING SYSTEMS*

An interactive debugging system provides programmers with facilities that aid in the testing and debugging of programs. Some interactive debuggers are now available; however, there are relatively few such systems in wide use. There are many conflicting points of view regarding the value of such products. It is unclear to what extent the current low usage of debugging products is a result of the functions and behavior of the particular products, and to what extent users are unaware of the potential benefits of such tools. It is, however, clear that there is a major need among nearly all users for a useful interactive debugger. This section discusses some of the most important requirements as perceived by the user community and some basic system considerations involved in providing such services.

The material presented here is the result of a collaboration between the GUIDE/SHARE Language Futures Task Force (LFTF) and IBM. The LFTF was formed at meetings of SHARE and GUIDE (two major computer users' groups) in 1979. The objective of the task force was to provide IBM with the views of the user community on the future of the application-development environment. One of the areas of concentration was interactive debugging. The discussion in Seidner (1983) presents IBM's understanding of the debugging requirements as pre-

* Adapted from Rich Seidner and Nick Tindall, "Interactive Debug Requirements." *SOFTWARE ENGINEERING Notes* and *SIGPLAN Notices*, August 1983. Copyright 1983, Association for Computing Machinery, Inc.

sented by the LFTF. The definition is deliberately broad in scope, and is not limited to System/370 architecture or to any current program products or operating systems.

Section 7.3.1 presents a brief introduction to the most important functions and capabilities of an interactive debugging system and discusses some of the problems involved. Section 7.3.2 describes how the debugging tool should be related to other parts of the system. Section 7.3.3 discusses the nature of a user interface for such a debugging system.

7.3.1 Debugging Functions and Capabilities

This section describes some of the most important capabilities of an interactive debugging system. Certain of these functions are much more difficult to implement than others, and in some cases, the best form of solution is not yet clear.

The most obvious requirement is for a set of *unit test* functions that can be specified by the programmer. One important group of such functions deals with *execution sequencing,* which is the observation and control of the flow of program execution. For example, the programmer may define *breakpoints* which cause execution to be suspended when a specified point in the program is reached. After execution is suspended, other debugging commands can be used to analyze the progress of the program and to diagnose errors detected; then execution of the program can be resumed. As an alternative, the programmer can define conditional expressions that are continually evaluated during the debugging session. Program execution is suspended when any of these conditions becomes true. Similar functions may provide for halting the program after a fixed number of instructions so the programmer can observe the state of execution via other debugging commands. Given a good graphic representation of program progress, it may even be useful to enable the program to run at various speeds, called *gaits*.

A debugging system should also provide functions such as tracing and traceback. *Tracing* can be used to track the flow of execution logic and data modifications. The control flow can be traced at different levels of detail: module, subroutine, branch instruction, and so on. This tracing can also be based on conditional expressions as just described. *Traceback* can show the path by which the current statement was reached. For a given variable or parameter, traceback can show which statements modified it. Such information should be displayed symbolically. For example, statement numbers rather than hexadecimal displacements should be displayed.

It is also important for a debugging system to have good *program-*

display capabilities. It must be possible to display the program being debugged, complete with statement numbers. The user should be able to control the level at which this display occurs. For example, the program may be displayed as it was originally written, after macro expansion, and so on. It is also useful to be able to modify and incrementally recompile the program during the debugging session. The system should save all the debugging specifications (breakpoint definitions, display modes, etc.) across such a recompilation, so the programmer does not need to reissue all of these debugging commands. It should be possible to symbolically display or modify the contents of any of the variables and constants in the program, and then resume execution. The intent is to give the appearance of an interpreter, regardless of the underlying mechanism that is actually used.

Many other functions and capabilities are commonly found in interactive debugging systems. Further descriptions of such features can be found in Seidner (1983).

In providing functions such as those just described, a debugging system should consider the language in which the program being debugged is written. Most user environments, and many applications systems, involve the use of several different programming languages. What is needed is a single debugging tool that is applicable to such multilingual situations. Debugger commands that initiate actions and collect data about a program's execution should be common across languages. However, a debugging system must be sensitive to the specific language being debugged so that procedural, arithmetic, and conditional logic can be coded in the syntax of that language.

These requirements have a number of consequences for the debugger and for other software. When the debugger receives control, the execution of the program being debugged is temporarily suspended. The debugger must then be able to determine the language in which the program is written and set its *context* accordingly. Likewise, the debugger should be able to switch its context when a program written in one language calls a program written in a different language. To avoid confusion, the debugger should inform the user of such changes in context.

The context being used has many different effects on the debugging interaction. For example, assignment statements that change the values of variables during debugging should be processed according to the syntax and semantics of the source programming language. A COBOL user might enter the debugging command MOVE 3.5 TO A, whereas a FORTRAN user might enter A = 3.5. Likewise, conditional expressions should use the notation of the source language. The condition that A be unequal to B might be expressed as IF A NOT EQUAL

TO B for debugging a COBOL program, and as IF (A .NE. B) for a FORTRAN program. Similar differences exist with respect to the form of statement labels, keywords, and so on.

The notation used to specify certain debugging functions varies according to the language of the program being debugged. However, the functions themselves are accomplished in essentially the same manner regardless of the source programming language. To perform these operations, the debugger must have access to information gathered by the language translator. However, the internal symbol dictionary formats vary widely between different language translators; the same is true for statement-location information. Future compilers and assemblers should aim toward a consistent interface with the debugging system. One approach is for the language translators to produce the needed information in a standard external form for the debugger regardless of the internal form used in the translator. Another possibility would be for the language translator to provide debugger interface modules that can respond to requests for information in a standard way regardless of the language being debugged.

A similar issue is related to the display of source code during the debugging session. Again, there are two main options. The language translator may provide the source code or source listing tagged in some standard way so that the debugger has a uniform method of navigating about it. Alternatively, the translator may supply an interface module that does the navigation and display in response to requests from the debugger.

It is also important that a debugging system be able to deal with optimized code. Application code used in production environments is usually optimized. It is not enough to return to unoptimized forms of the code, because the problem will often disappear. However, the requirement for handling optimized code may create many problems for the debugger. We briefly describe some of these difficulties. Further discussions can be found in Seidner (1983).

Many optimizations involve the rearrangement of segments of code in the program. For example, invariant expressions can be removed from loops. Separate loops can be combined into a single loop, or a loop may be partially unrolled into straight-line code. Redundant expressions may be eliminated; in some cases, this may cause entire statements to disappear. Blocks of code may be rearranged to eliminate unnecessary branch instructions, which provides more efficient execution. See Section 5.3.3 for examples of some of these transformations.

All these types of optimization create problems for the debugger. The user of a debugging system deals with the source program in its original form, before optimizations are performed. However, code rearrangement alters the execution sequence and may affect tracing, breakpoints, and even statement counts if entire statements are involved. If an error occurs, it may be difficult to relate the error to the appropriate location in the original source program.

A different type of problem occurs with respect to the storage of variables. When a program is translated, the compiler normally assigns a *home location* in main memory to each variable. However, as we discussed in Section 5.2.2, variable values may be temporarily held in registers at various times to improve speed of access. Statements referring to these variables use the value stored in the register, instead of taking the variable value from its home location. These optimizations present no problem for displaying the values of such variables. However, if the user changes the value of the variable in its home location while debugging, the modified value might not be used by the program as intended when execution is resumed. In a similar type of global optimization, a variable may be permanently assigned to a register. In this case, there may be no home location at all.

The debugging of optimized code requires a substantial amount of cooperation from the optimizing compiler. In particular, the compiler must retain information about any transformations that it performs on the program. Such information can be made available both to the debugger and to the programmer. Where reasonable, the debugger should use this information to modify the debugging request made by the user, and thereby perform the intended operation. For example, it may be possible to simulate the effect of a breakpoint that was set on an eliminated statement. Similarly, a modified variable value can be stored in the appropriate register as well as at the home location for that variable. However, some more complex optimizations cannot be handled as easily. In such cases, the debugger should merely inform the user that a particular function is unavailable at this level of optimization, instead of attempting some incomplete imitation of the function.

7.3.2 Relationship with Other Parts of the System

An interactive debugger must be related to other parts of the system in many different ways. The single most important requirement for any interactive debugger is that it always be available. This means at least that it must appear to be a part of the run-time environment and an integral part of the system. When an error is discovered, immediate debugging must be possible because it may be difficult or impossible to reproduce the program failure in some other environment or at some other time. Thus the debugger must communicate and cooperate with other operating system components such as interactive subsystems.

For example, users need to be able to debug in a production environment. Debugging is even more important at production time than it is at application-development time. When an application fails during a production run, work dependent on that application stops. Since the production environment is often quite different from the test environment, many program failures cannot be repeated outside the production environment.

The debugger must also exist in a way that is consistent with the security and integrity components of the system. It must not be possible for someone to use the debugger to access any data or code that would not otherwise be accessible to that individual. Similarly, it must not be possible to use the debugger to interfere with any aspect of system integrity. Use of the debugger must be subject to the normal authorization mechanisms and must leave the usual audit trails. One benefit of the debugger, at least by comparison with a storage dump, is that it controls the information that is presented. Whereas a dump may include information that happens to have been left in storage, the debugger presents information only for the contents of specific named objects.

The debugger must coordinate its activities with those of existing and future language compilers and interpreters, as described in the preceding section. It is assumed that debugging facilities in existing languages will continue to exist and be maintained. The requirements for a cross-language debugger assume that such a facility would be installed as an alternative to the individual language debuggers.

7.3.3 User-Interface Criteria

The behavior and presentation of an interactive system is crucial to its acceptance by users. Probably the most common complaint about debugging products is that they are something new to learn. The best way to overcome this difficulty is to have a system that is simple in its organization and familiar in its language. The facilities of the debugging system should be organized into a few basic categories of function, which should closely reflect common user tasks. This simple organization contributes greatly to ease of training and ease of use.

The user interaction should take advantage of full-screen terminal devices when they are available. The primary advantage offered by screen devices is that a great deal of information can be displayed and changed easily and quickly. With such techniques as menus and full-screen editors, the user has far less information to enter and remember. This can greatly contribute to the perceived friendliness of an interactive debugging system.

If the tasks a user needs to perform are reflected in the organization of menus, then the system will feel very natural to use. Menus should have titles that identify the task they help perform. Directions should precede any choices available to the user. Techniques such as indentation should be used to help separate portions of the menu. Often the most frustrating aspect of menu systems is their lack of direct routing. It should be possible to go directly to the menu that the user wants to select without having to retrace an entire hierarchy of menus.

The use of full-screen devices and techniques such as menus is highly desirable. However, a debugging system should also support interactive users when a full-screen terminal device is not present. Every action a user can take at a full-screen terminal should have an equivalent action in a linear debugging language. For example, there should be complete functional equivalence between commands and menus.

The command language should have a clear, logical, simple syntax. It should also be as similar to the programming language(s) as possible. Commands should be simple rather than compound and should require as few parameters as possible. There should be a consistent use of parameter names across the set of commands. Parameters should automatically be checked for errors in such attributes as type and range of values. Defaults should be provided for most parameters, and the user should be able to determine when such defaults have been applied.

Command formats should be as flexible as possible. The command language should minimize the use of such punctuation as parentheses, slashes, quotation marks, and other special characters. Where possible, information should be invoked through prompting techniques.

Any good interactive system should have an on-line HELP facility. Even a list of the available commands can provide valuable assistance for the inexperienced or occasional user. For more advanced users, the HELP function can be multi-level and quite specific. One powerful use of HELP with menus is to provide explanatory text for all options present on the screen. These can be selectable by option number or name, or by filling the choice slot with a question mark. HELP should be accessible from any state of the debugging session. The more difficult the situation, the more likely it is that the user will need such information.

CHAPTER 8

SOFTWARE ENGINEERING ISSUES

This chapter contains an introduction to software engineering concepts and techniques. A full treatment of such subjects is beyond the scope of this book. Therefore, our discussion will be focused primarily on techniques that might be most useful in designing and implementing a piece of system software such as an assembler. The presentation of this material is relatively independent of the rest of the text; this chapter can be read at any time after the introduction to assemblers in Section 2.1. You may find it useful to refer to the material in Chapter 8 as you read about the other types of system software in this book, and consider how the methods discussed could be applied in those situations as well.

Section 8.1 presents an introduction to software engineering concepts and terminology, in order to give a frame of reference for the material that follows. Section 8.2 discusses the writing of specifications to define precisely what a piece of software is to accomplish. Section 8.3 introduces data flow diagrams as a representation of the functioning of a system and illustrates the development of a data flow diagram for a simple assembler. Section 8.4 demonstrates how the data flow diagram can be used in designing an assembler as a set of relatively independent modules. Finally, Section 8.5 discusses strategies for testing the individual modules and the complete system.

8.1 INTRODUCTION TO SOFTWARE ENGINEERING CONCEPTS

This section contains a brief overview of software engineering terminology and ideas. Section 8.1.1 describes some of the problems that led to the development of software engineering techniques and presents several different definitions of the term "software engineering." Section 8.1.2 discusses the stages in the software development process and mentions some of the most important issues related to these stages. Section 8.1.3 continues this discussion into the important phase of software maintenance and evolution.

8.1.1 Background and Definitions

The development of software engineering tools and methods began in the late 1960's, largely in response to what many authors have termed "the software crisis." This crisis arose from the rapid increase in the size and complexity of computer applications. Systems became much too large and complicated to be programmed by one or two people; instead, large project teams were required. In some extremely large systems, it was difficult for any one individual even to have a full intellectual grasp of the entire project. The problems in managing such a large development project led to increases in development costs and decreases in productivity. Large software systems seemed always to be delivered late, to cost more than anticipated, and to have hidden flaws. For an excellent discussion of such problems, see Brooks (1975).

We can see evidence of the continuing problems today. The purchaser of a new automobile, television set, or personal computer usually expects that the product will correctly perform its intended function. On the other hand, the first releases of a new operating system, compiler, or other software product almost always contain major "bugs" and do not work properly for some users and in some situations. The software then goes through a series of different versions or "fixes" designed to resolve these problems. Even in later releases, however, it is usual to find new flaws from time to time.

The discipline now known as *software engineering* evolved gradually in response to the problems of cost, productivity, and reliability created by increasingly large and complex software systems. Software engineering has been defined in many different ways—for example,

> the establishment and use of sound engineering principles in order to obtain, economically, software that is reliable and works efficiently on real machines (Bauer, 1972);
>
> the process of creating software systems [using] techniques that reduce high software cost and complexity while increasing reliability and modifiability (Ramamoorthy and Siyan, 1983);
>
> the technological and managerial discipline concerned with systematic production and maintenance of software products that are developed on time and within cost estimates (Fairley, 1984);

Many useful tools and techniques have been developed to help with these problems of reliability and cost. For discussions of some of these methods, see Charette (1986), Gilbert (1983), Jensen and Tonies (1979), and Lamb (1988), In spite of the advances, however, the problems are far from completely solved. Parnas (1985) gives an interesting discussion of the present state of software engineering and his view of the prospects for the future.

8.1.2 The Software Development Process

This section briefly discusses the various steps in the software development process. Figure 8.1 shows the oldest and best-known model for this process—the so-called "waterfall" software life-cycle model. In this model, the software development effort is pictured as flowing through a sequence of different stages, much like a waterfall moving from one level to the next.

In the first stage, *requirements analysis,* the task is to determine what the software system must do. The focus of this stage is on the needs of the users of the system, not on the software solution. That is, the requirements specify *what* the system must do, not *how* it will be done. In some cases, it is necessary to do much analysis and consultation with the prospective users—there are often hidden assumptions or constraints that must be made precise before the system can be constructed. The result of the requirements analysis stage is a *requirements document* that states clearly the intent, properties, and constraints of the desired system in terms that can be understood by both the end user and the system developer.

The goal of the second stage, *system specification,* is to formulate a precise description of the desired system in software development terms. The information contained in the system specification is similar to that contained in the requirements document. The focus is still on what functions the system must perform, rather than on how these functions are to be accomplished. (Many authors, in fact, consider requirements analysis and system specification to be different aspects of

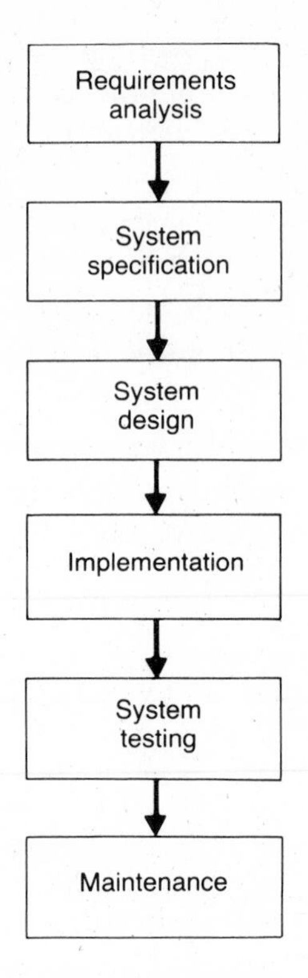

FIGURE 8.1 Software life cycle ("waterfall" model).

the same stage in the development process.) However, the approach is somewhat different. The requirements analysis step looks at the system from the point of view of the end user; the system specifications are written from the point of view of system developers and programmers, using software development terminology. Thus, the system specifications can be considered as a computer-oriented interpretation of the requirements document. We will consider examples of such specifications in Section 8.2.

The first two stages of the software development process are primarily concerned with understanding and describing the problem to be solved. The third stage, *system design,* begins to address the solution itself. The system design document outlines the most significant characteristics of the software to be developed. For example, it may describe the most important data structures and algorithms to be used and specify how the software is to be divided into modules. Smaller and less significant details are omitted—the goal is a high-level description of how the software will be constructed and how the various parts will work together to perform the desired function.

The system designer should attempt to make decisions that will minimize the problems of cost, complexity, and reliability that we have mentioned previously. For example, the software might be divided into a set of relatively independent modules in an effort to keep each module to a manageable size and complexity. The modular structure should also be designed so that the overall system is as easy as possible to implement, test, and modify. In Sections 8.3 and 8.4, we will consider an example of this process.

After the system design is complete, the fourth step, *implementation,* can begin. In this stage of the development process, the individual modules described by the design process are coded and preliminary testing is done. Although this step is what most people think of as "programming," it is actually a relatively small part of the software development effort. According to most estimates, coding should take only about 20% of the effort involved in building a system; system design and testing consume the rest of the time (Thibodeau and Dodson, 1980). One of the most common mistakes made by inexperienced system developers is to begin coding too soon, before adequate planning has been done.

The actual coding of the modules may use a variety of well-known techniques such as structured programming, stepwise refinement, and table-driven processes. Good discussions of these topics are given by Gilbert (1983). Thorough documentation for each module is also extremely important. This documentation should include (at the least) a precise description of the input and output parameters of the module, a basic description of how the module works, and important details concerning the algorithms and data structures used. The style used for programming and documentation should be consistent for all of the modules of the system.

The final phase, *system testing,* is usually the most expensive and time-consuming part of the software development process. According to most estimates, testing requires 40–50% of the total development effort (Thibodeau and Dodson, 1980). Actually, several different levels of testing are involved. Individual modules must be tested to be sure that they correctly perform their functions. The communications between the modules must be tested to be sure that they work properly together. And the entire system must be tested to ensure that it meets its overall specifications. Because the modules are related to each other in a variety of ways, these types of testing overlap to some extent. In Section 8.5, we will discuss different strategies for performing these testing tasks.

The "waterfall" model of software development treats each stage as though it were completed before the following stage begins. In reality, this is often not the case. For example, the system specification process may reveal that certain of the requirements are incomplete or inconsistent; it is then necessary to perform more requirements analysis. During implementation or testing, flaws in the design may be discovered. Thus, there is often a temporary "reverse" flow of information to earlier stages. However, the model does present a rational sequence of events in the system development process. Parnas and Clements (1986) suggest that we should attempt to follow this model as closely as possible and document the system development according to its stages, even though the actual software development may include the kind of backtracking just described.

Various authors have proposed other models of the software development process. For discussions of some of these, see Charette (1986).

8.1.3 Software Maintenance and Evolution

The last phase of the software life cycle model shown in Fig. 8.1 is *maintenance*. It is tempting to believe that most of the work is done after a system has been designed, implemented, and tested. However, this is often far from true. Systems that are used over a long period of time inevitably must change in order to meet changing requirements. According to some estimates, maintenance can account for as much as two-thirds of the total cost of software (Boehm, 1981).

Lamb (1988) identifies four major categories of software maintenance. *Corrective maintenance* fixes errors in the original system design and implementation—that is, cases in which the system does not satisfy the original requirements. *Perfective maintenance* improves the system by addressing problems that do not involve violations of the original requirements—for example, making the system more efficient or improving the user interface. *Adaptive maintenance* changes the system in order to meet changing environments and evolving user needs. *Enhancement* adds new facilities to the system that were not a part of the original requirements and were not planned for in the original design.

Maintenance can be made much easier and less costly by the presence of good documentation. The documents created during the system development (requirements, specifications, system design, and module documentation) should be kept throughout the lifetime of the system. It is very important to keep these documents updated as the system evolves (for example, when requirements change or new modules are added). There may be documentation that explicitly addresses questions of maintenance. For example, the designers of a system often plan for the likelihood of future change, identifying points where code or data items can easily be added or modified to meet changing requirements. Sample executions of the system should also be a part of the documentation. The test cases originally used during system development should be preserved, so that these tests can be repeated on the modified version of the system.

As the system is modified, it is extremely important to maintain careful control over all of the changes being made. A software system may go through many different versions or "releases." Each such version typically involves a set of related changes to requirements, specifications, design documents, code, and user manuals. Changes to one part of the system need to be carefully coordinated with changes to other parts. The process of controlling all of these changes to avoid confusion and interference is known as *configuration management*. For further discussions of configuration management issues, see Lamb (1988) and Babich (1986).

8.2 SYSTEM SPECIFICATIONS

In this section, we will examine system specifications in more detail. Section 8.2.1 discusses some of the properties that specifications should possess and examines the relationship of the specifications to other parts of the software development process. Section 8.2.2 describes several different types of specifications and gives examples of such specifications for a simple assembler. Section 8.2.3 discusses the important topic of error handling and shows how this subject is related to the system specifications.

8.2.1 Goals of System Specifications

As discussed in the preceding section, the system specification process lies between the steps of requirements analysis and system design. The requirements document is primarily concerned with the end user's view of the system and is usually written at a high level, with less important details omitted. During system specification, these details must be supplied to provide a basis for the system design to follow. The specifications must contain a complete description of all of the functions and constraints of the desired software system. They must be clearly and precisely written, consistent, and they must contain all of the information needed to write and test the software. In order to create such specifications, the developers must examine the purpose and goals of the system more closely. The process of formulating precise system specifications often reveals areas where the requirements are incomplete or ambiguous; this requires a temporary return to the requirements analysis stage.

Although the system specifications contain more detailed information than the requirements document, they are still concerned with *what* the system must do, rather than with *how* it will be done. The selection of algorithms, data representations, etc., belongs to the following phase, system design. However, it is important that the specifications explicitly address issues such as the performance needed from the system and how the software should interact with the users. During design, it is often necessary to make choices between conflicting goals. For example, one choice of data structures might lead to more efficient processing but consume more storage space; a different choice might save space but require more processing time. The selection of one type of user interface might optimize the speed of data entry but require more initial training time for the users. It is important that the system designers make such choices in a way that is consistent with the overall objectives of the system and the needs of the end users. Including such information in the requirements provides a basis for making the design decisions to follow.

The system specifications also form the basis for the system testing phase. Therefore, they must be written in a way that is objectively testable. In the case of "general" requirements such as efficiency and cost of operation, it is important to specify how these qualities will be measured and what will constitute acceptable performance.

As we can see, the system specifications are related closely to the other parts of the software development process. It is essential to maintain a record of these relationships so that the overall system documentation is coherent and consistent. For example, each specification should be explicitly connected with the appropriate item in the requirements document. Later in the process, design decisions may contain cross-references to the specifications that formed the basis for the decision. In the testing phase, each test case may refer explicitly to the specification that is being verified.

8.2.2 Types of Specifications

In this section, we will give examples of system specifications for an assembler. These examples are not intended to be a complete set of specifications for even a very simple assembler. Instead, they are intended to illustrate some of the possible kinds of specifications that may be written.

Figure 8.2 shows several different types of specifications. Specifications 1–6 give constraints on the input to the system—in this case, the source program. Such constraints describe the form and content of allowable inputs. A complete set of these specifications would precisely define the set of input conditions that the assembler must handle. Specification 1 deals with the *format* of the input, describing how some of the subfields of the source statement are positioned. Specification 2

Input specifications

1. The label on a source program statement, if present, must begin in column 1 of the statement. The Operation field is separated from the Label field by at least one blank; if no label is present, the Operation field may begin anywhere after column 1.
2. Labels may be from 1 to 6 characters in length. The first character must be alphabetic (A–Z); each of the remaining characters may be alphabetic or numeric (0–9).
3. The Operation field must contain one of the SIC mnemonic opcodes, or one of the assembler directives BYTE, WORD, RESB, RESW, START, END.
4. An instruction operand may be either a symbol (which appears as a label in the program) or a hexadecimal number that represents an actual machine address. Hexadecimal numbers used in this way must begin with a leading zero (to distinguish them from symbols) and must be between 0 and 0FFFF in value.
5. A hexadecimal string operand in a BYTE directive must be of the form X'hhh...', where each *h* is a character that represents a hexadecimal digit (0–9 or A–F). There must be an even number of such hex digits. The maximum length of the string is 32 hex digits (representing 16 bytes of memory).
6. The source program may contain as many as 500 distinct labels.

Output specifications

7. The assembly listing should show each source program statement (including any comments), together with the current location counter value, the object code generated, and any error messages.
8. The object program should occupy no address greater than hexadecimal FFFF.
9. The object program should not be generated if any assembly errors are detected.

Quality specifications

10. The assembler should be able to process at least 50 source statements per second of compute time.
11. Experienced SIC programmers using this assembler for the first time should be able to understand at least 90% of all error messages without assistance.
12. The assembler should fail to process source programs correctly in no more than 0.01% of all executions.

FIGURE 8.2 Sample program specifications.

gives rules for the formation of labels; such *lexical* rules determine the algorithm that must be used in scanning the input characters. (You may want to refer to the discussion of lexical analysis in Section 5.1.2.) Specification 3 describes the set of entries that are allowed to occur in a particular subfield, thus giving constraints on the *content* of the input. Specifications 4 and 5 give similar content restrictions that are also *context-dependent*. In this case, the allowable entries in a certain portion of the input (the Operand field) depend on the context in which the entry occurs—that is, the value in the Operation field of the same statement. Specification 6 describes a type of constraint commonly known as an *implementation restriction.* Such restrictions allow the system designer to define data structures that are large enough to accommodate the full range of anticipated inputs. Most system software involves a number of such assumptions—for example, the maximum number of concurrent jobs to be run by an operating system or the maximum nesting depth of blocks in a program being compiled.

Output specifications are intended to define precisely the results to be produced by the system. Such specifications may describe the form and content of the desired output (specification 7) or the conditions under which the output is to be generated (specification 9). They may also specify constraints on the values being output (as in specification 8). Although the output values are generally a consequence of the input, such restrictions are sometimes easier to state and check with respect to the output.

Specifications 10–12 are concerned with global characteristics of the software system, such as efficiency (specification 10), ease of use (specification 11), and reliability (specification 12). Such *quality specifications* describe overall expectations of the desired system and its operational properties. Other attributes that are often the subject of quality specifications are response time, portability, operating cost, maintainability, and training time for new users.

It is important to write quality specifications that can be objectively tested. Generalities like "efficient," "reliable," and "easy to use" may represent worthwhile goals for the system; however, without quantitative measures, it would be difficult to decide whether or not the specification had been met. In some cases, it is desirable to specify both a desired level of quality and a minimum acceptable level. For example, the desired frequency of system failures (as in specification 12) might be zero. However, it would be unrealistic to reject an entire system for a single failure. In some cases, the desired qualities for a system might be in conflict. For example, a system that is designed to be as efficient as possible might not be particularly easy to use. Features designed to make the system attractive for new users, such as detailed explanations and menus, might become cumbersome and annoying for experienced users. Therefore, it is often desirable to specify the relative importance of the various qualities desired in the system. This provides the system designer with the information needed to make intelligent choices that are in line with the overall goals for the system. Further discussions and examples of quality specifications may be found in Gilbert (1983).

Specifications often involve conditions or combinations of conditions that cannot conveniently be expressed in simple narrative sentences like those in Fig. 8.2. For example, a specification like number 4 in Fig. 8.2 might adequately describe the contents of an operand field for a SIC (standard version) program. In this case, an operand can be only a symbol or a hexadecimal constant, and issues such as relative addressing and program relocation do not arise. For a SIC/XE assembler, on the other hand, the set of conditions to be tested in processing an operand field are much more complex.

Figure 8.3 shows a sample specification for a SIC/XE assembler, expressed in *decision table* form. The upper portion of this table gives a set of *conditions,* and the lower portion describes a related set of *actions* to be taken, based on the conditions. The first column of the table lists the conditions and actions. Each column after the first describes a *rule* that specifies which actions to take under a particular combination of conditions. (You may want to refer to Section 2.2 in order to understand the logic being expressed in this specification.)

Thus, for example, Rule R5 states that if (1) the operand value type is "rel" (relative), (2) extended format is not specified in the instruc-

FIGURE 8.3 Sample decision table specification.

	Rule								
Conditions/actions	**R1**	**R2**	**R3**	**R4**	**R5**	**R6**	**R7**	**R8**	**R9**
Operand value type	abs	abs	abs	abs	rel	rel	rel	rel	neither abs nor rel
Operand value < 4096?	Y	Y	N	N	—	—	—	—	—
Extended format specified?	N	Y	N	Y	N	N	N	Y	—
Operand in range for PC-relative?	—	—	—	—	Y	N	N	—	—
Operand in range for base-relative?	—	—	—	—	—	Y	N	—	—
Set bit P to	0	0		0	1	0		0	
Set bit B to	0	0		0	0	1		0	
Set bit E to	0	1		1	0	0		1	
Flag for relocation								X	
Error			X				X		X

tion, and (3) the operand is in range for PC-relative addressing, then the assembler should (1) set bit P to 1, (2) set bit B to 0, and (3) set bit E to 0. Rule R7 states that if (1) the operand value type is "rel", (2) extended format is not specified, (3) the operand is not in range for PC-relative addressing, and (4) the operand is not in range for base-relative addressing, then an error has been detected. The entry "—" as part of a rule indicates that the corresponding condition is irrelevant or does not apply to that rule. Of course, if this decision table were used as a part of the system specification for an assembler, other specifications would be needed to define precisely what is meant by such terms as "absolute operand," "relative operand," and "in range for PC-relative addressing." Further information about the construction and use of decision tables may be found in Gilbert (1983) and Fergus (1974).

8.2.3 Error Conditions

One of the most frequently overlooked aspects of software writing is the handling of error conditions. It is much easier to write a program that simply processes valid inputs properly than it is to write one that detects and processes erroneous input as well. However, effective handling of error conditions is essential to the creation of a usable software product.

Properly written specifications implicitly define what classes of inputs are not acceptable. For example, specification 2 in Fig. 8.2 implies that a label should be considered invalid if it (1) is longer than six characters, (2) does not begin with an alphabetic character, or (3) contains any character that is not either alphabetic or numeric. Figure 8.4 shows a number of input errors derived from the input specifications in Fig. 8.2. Other erroneous input conditions may be explicitly defined by the specifications, as in the decision table in Fig. 8.3. It is extremely important that such error conditions be a part of the testing of the overall software system.

Software may take many different actions when faced with erroneous input. Sometimes a program simply aborts with a run-time error (such as a subscript out of range) or halts with no output. Obviously, these are unacceptable actions—they leave the user of the program with little or no help in finding the cause of the problem. An even worse alternative is simply to ignore the error and continue normally. This may deceive the user into thinking that everything is correct, which may lead to confusion when the output of the system is not as expected.

The preferred response to an error condition is to issue an error message and continue processing. The program should *not* terminate when an error is found, except in very unusual situations (such as

Error number	Violates specification	Statement		
1	1	ALPHA	LDA	BETA
2	1	ALPHALDA		BETA
3	1	LDA		BETA
4	2	ALPHAXX	LDA	BETA
5	2	1LPHA	LDA	BETA
6	2	ALP*A	LDA	BETA
7	3	ALPHA	XXX	BETA
8	4	ALPHA	LDA	7FD3
9	4	ALPHA	LDA	010000
10	5	BETA	BYTE	XA3B2
11	5	BETA	BYTE	X'01G9'
12	5	BETA	BYTE	X'A3B'

FIGURE 8.4 Sample input errors derived from specifications.

running out of internal storage) when it is impossible to continue. Instead, it should process the rest of the input as completely as possible in order to detect and flag any other errors that exist. Sometimes this may involve discarding a small amount of erroneous input (for example, skipping to the beginning of the next statement to be assembled), or taking a default action (for example, substituting 00 for an invalid operation code or address).

Effective handling of error conditions is one mark of a well written piece of software. A program should not "crash" when presented with any input (no matter how erroneous). The philosophy "Garbage In—Garbage Out" has no place in software development—a more appropriate maxim would be "Garbage In—Meaningful Error Messages Out."

8.3 DATA FLOW DIAGRAMS

This section introduces data flow diagrams as a way of representing the procedures, data, and information flows of a system to be designed. Section 8.3.1 gives a brief introduction to the concepts and notation to be used, and Section 8.3.2 illustrates the development of a data flow diagram for a simple assembler. In Section 8.4, we show how such data flow diagrams can be used in the software design process.

8.3.1 Data Flow Concepts and Notation

A *data flow diagram* is a representation of the movement of information between storage and processing steps within a software system. The diagram shows the major *data objects* of the system, such as files, variables, and data structures. It also displays the major *processing actions* that move, create, or transform data, and the flow of data between objects and actions.

Figure 8.5(a) shows the basic notation used in a data flow diagram. Processing actions are represented by rectangular boxes and data objects by circles. Arrows show the flow of information between objects and actions. Thus, a procedure that simply copies one file to another could be represented as shown in Fig. 8.5(b). In a more complicated situation, the diagram might include the action or actions that produced File 1 and the action or actions that use information from File 2.

FIGURE 8.5 Sample data flow diagrams.

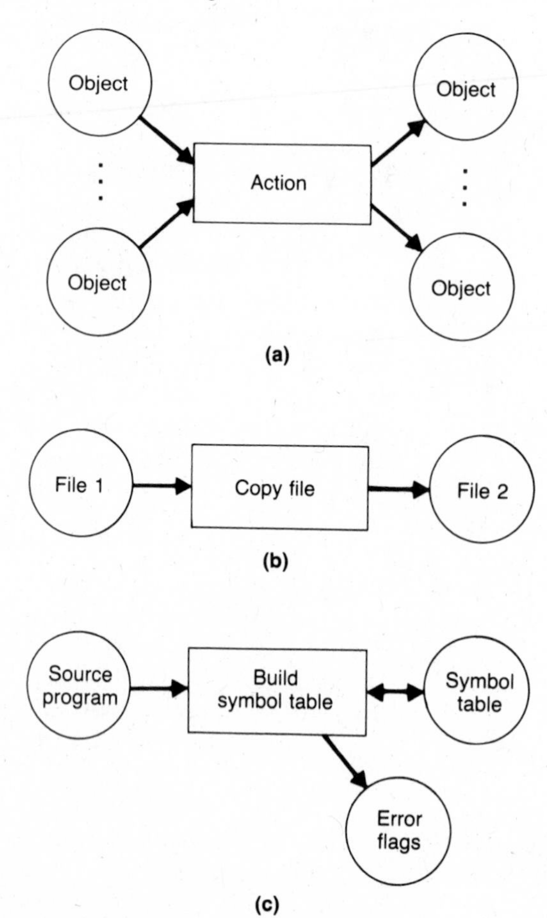

Figure 8.5(c) shows an action that reads a source program and produces a symbol table that contains the labels defined in the program and their associated addresses. The action may also set flags to indicate errors that were detected in the source program. Notice that the symbol table is both an input object and an output object for this action. This reflects the fact that the action must first search the table before adding a symbol, in order to detect duplicate symbol definitions.

Data flow diagrams usually represent the most important actions and objects in a system, with less-important details omitted. For example, the action shown in Fig. 8.5(c) probably uses several other data objects, such as a location counter and working-storage variables. Likewise, the action itself could be divided into smaller actions, such as updating the location counter or scanning the source statement for a label. However, the high-level representation shown in Fig. 8.5(c) conveys the overall approach being taken.

8.3.2 Refinement of the Data Flow Diagram

During the system design process, data flow diagrams may initially be drawn at a relatively high level. The diagrams are then refined and made more detailed as the design progresses. As an illustration of this, let us consider a simple assembler for a SIC machine (standard version). You may want to refer to Sections 1.3.1 and 2.1 as you read this material.

Figure 8.6 shows a very-high-level data flow diagram for an assembler. There is only one action, "assemble program," and the only data objects are the source and object programs and the assembly listing. Obviously, this representation is of little value in designing the assembler. However, it may serve as a starting point for the process of developing and refining the data flow diagram.

FIGURE 8.6 High-level data flow diagram for assembler.

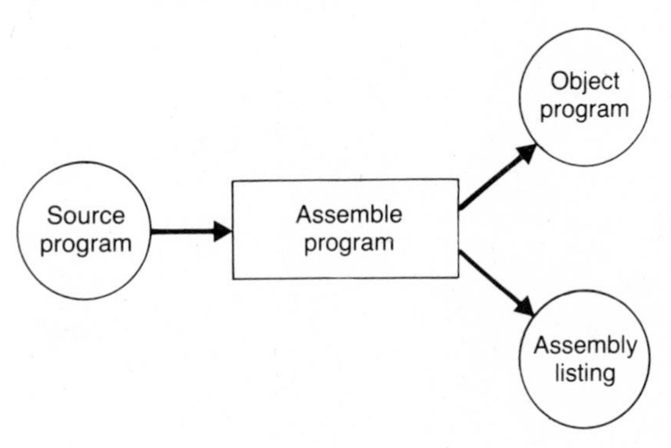

One method for refining the diagram is to begin with the desired outputs from the system. We then add to the diagram a relatively simple action that produces each required output. (The definition of "relatively simple" obviously depends on the level of detail we wish to represent in the diagram.) If the new action does not operate directly on a primary input to the system, we must define new "intermediate" data objects to provide the required input. We then add new actions that produce these intermediate data objects, and we continue in this way until the refined diagram is complete.

Figure 8.7 illustrates this process. In Fig. 8.7(a), we have created a new action whose purpose is to format and write the object program. The object program contains all of the assembled instructions and data items, together with the addresses where they are to be loaded in

FIGURE 8.7 Refinement of data flow diagram for assembler.

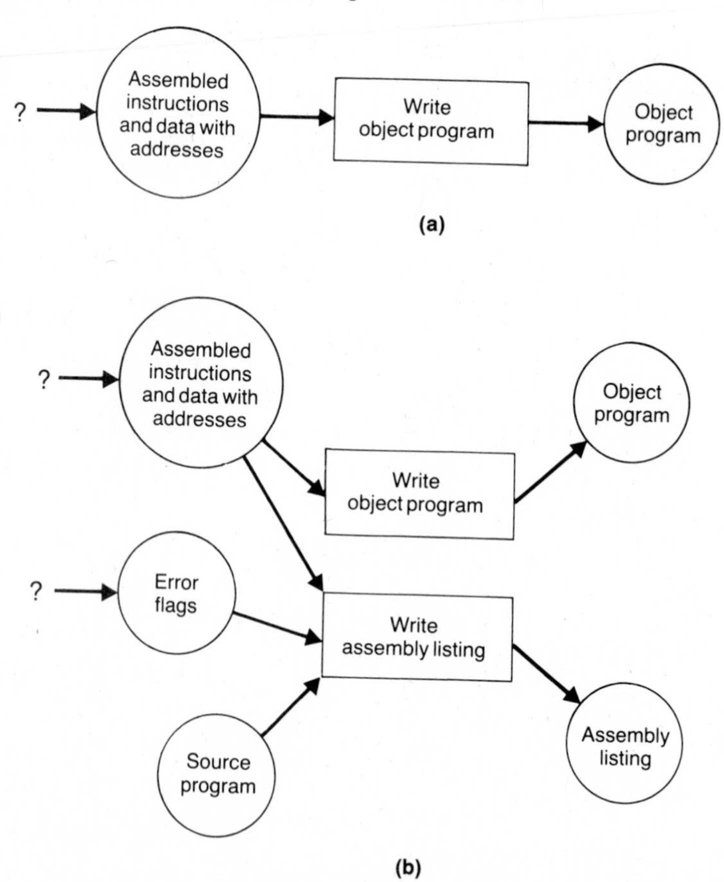

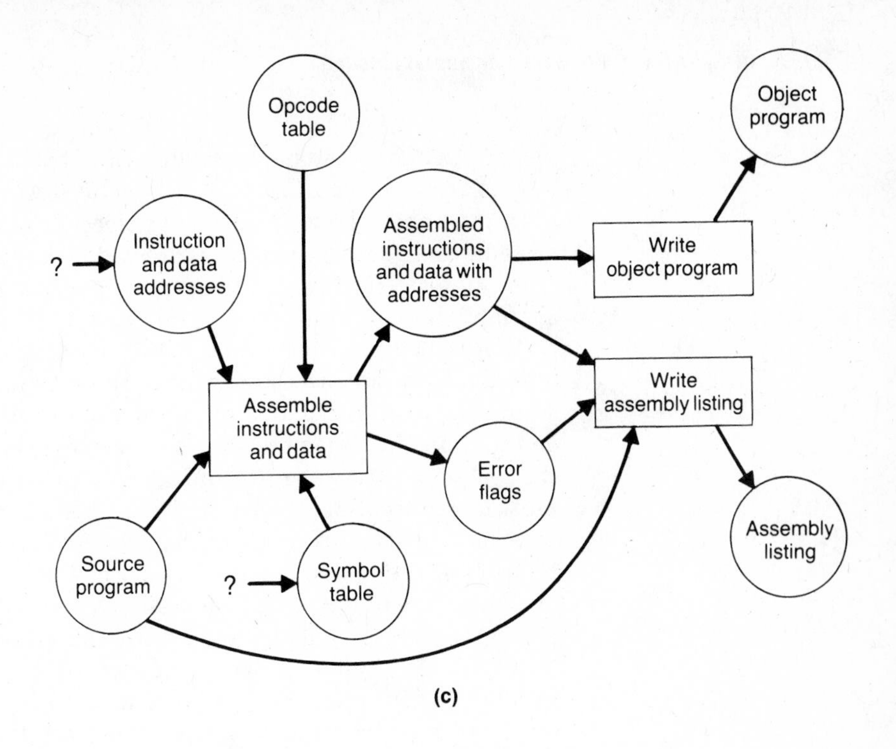

(c)

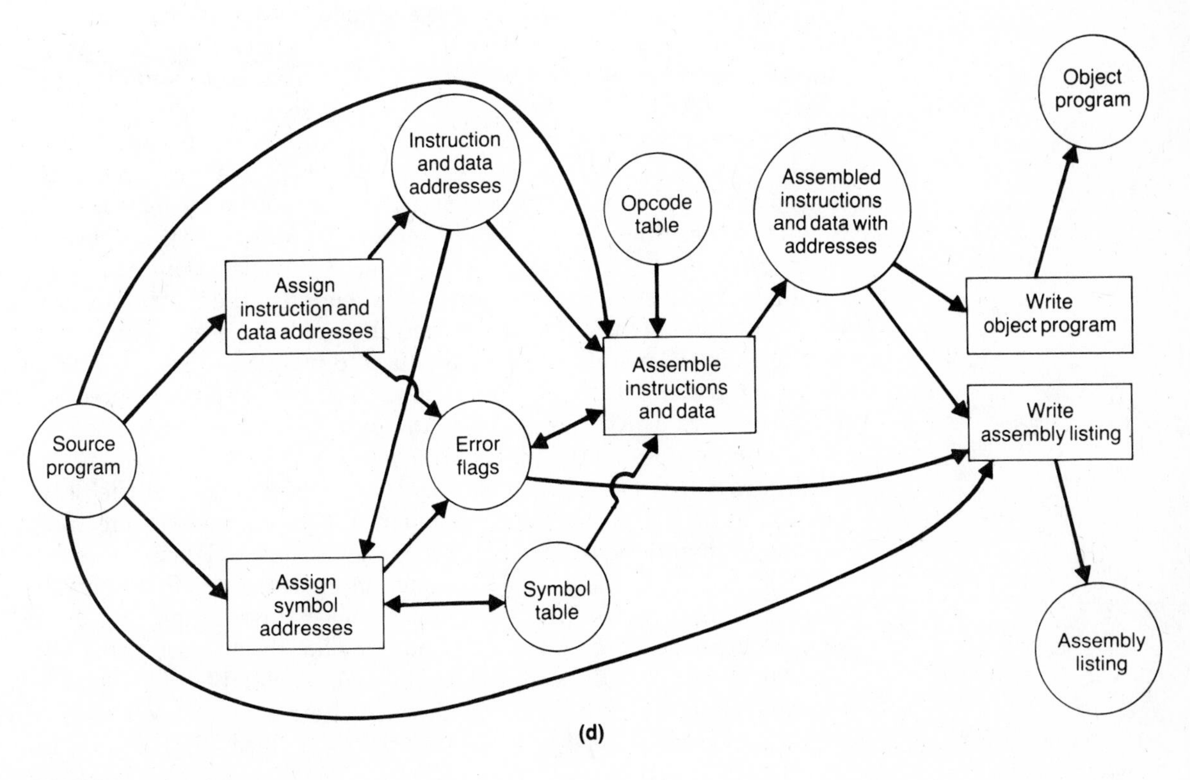

(d)

FIGURE 8.7 *(cont'd)*

memory. Thus, we define a new data object that contains this information to serve as input for the new action. At this stage, we are not concerned about how the new data object is actually produced by the system; this question will be addressed at a later step in the process.

Similarly, in Fig. 8.7(b) we have created an action to write the assembly listing. This action requires as input some of the same information that was needed for the object program. However, it also requires the source program (so that the original assembler language statements can be listed) and information about any errors that were detected during assembly. The source program is a primary input to the system, so it can be used directly by the new action. However, we must introduce a new object that contains the required error flags.

The process continues in Fig. 8.7(c). At this stage, we consider how to produce the intermediate data object that contains the assembled instructions and data with addresses. Obviously, the new action that creates this object must have the source program as one of its inputs. It also requires a table of operation codes (to translate the mnemonic instructions into machine opcodes) and a symbol table (to translate symbolic operands into machine addresses). In addition, we define a new object that contains the address assigned to each instruction and data item to be assembled. We prefer to separate the assignment of addresses from the translation process itself in order to simplify the translation action and also to make the addresses available to other actions that may be defined.

During the translation of instructions, certain error conditions may be detected. Thus, the data object that contains error flags is also an output from the new action in Fig. 8.7(c). (As we shall see, other actions may also set error flags in this object.) The operation code table contains only constant information that is related to the instruction set of the machine. This information may be predefined as part of the assembler itself, instead of being produced during each assembly of a source program. Thus, the operation code table may be treated as though it were a primary input to the system—we will not need to create any new action to produce this data object.

Figure 8.7(d) shows the final step in the development of the data flow diagram for our simple assembler. We have created one new action to compute the addresses for the instructions and data items for the program being assembled. This action operates by scanning the source program and noting the length of each instruction and data item, as described in Section 2.1. Another new action uses these addresses to make entries in the symbol table for each label in the source program. Both of the new actions may detect errors in the source program, so they may need to write into the data object that is used to store error flags. With these new actions defined, there are no "disconnected" actions or objects. Thus, the data flow diagram in Fig. 8.7(d) is complete.

The data flow diagram is intended to represent the flow and transformation of information from the primary inputs through intermediate data objects to the final outputs of the system. As the diagram is developed, it is important to write down documentation for the data objects and processing actions that are being defined. For example, the documentation should describe what data are stored in each object and how each action transforms the data with which it deals. However, the data structures being used to store the information and the algorithms used to access this information are not a part of the data flow representation. Likewise, the mechanisms by which data are passed from one processing action to another are not specified by the representation. Such implementation details are a part of the modular design process that we discuss in the following section. Thus, the data flow diagram and associated documentation can be considered as an intermediate step between the specifications (which describe *what* is to be done) and the system design (which describes *how* the tasks are to be accomplished).

8.4 MODULAR DESIGN

In this section, we discuss methods for designing a piece of software as a set of modules. The data flow representation of the problem is used as a starting point for this process. Section 8.4.1 discusses the general principles and goals of modular design—that is, what the system designer attempts to accomplish. Section 8.4.2 describes ways in which the data flow diagram can be partitioned into modules. The diagram for a simple assembler that was developed in Section 8.3 is used to illustrate this partitioning. Section 8.4.3 considers how the modules interact with each other and with the data objects being processed, and how these interfaces should be documented.

8.4.1 General Principles

The data flow diagram for a system represents the flow and transformation of information from the primary inputs through intermediate data objects to the final outputs of the system. However, there are many different ways in which these flows and transformations could be implemented in a piece of software. For example, consider the action in Fig. 8.7(d) that assigns instruction and data addresses. This action might be implemented as a separate pass over the source program, computing all addresses before any other processing is done. In that case, the object that contains the addresses might be a data structure with one entry for each line of the source program. This structure would then be used by the other actions of the assembler.

On the other hand, the action that assigns addresses might deal with the source program one line at a time. It might compute the address for each instruction or data item and then pass this address to the other parts of the assembler that require it. In that case, the data object created might be a simple variable, containing only the address for the line currently being processed.

Similar options exist for many of the other actions and data objects in Fig. 8.7(d). Thus, this data flow diagram could describe an ordinary two-pass assembler. However, it could equally well describe a one-pass assembler or a multi-pass assembler (see Section 2.4). The choices between such alternatives are made by the system designer as the data flow diagram is converted into a modular design for a piece of software.

Obviously, the goal of the modular design process is a software design that meets the system specifications. However, it is almost equally important to design systems that are easy to implement, understand, and maintain. Thus, for example, the modules should be small enough that each could be implemented by one person within a relatively short time period. Modules should have high *cohesion*—that is, the functions and objects within a module should be closely related to each other. At the same time, the modules in the system should have low *coupling*—that is, each module should be as independent as possible of the others. Systems organized into modules according to these "divide and conquer" principles tend to be much easier to understand. They are also easier to implement, because a programmer needs to understand and remember fewer details in order to code each module. The resulting system is easier to maintain and modify, because the changes that need to be made are usually isolated within one or two modules, not distributed throughout the entire system. In the remainder of this section, we will see examples of the application of these general principles to the design of an assembler.

8.4.2 Partitioning the Data Flow Diagram

The modular design process may be thought of as a partitioning of the data flow diagram into modules. As the modules are defined, the interfaces between modules and the data structures to be used for objects are also specified. One common approach to this partitioning, called *top-down design,* begins by dividing the data flow diagram into a relatively small set of major processing units. Each such unit may then be divided into subunits, and the process continues until the design is complete.

The division of the diagram (or a portion of the diagram) into units may be based on a variety of factors. Two of the most common criteria are the sequence in which functions are performed and the type of function being performed. For example, Fig. 8.8 shows the data flow diagram for our simple assembler partitioned into two passes. This division is based on the processing sequence to be used—assigning addresses to labels in Pass 1 and then assembling the source statements in Pass 2. (You may want to review the discussion in Section 2.1 for further explanation of this division into passes.) At a later stage in the design process, we might use divisions based on type of function. For example, the task of assembling a single line from the source program might be divided into one module that assembles machine instructions, one module that assembles data constants, and so on.

Another important factor to consider in modular design is the desirability of minimizing the coupling between modules. Consider, for

FIGURE 8.8 Division of assembler data flow diagram into passes.

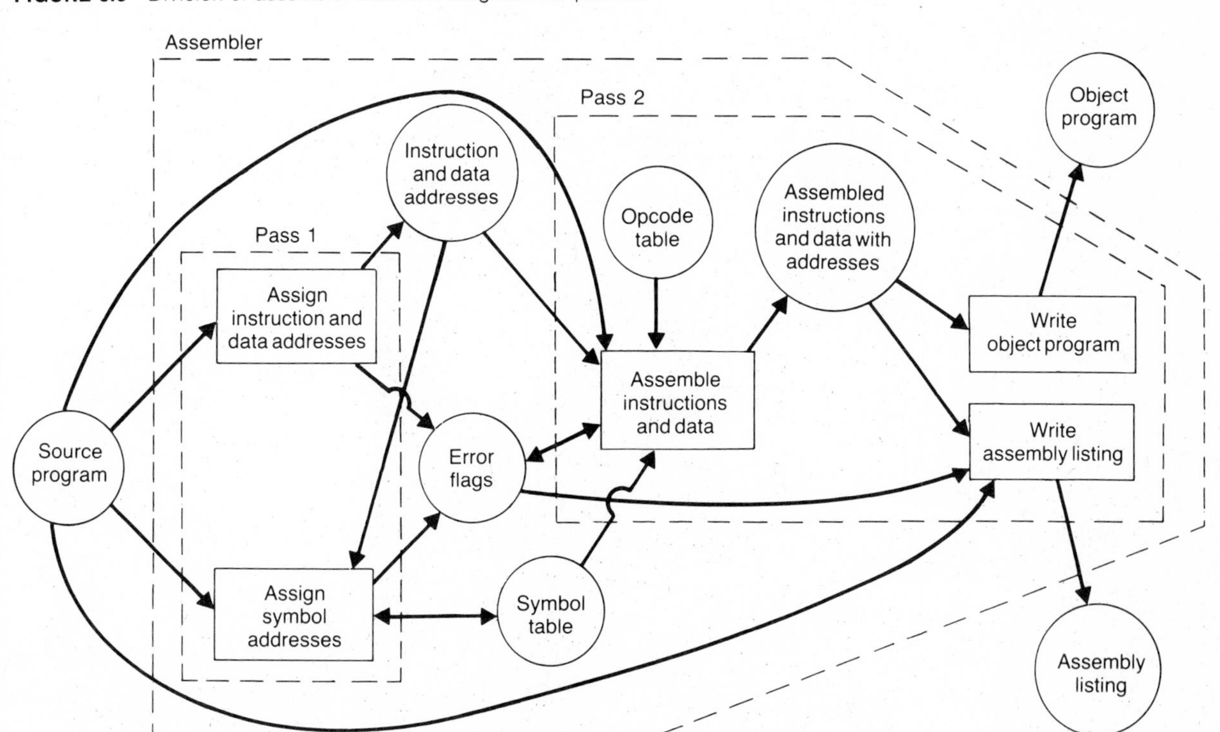

example, the portion of the data flow diagram that is shown in Fig. 8.9(a). Suppose that each of the two actions shown is implemented as a separate module. These two modules access the symbol table directly (to add new entries and to search the table). Thus, both modules must know the internal structure of the symbol table. For example, if the

FIGURE 8.9 Isolation of symbol table design.

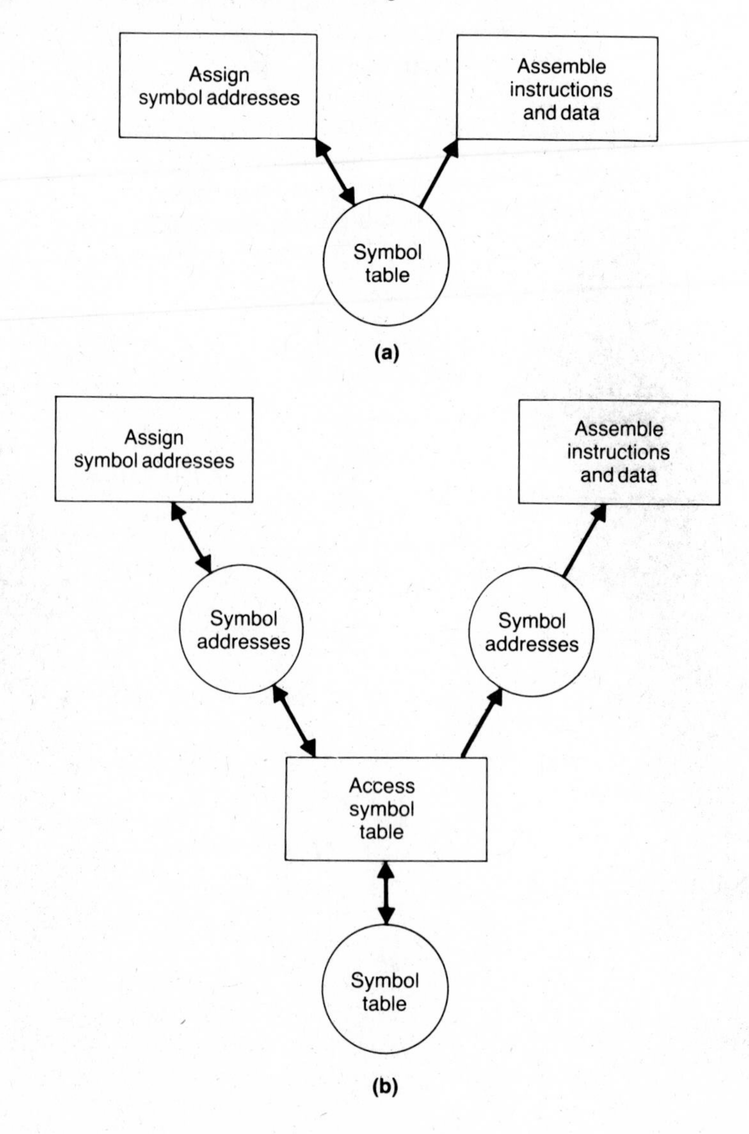

symbol table is a hash table, then both modules must know the size of the table, the hashing function, and the methods used for resolving collisions. This creates more work for the programmers who implement the modules. It also leads to duplication of effort, because the same code is written twice, and it creates additional possibilities for errors in the implementation.

Similar problems may occur during the maintenance phase. If the organization of the table or the methods for accessing it are changed, then both of the processing modules must be modified. As before, this requires more work and may lead to errors if one module is updated and the other is not.

The difficulties just described are a consequence of the undesirable coupling between the two modules—closely related items of information and processing logic occur in both modules. The modules also exhibit relatively poor cohesion—each module must contain information and logic that is related to the design of the symbol table, instead of focusing only on the logical requirements of the specific processing task being performed.

A better design, with increased cohesion and reduced coupling, is shown in Fig. 8.9(b). We have defined a new module whose sole purpose is to access the symbol table. This module is called by the other two whenever they need to perform any operation on the table. Thus, the two original modules need only know the calling interface (parameters, etc.) used to invoke the "access" module. The internal structure of the symbol table—size, organization, algorithms for access, etc.—are of concern only within the new module that performs the actual access. This reduces the amount of knowledge that must be included in the two main processing modules. It also simplifies the maintenance of the system in case the internal details of the table structure need to be changed.

In the process just illustrated, the effect of a design decision (i.e., the internal structure and representation of the symbol table) was "isolated" within a single module. This design principle is sometimes referred to as *isolation of design factors,* or simply *factor isolation*. The same general concept is also often called *information hiding* (because a module "hides" some design decision), or *data abstraction* (because the rest of the system deals with the data as an "abstract" entity, separated from its actual representation). Further discussions of these topics can be found in Gilbert (1983) and Lamb (1988).

Figure 8.10 shows a modularization of the data flow diagram from Fig. 8.8 according to the principles just described. This design includes the new module introduced in Fig. 8.9 (Access_symtab); there are also several other similar changes. The source statements, instruction and

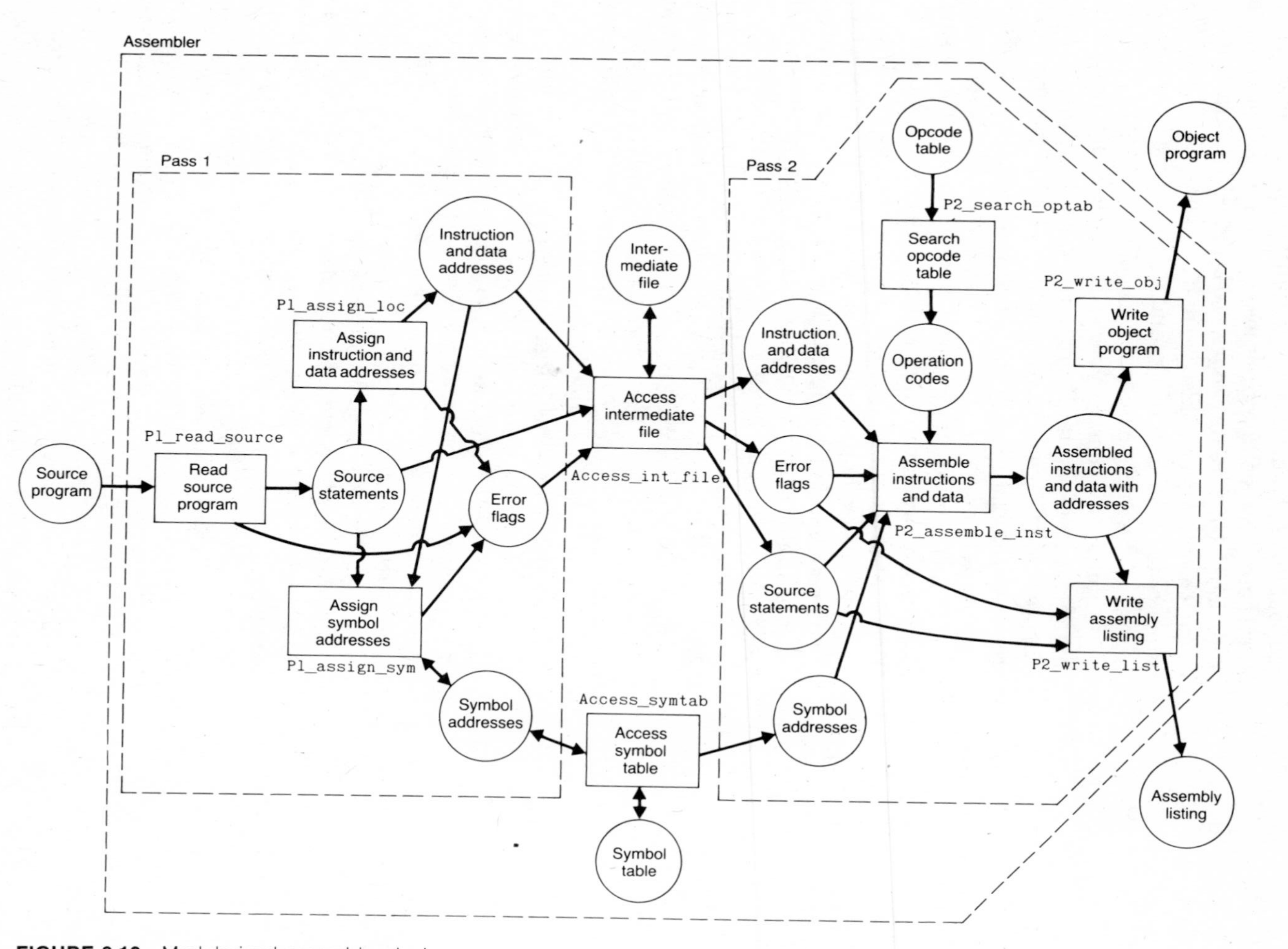

FIGURE 8.10 Modularized assembler design.

data addresses, and error flags that are communicated from Pass 1 to Pass 2 are included in an intermediate file. (A discussion of the reasons for this design decision may be found in Section 2.1.) A new module (Access_int_file) has been defined to handle all of the reading and writing of this intermediate file. The reasons for including this module are essentially the same as those discussed above—all of the details concerning the structure and access techniques for the intermediate file are isolated within a single module and removed from the rest of the system.

Likewise, we have defined a module (P2_search_optab) whose sole purpose is to access the operation code table. This design decision is somewhat different from the two just discussed, because it does not materially reduce the coupling between modules. (Whether or not the new module is defined, there is still only one place in the system where the structure of the table must be known.) However, the decision does make the module that assembles instructions smaller and less complex. It also improves module cohesion, by separating two logically unrelated functions that were previously part of the same module. Thus, it leads to a modular structure that is easier to implement, understand, and modify. For similar reasons, we have introduced a module (P1_read_source) that reads the source program and passes the source statements to the rest of the assembler. This module could, for example, handle details such as scanning for the various subfields in a free-format source program.

Figure 8.10 represents one possible stage in the modular design of our assembler. Depending upon the detailed specifications for the assembler, however, some of these modules may still be larger and more complex than is desirable. In that case, the decomposition could be carried further. For example, module P2_assemble_inst could be divided into several submodules according to the type of statement being processed—one to assemble Format 3 instructions, one to process BYTE assembler directives, etc. There may also be a number of "common" or "utility" functions that are used by more than one module in the system. These functions might include, for example, conversion of numeric values between internal (integer) and external (hexadecimal character string) representations. Each such function could be isolated within its own module and called by the other modules as needed. This isolation would improve module cohesion and reduce module coupling, resulting in the benefits previously described.

8.4.3 Module Interfaces

A diagram like Fig. 8.10 describes the decomposition of a problem into a set of modules. However, it does not specify the sequence in which these modules are to be executed or the interfaces between the modules. These are questions that must be addressed by the system designer before the implementation can begin.

There are often many ways in which a given set of modules can be organized into a system. For example, consider the three modules that make up Pass 1 in Fig. 8.10. In one possible organization, module P1_read_source would be called by the main procedure for Pass 1. For each call, P1_read_source would read a line from the source program. It would then call P1_assign_loc (for noncomment lines) to assign an address to the current statement. For statements containing a label, P1_read_source would call P1_assign_sym to make the required entry in the symbol table. A similar organization would have P1_read_source call P1_assign_loc for every noncomment line read. After calculating the appropriate address, P1_assign_loc would then call P1_assign_sym itself (instead of returning immediately to P1_read_source).

In the organizations just described, the processing of Pass 1 is "driven" by P1_read_source. On the other hand, the processing could also be controlled by P1_assign_loc. This module could call P1_read_source whenever it needs another line from the source program. It could assign an address to the line returned by P1_read_source and then call P1_assign_sym. Similarly, the processing could be driven by P1_assign_sym.

Yet another possibility is for the main procedure of Pass_1 to call all three of the other modules directly. Thus, Pass_1 could call P1_read_source to read each line from the source program. Pass_1 would then call P1_assign_loc and P1_assign_sym in turn, passing as a parameter the source line just read. Because of its simplicity, we have chosen this form of organization for the remainder of our discussions. We have also chosen to follow a similar organization in Pass 2. Figure 8.11(a) summarizes this calling structure for the modules of our assembler. In this diagram, an arrow from one module to another indicates that the first module may call the second.

A closely related issue is the placement of data objects within the modular structure. If an object is used by only one module, it is natural to place that object within the module that uses it. This is especially true in the case of modules whose purpose is to hide the internal structure of the object. For example, the symbol table in our assembler design should probably be declared within module Access_symtab. If an object is required by more than one module, the data can be shared either via parameter passing or through the use of global variables. In the first approach, the data object itself is located within one module, which passes the information in the form of parameters to other modules as needed. For example, the data object that contains the source statement currently being processed could be declared in the module

Pass_1 and passed as a parameter to P1_read_source, P1_assign_loc, and P1_assign_sym. In the second approach, the data object is made global or common to the modules that require it. For example, suppose that the modules P1_read_source, P1_assign_loc, and P1_assign_sym are contained within the module Pass_1. If the data object containing the source statement were also declared within Pass_1, it could be directly accessible to the other three modules, without the need for parameter passing.

Although these two approaches to data placement—parameter passing and global variables—can be viewed as equivalent in their effect, the choice of one over the other may be influenced by a variety of factors. Some of these factors are related to the programming language being used—for example, the amount of overhead involved in parameter passing and the mechanisms for allocating and using local and global variables. Other factors are related to the structure of the software system itself. The use of parameters provides a clearly defined and limited interface between modules. However, it may be inconvenient if the need for data transmission does not closely match the calling structure of the system. For example, a particular item of information might have to be passed through a chain of calls before reaching the module that requires it. On the other hand, the use of global variables provides a simple and efficient means of sharing data. However, it can increase the coupling between modules, because any module in the entire system could potentially use or modify any global data object.

In general, it seems desirable to avoid the use of global variables unless there is a clear reason for preferring to use them in a specific situation. If global variables are used, it is important to document clearly the use of these variables in all of the modules affected. Further discussions of module interfaces and the implications of data placement may be found in Gilbert (1983).

Figure 8.11(b) shows a high-level description of the calling sequence and parameters for our modular assembler design. In this description, we have assumed that all data are passed between modules in the form of parameters. This type of documentation is an extremely important part of the design process, because it forms the basis for the implementation phase to follow. An actual system design, of course, would include much more detailed information than is present in this example. The parameters for each module would be described completely, with data types specified and detailed descriptions of the contents and use of each parameter. The processing to be performed by each module would also be described carefully and precisely. You may want to write down some of these details for yourself, in order to gain further insight into the process of specifying a software design.

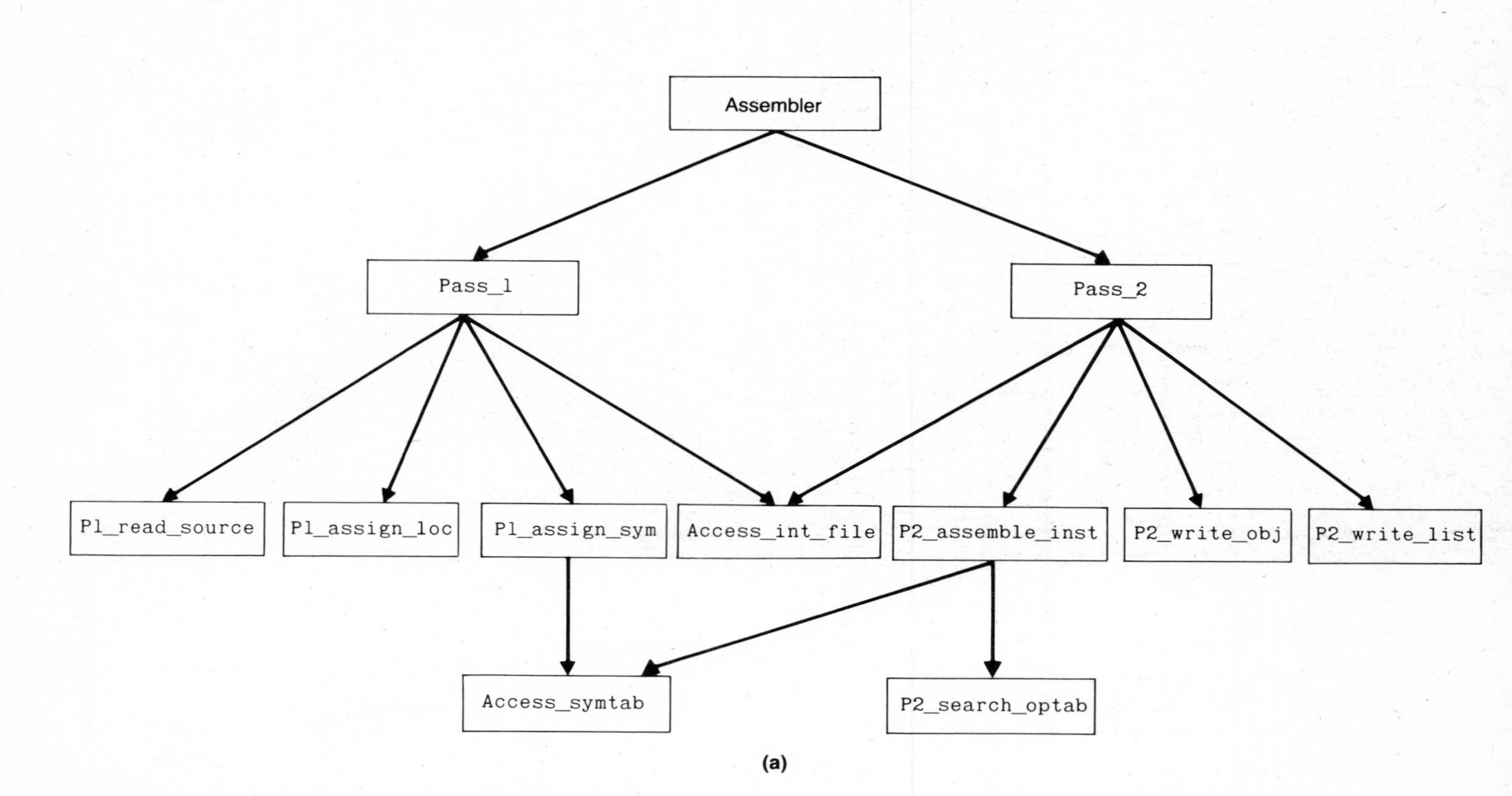

FIGURE 8.11 (a) Calling structure for modules of assembler. (b) Parameters and calling sequence for modules of assembler.

Module	Parameters	Called by	Calls
Assembler	—	User	Pass_1 (once) Pass_2 (once)
Pass_1	—	Assembler	P1_read_source (for each source line) P1_assign_sym (for each noncomment line) P1_assign_loc (for each noncomment line) Access_int_file (for each source line)
Pass_2	—	Assembler	Access_int_file (for each source line) P2_assemble_inst (for each noncomment line) P2_write_obj (for each noncomment line) P2_write_list (for each source line)
Access_int_file	Request code (I) Return code (O) Source statement (I/O) Current location counter (I/O) Error flags (I/O)	Pass_1 Pass_2	—
Access_symtab	Request code (I) Return code (O) Symbol(I) Address (I/O)	P1_assign_sym P2_assemble_inst	—
P1_read_source	Return code (O) Source statement (O) Error flags (O)	Pass_1	—
P1_assign_loc	Source statement (I) Error flags (O) Current location counter (I) Next location counter (O)	Pass_1	—
P1_assign_sym	Source statement (I) Error flags (O) Current location counter (I)	Pass_1	Access_symtab (for each label)
P2_assemble_inst	Source statement (I) Error flags (I/O) Current location counter (I) Object code (O)	Pass_2	P2_search_optab (for each instruction) Access_symtab (for each operand symbol)
P2_search_optab	Mnemonic opcode (I) Return code (O) Machine opcode (O)	P2_assemble_inst	—
P2_write_obj	Current location counter (I) Object code (I)	Pass_2	—
P2_write_list	Source statement (I) Current location counter (I) Error flags (I) Object code (I)	Pass_2	—

FIGURE 8.11 *(cont'd)* **(b)**

8.5 SYSTEM TESTING STRATEGIES

This section provides an introduction to system testing techniques. Many books and papers have been written on software testing methodologies; a detailed discussion of such topics is beyond the scope of this chapter. Instead, we consider general strategies that can be used in testing a set of related modules that make up a software system. In particular, we focus on alternatives for the sequence in which the modules of a system are coded and tested. Further information about software testing in general can be found in Evans (1984), Gilbert (1983), and Myers (1979).

Section 8.5.1 introduces some terminology and briefly mentions the various levels of system testing. Sections 8.5.2 and 8.5.3 discuss two commonly used sequences for testing a collection of modules.

8.5.1 Levels of Testing

A large software system, composed of many thousands of lines of source code, is almost always too complex to be debugged effectively all at once—when errors occur, there are too many places to look for them. Most such systems are tested in a series of stages. At each stage, the amount of code actually being debugged is kept relatively small, so that it can be tested thoroughly and efficiently.

In the *unit test* phase, individual modules are tested in isolation from the rest of the system. During this process, it is usually necessary to simulate the interaction of the module being tested with the other parts of the system. Methods for doing this are discussed in the following sections.

The unit test of an individual module is usually governed by the module specifications. That is, test cases are generated from the specifications for the module, without considering the code itself. (Obviously, the code must be considered in finding and correcting errors.) This is known as *black box* testing. Usually, a module is small enough to allow some additional testing based on logic paths in the code. Test cases might be designed that force the module through certain statements or sequences of statements. For example, the tester could make sure that both the *then* and the *else* clause of an *if* statement are executed. This is called *white box* testing.

Unit tests are often conducted by the programmer who writes each module. However, there are many advantages to having other people involved in the unit testing. From the programmer's point of view, a test is successful if it shows that the program works correctly. However, the purpose of testing is to reveal bugs in the software under test—thus, a test *should* be considered successful if it discloses a flaw.

The programmer who writes a piece of code may be too close to it, and too psychologically invested in its success, to design rigorous tests.

After modules are tested individually, they must be tested in combination with each other to be sure that the interfaces are correct. This is known as *integration testing*. Except in relatively simple systems, the integration testing is usually done using an *incremental* approach. The system is built up by adding one module (or a small number of closely related modules) at a time, and the partial system is tested at each stage. Modules are usually added to the partial system using either a bottom-up or top-down ordering; these strategies are described in the following sections.

After all of the modules have been integrated into a complete system, the final phase, *system testing,* can begin. The goal of this testing is to verify that the entire system meets all of its specifications and requirements. System testing often takes place in two or more stages. *Alpha testing* is performed by the organization that developed the system, before it is released to any outside users. *Beta testing* or *field testing* involves placing the system into actual use in a limited number of environments. This sort of customer-performed testing often turns up problems that were missed by the system developers. Finally, customers may perform their own *acceptance testing* to decide whether or not to accept delivery of the system.

8.5.2 Bottom-Up Testing

The term *bottom-up testing* describes one common sequence in which the modules of a system undergo unit testing and are integrated into a partial system. *Bottom-up* refers to the hierarchical calling structure of the system, such as the one shown in Fig. 8.11(a). The modules at the lowest level of the hierarchy (i.e., farthest from the root) are tested first, then the modules at the next higher level, and so on.

Thus, in the structure of Fig. 8.11(a), we might first perform unit testing on the modules Access_symtab and P2_search_optab. Then we might unit test module P2_assemble_inst. After this unit testing, we could combine these three modules into a partial system and perform integration testing on them. The other modules at level 3 of the hierarchy would also be unit tested individually. Then they would be integrated together to form Pass_1 and Pass_2, and these larger subsystems would be tested. Finally, the "driver" routine for the assembler would be unit tested, and all of the modules would be combined for system testing.

During the unit testing of individual modules and the integration testing of partial systems, it is necessary to simulate the presence of the remainder of the system. This is done by writing a *test driver* for each module. Figure 8.12 shows the outline of a simple test driver for Access_symtab. This driver reads test cases (i.e., sets of calling parameters) that are supplied by the person performing the test and calls Access_symtab with these parameters. It then writes out the results returned by the module, so that these can be compared with the correct input.

Bottom-up testing is the most frequently used strategy for unit testing and module integration. Test cases are delivered directly to the module, instead of being passed through the rest of the system. Thus, it is relatively easy to test a large variety of different conditions. Bottom-up testing also allows for the simultaneous unit testing and integration of many different low-level modules in the system. This can be of real benefit in meeting project deadlines.

However, bottom-up testing has been criticized by a number of authors. The most frequent objection is that, with bottom-up testing, design errors that involve interfaces between modules are not discovered until the later stages of testing. When such errors are discovered, fixing them can be very expensive and time-consuming. In some complex systems, it can also be difficult to write drivers that exactly simulate the environment of the unit or partial system being tested.

8.5.3 Top-Down Testing

In *top-down testing,* modules are unit tested and integrated into partial systems beginning at the highest level of the hierarchical structure. Thus, in the structure shown in Fig. 8.11(a) the first module to be unit tested would be the main routine for the assembler (the root node in the tree structure). Next, the main routines for Pass_1 and Pass_2 would be tested individually, and these three routines would be inte-

FIGURE 8.12 Test driver for Access_symtab module.

```
program test_symtab (input, output);
   var
      ...
   begin
      while  not eof(input) do
         begin
            readln(request_code, symbol, address);
                                          { read test case from input }
            access_symtab(request_code, return_code, symbol, address);
                                          { call access_symtab with test case }
            writeln(request_code, return_code, symbol, address);
                                            { print result }
         end;
      ...
   end.
```

grated together. The modules at the next level would be individually tested and integrated into the partial system being developed, and this process would continue until the system is complete.

During the top-down testing of modules and partial systems, we must simulate the presence of lower-level modules. This is done by writing *stubs* for the modules to be simulated. A stub contains only enough code to allow communication with the higher-level module being tested. For example, the stub for a module that computes a data value might return a constant value or a random value in some range, regardless of its input. A stub might also write a message that indicates that the module has been executed and that shows the parameters it was passed. In some cases, the stub need do nothing except exit back to the calling procedure. Figure 8.13 shows sample stubs for some of the modules in Fig. 8.11.

FIGURE 8.13 Sample stubs for modules of assembler.

```
procedure p1_assign_loc (.....);
   var
      ...
   begin
      next_locctr := curr_locctr + 3;
   end;


procedure p2_search_optab (.....);
   var
      ...
   begin
      if mnemonic = 'LDA' then
         begin
            return_code := 0;          { mnemonic found in opcode table }
            opcode := '00';            { machine opcode = hex 00 }
         end
      else
         begin
            return_code := 1;          { mnemonic not found in table }
            opcode := 'FF';            { set machine opcode to hex FF }
         end;
   end;


procedure p2_write_obj (.....);
   var
      ...
   begin
      writeln('*** p2_write_obj executed ***');
   end;
```

Top-down testing detects design errors that involve module interfaces at an earlier stage than bottom-up testing. Thus, such errors can be fixed with less wasted time and effort. However, the testing process is more difficult to plan and execute. In some systems, it is very hard to devise input data that, when passed down through the hierarchical structure, will adequately test all situations in a low-level module. In addition, top-down testing is very sensitive to delays in the planned schedule of tests. If one high-level module takes longer than anticipated to debug, the testing of many lower-level modules is also delayed.

In many cases, a combination of the bottom-up and top-down approaches is used in practice. For example, some of the most critical low-level modules could be tested first, and then the rest of the system could be integrated from the top down. Another approach uses both stubs and test drivers for the unit testing process; the resulting modules can be integrated together in any convenient sequence. Further discussions of bottom-up and top-down testing strategies can be found in Evans (1984) and Gilbert (1983).

EXERCISES

Section 8.2

1. Write a complete set of input specifications for a SIC (standard version) assembler, as described in Section 2.1. You may make any decisions about requirements that you feel are appropriate.
2. Write a complete set of specifications for an absolute loader, as described in Section 3.1.
3. List implementation restrictions that would be appropriate for a SIC/XE assembler that incorporates all of the features described in Sections 2.2 and 2.3.
4. What quality specifications might be appropriate for an operating system for a personal computer?
5. Write a decision table that specifies how bits n and i should be set in a SIC/XE Format 3 instruction (see Section 2.2).
6. Write a decision table specification for determining whether an expression specifies an absolute or relative value (see Section 2.3.3).
7. List input error conditions derived from the specifications that you wrote in Exercises 1 and 2.

Section 8.3

1. Draw a data flow diagram for an absolute loader (see Section 3.1).
2. Draw a data flow diagram for a linking loader (see Section 3.2).
3. Draw a data flow diagram for a linkage editor (see Section 3.4.1).
4. Draw a data flow diagram for a one-pass macro processor (see Section 4.1).

5. Draw a data flow diagram for a one-pass assembler that generates the object program in memory (see Section 2.4.2).
6. Modify the data flow diagram in Fig. 8.7(d) for a SIC/XE assembler that supports literals (see Sections 2.2 and 2.3.1).
7. Modify the data flow diagram in Fig. 8.7(d) for a SIC/XE assembler that supports program blocks (see Sections 2.2 and 2.3.4).
8. Write descriptions, at a suitable level of detail, for the objects and actions in Fig. 8.7(d).

Section 8.4

1. What information is "hidden" by module P1_read_source in Fig. 8.10?
2. Divide module P2_assemble_inst in Fig. 8.10 into a set of smaller modules, and specify the interfaces between these new modules.
3. Complete the interface description for P1_read_source that is given in Fig. 8.11 by specifying the representation, contents, and use of all of the parameters involved.
4. Complete the interface description for Access_symtab that is given in Fig. 8.11 by specifying the representation, contents, and use of all of the parameters involved.
5. Suppose that the execution of Pass_1 is "driven" by P1_read_source as described in the text. Specify the calling structure and parameters for this new organization, at the same level of detail as shown in Figs. 8.10 and 8.11.
6. Write a complete set of specifications for module P1_assign_sym.
7. Write a complete set of specifications for module Access_symtab.
8. Which of the parameters in Fig. 8.11 would you consider making global or common to two or more modules? Justify your choice.
9. Design a set of modules for a linking loader (see Section 3.2).
10. Design a set of modules for a one-pass macro processor (see Section 4.1).
11. Design a set of modules for a one-pass assembler that generates the object program in memory (see Section 2.4.2).

Section 8.5

1. Outline a test driver for module P1_read_source (see Figs. 8.10 and 8.11).
2. Devise a set of test cases to be used in the unit testing of P1_read_source with the test driver you outlined in Exercise 1.
3. Outline a test driver for module P2_write_obj (see Figs. 8.10 and 8.11).
4. Devise a set of test cases to be used in the unit testing of P2_write_obj with the test driver you outlined in Exercise 3.
5. Write a stub for Access_symtab to be used in the top-down testing of the modular structure in Figs. 8.10 and 8.11.
6. Write a stub for P1_read_source to be used in the top-down testing of the modular structure in Figs. 8.10 and 8.11.

APPENDIXES

APPENDIX A: SIC/XE INSTRUCTION SET AND ADDRESSING MODES

Instruction Set

In the following descriptions, uppercase letters refer to specific registers. The notation *m* indicates a memory address, *n* indicates an integer between 1 and 16, and *r*1 and *r*2 represent register identifiers. Parentheses are used to denote the contents of a register or memory location. Thus A $\leftarrow (m..m+2)$ specifies that the contents of the memory locations *m* through $m + 2$ are loaded into register A; $m..m+2 \leftarrow$ (A) specifies that the contents of register A are stored in the word that begins at address *m*.

The letters in the Notes column have the following meanings:

- P Privileged instruction
- X Instruction available only on XE version
- F Floating-point instruction
- C Condition code CC set to indicate result of operation (<, =, or >)

The Format column indicates which SIC/XE instruction format is to be used in assembling each instruction; 3/4 means that either Format 3 or

Mnemonic	Format	Opcode	Effect	Notes
ADD m	3/4	18	A ← (A) + (m..m+2)	
ADDF m	3/4	58	F ← (F) + (m..m+5)	X F
ADDR r1,r2	2	90	r2 ← (r2) + (r1)	X
AND m	3/4	40	A ← (A) & (m..m+2)	
CLEAR r1	2	B4	r1 ← 0	X
COMP m	3/4	28	(A) : (m..m+2)	C
COMPF m	3/4	88	(F) : (m..m+5)	X F C
COMPR r1,r2	2	A0	(r1) : (r2)	X C
DIV m	3/4	24	A ← (A) / (m..m+2)	
DIVF m	3/4	64	F ← (F) / (m..m+5)	X F
DIVR r1,r2	2	9C	r2 ← (r2) / (r1)	X
FIX	1	C4	A ← (F) [convert to integer]	X F
FLOAT	1	C0	F ← (A) [convert to floating]	X F
HIO	1	F4	Halt I/O channel number (A)	P X
J m	3/4	3C	PC ← m	
JEQ m	3/4	30	PC ← m if CC set to =	
JGT m	3/4	34	PC ← m if CC set to >	
JLT m	3/4	38	PC ← m if CC set to <	
JSUB m	3/4	48	L ← (PC); PC ← m	
LDA m	3/4	00	A ← (m..m+2)	
LDB m	3/4	68	B ← (m..m+2)	X
LDCH m	3/4	50	A [rightmost byte] ← (m)	
LDF m	3/4	70	F ← (m..m+5)	X F
LDL m	3/4	08	L ← (m..m+2)	
LDS m	3/4	6C	S ← (m..m+2)	X
LDT m	3/4	74	T ← (m..m+2)	X
LDX m	3/4	04	X ← (m..m+2)	
LPS m	3/4	D0	Load processor status from information beginning at address m (see Section 6.2.1)	P X
MUL m	3/4	20	A ← (A) * (m..m+2)	
MULF m	3/4	60	F ← (F) * (m..m+5)	X F
MULR r1,r2	2	98	r2 ← (r2) * (r1)	X
NORM	1	C8	F ← (F) [normalized]	X F
OR m	3/4	44	A ← (A) | (m..m+2)	
RD m	3/4	D8	A [rightmost byte] ← data from device specified by (m)	P

Mnemonic	Format	Opcode	Effect	Notes
RMO r1,r2	2	AC	r2 ← (r1)	X
RSUB	3/4	4C	PC ← (L)	
SHIFTL r1,n	2	A4	r1 ← (r1); left circular shift n bits. {In assembled instruction, r2=n−1}	X
SHIFTR r1,n	2	A8	r1 ← (r1); right shift n bits with vacated bit positions set equal to leftmost bit of (r1). {In assembled instruction, r2=n−1}	X
SIO	1	F0	Start I/O channel number (A); address of channel program is given by (S)	P X
SSK m	3/4	EC	Protection key for address m ← (A) (see Section 6.2.4)	P X
STA m	3/4	0C	m..m+2 ← (A)	
STB m	3/4	78	m..m+2 ← (B)	X
STCH m	3/4	54	m ← (A) [rightmost byte]	
STF m	3/4	80	m..m+5 ← (F)	X F
STI m	3/4	D4	Interval timer value ← (m..m+2) (see Section 6.2.1)	P X
STL m	3/4	14	m..m+2 ← (L)	
STS m	3/4	7C	m..m+2 ← (S)	X
STSW m	3/4	E8	m..m+2 ← (SW)	P
STT m	3/4	84	m..m+2 ← (T)	X
STX m	3/4	10	m..m+2 ← (X)	
SUB m	3/4	1C	A ← (A) − (m..m+2)	
SUBF m	3/4	5C	F ← (F) − (m..m+5)	X F
SUBR r1,r2	2	94	r2 ← (r2) − (r1)	X
SVC n	2	B0	Generate SVC interrupt. {In assembled instruction, r1=n}	X
TD m	3/4	E0	Test device specified by (m)	P C
TIO	1	F8	Test I/O channel number (A)	P X C
TIX m	3/4	2C	X ← (X) + 1; (X) : (m..m+2)	C
TIXR r1	2	B8	X ← (X) + 1; (X) : (r1)	X C
WD m	3/4	DC	Device specified by (m) ← (A) [rightmost byte]	P

Format 4 can be used. All instructions for the standard version of SIC are assembled using the format described in Section 1.3.1 (which is compatible with Format 3.) Instruction subfields that are not required, such as the address field for an RSUB instruction, are set to zero.

Instruction Formats

Format 1 (1 byte):

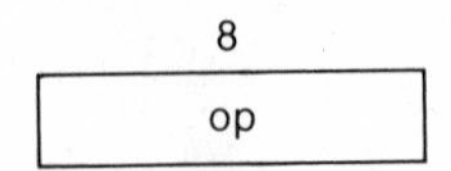

Format 2 (2 bytes):

8	4	4
op	r1	r2

Format 3 (3 bytes):

6	1	1	1	1	1	1	12
op	n	i	x	b	p	e	disp

Format 4 (4 bytes):

6	1	1	1	1	1	1	20
op	n	i	x	b	p	e	address

Addressing Modes

The following addressing modes apply to Format 3 and 4 instructions. Combinations of addressing bits not included in this table are treated as errors by the machine. In the description of assembler language notation, *c* indicates a constant between 0 and 4095 (or a memory address known to be in this range); *m* indicates a memory address or a constant value larger than 4095. Further information can be found in Section 1.3.2.

The letters in the Notes column have the following meanings:

- 4 Format 4 instruction
- D Direct-addressing instruction
- A Assembler selects either program-counter relative or base-relative mode
- S Compatible with instruction format for standard SIC machine. Operand value can be between 0 and 32,767 (see Section 1.3.2 for details).

Addressing type	Flag bits n	i	x	b	p	e	Assembler language notation	Calculation of target address TA	Operand	Notes
Simple	1	1	0	0	0	0	op c	disp	(TA)	D
	1	1	0	0	0	1	+op m	addr	(TA)	4 D
	1	1	0	0	1	0	op m	(PC) + disp	(TA)	A
	1	1	0	1	0	0	op m	(B) + disp	(TA)	A
	1	1	1	0	0	0	op c,X	disp + (X)	(TA)	D
	1	1	1	0	0	1	+op m,X	addr + (X)	(TA)	4 D
	1	1	1	0	1	0	op m,X	(PC) + disp + (X)	(TA)	A
	1	1	1	1	0	0	op m,X	(B) + disp + (X)	(TA)	A
	0	0	0	–	–	–	op m	b/p/e/disp	(TA)	D S
	0	0	1	–	–	–	op m,X	b/p/e/disp + (X)	(TA)	D S
Indirect	1	0	0	0	0	0	op @c	disp	((TA))	D
	1	0	0	0	0	1	+op @m	addr	((TA))	4 D
	1	0	0	0	1	0	op @m	(PC) + disp	((TA))	A
	1	0	0	1	0	0	op @m	(B) + disp	((TA))	A
Immediate	0	1	0	0	0	0	op #c	disp	TA	D
	0	1	0	0	0	1	+op #m	addr	TA	4 D
	0	1	0	0	1	0	op #m	(PC) + disp	TA	A
	0	1	0	1	0	0	op #m	(B) + disp	TA	A

APPENDIX B: ASCII CHARACTER CODES

Hex code	ASCII character	Hex code	ASCII character	Hex code	ASCII character	Hex code	ASCII character
00	NUL	20	SP	40	@	60	`
01	SOH	21	!	41	A	61	a
02	STX	22	"	42	B	62	b
03	ETX	23	#	43	C	63	c
04	EOT	24	$	44	D	64	d
05	ENQ	25	%	45	E	65	e
06	ACK	26	&	46	F	66	f
07	BEL	27	'	47	G	67	g
08	BS	28	(	48	H	68	h
09	HT	29	)	49	I	69	i
0A	LF	2A	*	4A	J	6A	j
0B	VT	2B	+	4B	K	6B	k
0C	FF	2C	,	4C	L	6C	l
0D	CR	2D	–	4D	M	6D	m
0E	SO	2E	.	4E	N	6E	n
0F	SI	2F	/	4F	O	6F	o
10	DLE	30	0	50	P	70	p
11	DC1	31	1	51	Q	71	q
12	DC2	32	2	52	R	72	r
13	DC3	33	3	53	S	73	s
14	DC4	34	4	54	T	74	t
15	NAK	35	5	55	U	75	u
16	SYN	36	6	56	V	76	v
17	ETB	37	7	57	W	77	w
18	CAN	38	8	58	X	78	x
19	EM	39	9	59	Y	79	y
1A	SUB	3A	:	5A	Z	7A	z
1B	ESC	3B	;	5B	[	7B	{
1C	FS	3C	<	5C	\	7C	|
1D	GS	3D	=	5D	]	7D	}
1E	RS	3E	>	5E	^	7E	~
1F	US	3F	?	5F	_	7F	DEL

APPENDIX C: SIC/XE REFERENCE MATERIAL

Status Word Contents

Bit position	Field name	Use
0	MODE	0=user mode, 1=supervisor mode
1	IDLE	0=running, 1=idle
2–5	ID	Process identifier
6–7	CC	Condition code
8–11	MASK	Interrupt mask
12–15		Unused
16–23	ICODE	Interruption code

Interrupts

Class	Interrupt type	Address of work area (hex)	Interruption code
I	SVC	100	Code from SVC instruction
II	Program	130	Condition (see below)
III	Timer	160	None
IV	I/O	190	Channel number

SVC Codes

Code	Mnemonic	Register parameters
0	`WAIT`	(A) = address of ESB for event
1	`SIGNAL`	(A) = address of ESB for event
2	`I/O`	(A) = address of channel program (S) = channel number (T) = address of ESB for I/O operation
3	`REQUEST`	(T) = address of resource name
4	`RELEASE`	(T) = address of resource name

Program Interrupt Codes

Code (hex)	Meaning
00	Illegal instruction
01	Privileged instruction in user mode
02	Address out of range
03	Memory-protection violation
04	Arithmetic overflow
10	Page fault
11	Segment fault
12	Segment-protection violation
13	Segment length exceeded

Channel Command Format

Bit positions	Contents
0–3	Command code (see below)
4–7	Device code
8–23	Number of bytes to transfer
24–27	Unused
28–47	Memory address for start of data transfer

Channel Command Codes

Code (hex)	Meaning
0	Halt device
1	Read data
2	Write data
3–F	Device-dependent; assigned individually for each specific type of I/O device

Channel Work Areas

Bytes	Contents
0–2	Address of current channel program
3–5	Address of ESB for current I/O operation
6–8	Address of I/O request queue for channel
9–B	Status flags
C–F	Reserved

The work area for channel n begins at hexadecimal memory address $2n0$.

REFERENCES

Aho, Alfred V. and Jeffrey D. Ullman, *Principles of Compiler Design,* Addison-Wesley Publishing Co., Reading, Mass., 1977.

Ammann, Urs, "The Zurich Implementation," in *Pascal—The Language and Its Implementation,* edited by D. W. Barron, John Wiley & Sons, New York, 1981.

Baase, Sara, *VAX-11 Assembly Language Programming,* Prentice-Hall, Inc., Englewood Cliffs, N.J., 1983.

Babich, Wayne A., *Software Configuration Management: Coordination for Team Productivity,* Addison-Wesley Publishing Co., Reading, Mass., 1986.

Bauer, F. L., "Software Engineering," *Information Processing 71,* North-Holland Publishing Co., Amsterdam, 1972.

Boehm, Barry W., *Software Engineering Economics,* Prentice-Hall, Englewood Cliffs, N.J., 1981.

Bourne, Stephen R., *The UNIX System,* Addison-Wesley Publishing Co., Reading, Mass., 1983.

Brooks, Frederick P., *The Mythical Man-Month: Essays on Software Engineering,* Addison-Wesley Publishing Co., Reading, Mass., 1975.

Brown, Peter J., *Macro Processors and Techniques for Portable Software,* John Wiley & Sons, New York, 1974.

Campbell–Kelley, M., *An Introduction to Macros,* MacDonald, 1973.

Charette, Robert N., *Software Engineering Environments: Concepts and Technology,* Intertext Publications, New York, 1986.

Control Data Corporation, *CYBER 170 Computer Systems Hardware Reference Manual,* 1981a.

——*COMPASS Version 3 Reference Manual,* 1982a.

——*CYBER Loader Version 1 Reference Manual,* 1982b.

——*NOS Internal Maintenance Specification,* 1979.

——*NOS Reference Manual,* 1981b.

Cole, A. J., *Macro Processors,* Cambridge University Press, New York, 1981.

Date, C. J., *An Introduction to Database Systems,* 3rd ed., Addison-Wesley Publishing Co., Reading, Mass., 1981.

Digital Equipment Corporation, *VAX Architecture Handbook,* 1981.

——*VAX-11 Linker Reference Manual,* 1978.

——*VAX-11 MACRO Language Reference Manual,* 1979.

——*VAX Software Handbook,* 1982.

Deitel, Harvey M., *An Introduction to Operating Systems,* rev. ed., Addison-Wesley Publishing Co., Reading, Mass., 1984.

Denning, Dorothy E., *Cryptography and Data Security,* Addison-Wesley Publishing Co., Reading, Mass., 1982.

Donovan, John, *Systems Programming,* McGraw-Hill Book Co., New York, 1972.

Evans, Michael W., *Productive Software Test Management,* John Wiley & Sons, New York, 1984.

Fairley, R., *Software Engineering Concepts,* McGraw-Hill, New York, 1984.

Fergus, Raymond M., "Decision Tables—What, Why, and How," in J. D. Couger and R. W. Knapp (ed.), *System Analysis Techniques,* John Wiley & Sons, New York, 1974.

Fernandez, Eduardo B., R. C. Summers, and C. Wood, *Database Security and Integrity,* Addison-Wesley Publishing Co., Reading, Mass., 1981.

Gear, C. William, *Computer Organization and Programming,* 3rd ed., McGraw-Hill Book Co., New York, 1981.

Gilbert, Philip, *Software Design and Development,* Science Research Associates, Chicago, 1983.

Gries, David, *Compiler Construction for Digital Computers,* John Wiley & Sons, New York, 1971.

Grishman, Ralph, *Assembly Language Programming for the Control Data 6000 Series and the CYBER 70 Series,* 2nd ed., Algorithmics Press, New York, 1974.

Holt, Richard C., G. S. Graham, E. D. Lazowska, and M. A. Scott, *Structured Concurrent Programming with Operating Systems Applications,* Addison-Wesley Publishing Co., Reading, Mass., 1978.

Hopgood, Frank R. A., *Compiling Techniques,* MacDonald/American Elsevier, London, 1969.

Hunter, Robin, *The Design and Construction of Compilers,* John Wiley & Sons, New York, 1981.

International Business Machines Corporation, *IBM System/370 Principles of Operation,* 1983.

——*OS/VS DOS/VSE VM/370 Assembler Language,* 1979.

——*OS/VS VM/370 Assembler PLM,* 1974.

——*OS/VS VM/370 Assembler Programmer's Guide,* 1982.

——*OS/VS Linkage Editor and Loader,* 1978.

——*OS/VS Linkage Editor Logic,* 1972a.

——*FORTRAN IV (H) Compiler Program Logic Manual,* 1972b.

Jensen, Kathleen and Niklaus Wirth, *PASCAL User Manual and Report,* Springer-Verlag, New York, 1974.

Jensen, Randall W. and Charles C. Tonies, *Software Engineering,* Prentice-Hall, Englewood Cliffs, N.J., 1979.

Johnson, Stephen C., "YACC—Yet Another Compiler-Compiler," Computing Science Technical Report 32, Bell Laboratories, Murray Hill, N.J., 1975.

——"Language Development Tools on the Unix System," *Computer 13:* 16–21, August 1980.

Kernighan, Brian W. and P. J. Plauger, *Software Tools,* Addison-Wesley Publishing Co., Reading, Mass., 1976.

Knuth, Donald E., *The Art of Computer Programming, Vol. 1: Fundamental Algorithms,* 2nd ed., Addison-Wesley Publishing Co., Reading, Mass., 1973a.

——*The Art of Computer Programming, Vol. 3: Sorting and Searching,* Addison-Wesley Publishing Co., Reading, Mass., 1973b.

Lamb, David A., *Software Engineering: Planning for Change,* Prentice-Hall, Englewood Cliffs, N.J., 1988.

Lesk, M. E., "Lex—A Lexical Analyzer Generator," Computing Science Technical Report 39, Bell Laboratories, Murray Hill, N.J., 1975.

Lewis, Philip M., D. J. Rosenkrantz, and R. E. Stearns, *Compiler Design Theory,* Addison-Wesley Publishing Co., Reading, Mass., 1976.

Loomis, Mary E. S., *Data Management and File Processing,* Prentice-Hall, Inc., Englewood Cliffs, N.J., 1983.

Lorin, Harold and Harvey M. Deitel, *Operating Systems,* Addison-Wesley Publishing Co., Reading, Mass., 1981.

Madnick, Stuart E. and John J. Donovan, *Operating Systems,* McGraw-Hill Book Co., New York, 1974.

Martin, James, *Computer Data-Base Organization,* 2nd ed., Prentice-Hall, Inc., Englewood Cliffs, N.J., 1977.

Meyrowitz, Norman and A. van Dam, "Interactive Editing Systems: Part I," *ACM Computing Surveys 14:* 321–352, September 1982.

——"Interactive Editing Systems: Part II," *ACM Computing Surveys 14:* 353–416, September 1982.

Myers, G. J., *The Art of Software Testing,* John Wiley & Sons, New York, 1979.

Nori, K. V., U. Ammann, K. Jensen, H. H. Nageli, and C. Jacobi, "Pascal-P Implementation Notes," in *Pascal—The Language and Its Implementation,* edited by D. W. Barron, John Wiley & Sons, New York, 1981.

Overgaard, M., "UCSD Pascal: A Portable Software Environment for Small Computers," *Proceedings, 1980 National Computer Conference 49:* 747–754.

Parnas, David L., "Software Aspects of Strategic Defense Systems," *Communications of the ACM 28:* 1326–1335, December 1985.

Parnas, David L. and Paul C. Clements, "A Rational Design Process: How and Why to Fake It," *IEEE Transactions on Software Engineering SE-12;* 251–257, February 1986.

Peterson, James L. and Abraham Silberschatz, *Operating System Concepts,* Addison-Wesley Publishing Co., Reading, Mass., 1983.

Pfleeger, Charles P., *Machine Organization: An Introduction to the Structure and Programming of Computer Systems,* John Wiley & Sons, New York, 1982.

Ramamoorthy, C. V. and K. Siyan, "Software Engineering," in *Encyclopedia of Computer Science and Engineering,* 2nd ed., Van Nostrand Reinhold, New York, 1983.

Ritchie, Dennis M. and K. Thompson, "The UNIX Time-Sharing System," *Bell System Technical Journal 57,* no. 6, pt. 2:1905–1930, July–August 1978.

Sassa, Masataka, "A Pattern Matching Macro Processor," *Software: Practice and Experience,* pp. 439–456, June 1979.

Seawright, L. H. and R. A. MacKinnon, "VM/370—A Study of Multiplicity and Usefulness," *IBM Systems Journal 18,* no. 1:4–17, 1979.

Seidner, Rich and Nick Tindall, "Interactive Debug Requirements," *Proceedings of the ACM SIGSOFT/SIGPLAN Software Engineering Symposium on High-Level Debugging,* March 1983, appearing in *SOFTWARE ENGINEERING Notes* and *SIGPLAN Notices,* August 1983, pp. 9–22.

SofTech Microsystems, *UCSD Pascal Users Manual,* 1980.

——*P-System Internal Architecture Reference Manual,* 1983.

Standish, Thomas A., *Data Structure Techniques,* Addison-Wesley Publishing Co., Reading, Mass., 1980.

Struble, George W., *Assembler Language Programming: The IBM System 370,* 3rd ed., Addison-Wesley Publishing Co., Reading, Mass., 1983.

Tannenbaum, A. S., *Structured Computer Organization,* Prentice-Hall, Inc., Englewood Cliffs, N.J., 1984.

Thibodeau, R. and E. N. Dodson, "Life Cycle Phase Interrelationships," *Journal of Systems and Software 1:* 203–211, 1980.

Tremblay, Jean-Paul and P. G. Sorenson, *An Introduction to Data Structures with Applications,* 2nd ed., McGraw-Hill Book Co., New York, 1984.

Ullman, Jeffrey D., *Principles of Database Systems,* Computer Science Press, Princeton, N.J., 1980.

Wiederhold, Gio, *Database Design,* 2nd ed., McGraw-Hill Book Co., New York, 1983.

INDEX